THE TORAH AND THE STOICS ON HUMANKIND AND NATURE

THE TORAH AND THE STOICS ON HUMANKIND AND NATURE

A Contribution to the Debate on Sustainability and Quality

BY

JAN J. BOERSEMA

BRILL
LEIDEN · BOSTON · KÖLN
2001

This book is printed on acid-free paper.

Cover illustration

Edward Hicks was an American artist who for years travelled the land to spread the word of the Quakers. The preferred subject of his paintings was paradise, a world in which lion and lamb lie together in peace. The illustration shows *The Peaceable Kingdom (c.* 1833), his rendering of the vision of Isaiah 11 and 65. It is noteworthy that Hicks has included in the scene William Penn's signing of a treaty with the Indians, in 1681, an act that showed Penn to be far ahead of his time. Hicks, for his part, was articulating in this work his vision of the peaceable kingdom of Isaiah as a model for peace among nations.

Deutsche Bibliothek – CIP-Einheitsaufnahme

Boersema, J.J.
The Torah and the stoics on humankind and nature : a contribution to the debate on sustainability and quality. / by Jan J. Boersema
– Leiden ; Boston ; Köln : Brill, 2001
ISBN 90-04-11886-1

Library of Congress Cataloging-in Publication data

Library of Congress Cataloging-in Publication data is also available

ISBN 90 04 11886 1

PRINTED IN THE NETHERLANDS

CONTENTS

PREFACE

This book is an adaptation of my doctoral thesis, originally published in Dutch in 1997. In terms of substance, the changes go no further than an update in the light of recent literature and the comments made by an anonymous reviewer. The English translation is by Nigel Harle, who is not only skilled in his art but has also joined me in elaborating some of the nuances of the text, both linguistically and in terms of substance. The book has improved enormously as a result, I feel, and it was indeed most stimulating to work with him. The translation was funded by the Dutch Ministry of Housing, Spatial Planning and the Environment (VROM), at the instigation of former minister Margaretha de Boer, and in thanking her I thank the ministry for its generous support.

I would also like to express my gratitude to a number of friends and colleagues who have contributed in numerous ways in the various phases of this study:

To the late Adam van der Woude for passing on to me some of his vast knowledge of the Old Testament, Ton Schoot Uiterkamp for his ability to create pleasant professional environments, Karel van der Toorn for stimulating me to enhance the comparative religious perspective, David Runia for his useful comments on chapter 4, Helias Udo de Haes for hosting me at his institute and fruitful discussions on chapter 1, Dineke Houtman for teaching me Hebrew and Inge Sievers for her enduring support.

Finally, I would like to dedicate this book to my *ʿēzer kᵉnegdô*, Toos Bremmer, and to my late parents.

Leiden, June 2001

CHAPTER ONE

THE ENVIRONMENTAL PROBLEMATIQUE IN WESTERN CULTURE

1.1 Introduction

Is the environmental problematique of recent date and characteristic of our modern western culture? Those aware that an air pollution act was introduced in England as long ago as 1273 will answer the first question in the negative, and those familiar with the salinization problems in ancient Mesopotamia and with the history of Easter Island the second likewise.[1] In view of the steady escalation of environmental problems since the industrial revolution and their burgeoning scale and intensity in the latter half of this century, though, both questions may perhaps be more appropriately answered in the affirmative. This is also borne out by the rapid profusion of environmental terminology in recent decades and by the far greater public concern about environmental quality as we move into the twenty-first century.[2] Compared with the past, there also seems to be a difference in how the problem is now appraised. Many people are coming to realize that we are faced with a very complex phenomenon that is closely intertwined with our culture. Behind everyday appearances, underlying mechanisms and causes are beginning to emerge. The environmental problematique has deep historical

[1] For the 1273 act see P. Brimblecombe 1975; for Mesopotamia, see D. Hillel 1994 and for Easter Island C. Ponting 1991, P. Bahn & J. Flenley 1992.

[2] A comparison of the number of environment-related terms in editions of the standard dictionaries from the fifties to the nineties shows an enormous, almost exponential increase. This has even spawned complete environmental encyclopaedic reference works (*Encyclopedie van het milieu*, Oosthoek, Utrecht 1984; A. Porteous *Dictionary of Environmental Science and Technology*, Wiley, Chichester 2000; *Umwelt Lexicon*, Springer Verlag, Berlin 1994, to name just a few). For a description of social developments from the political perspective and the role of the west vis-à-vis other parts of the world, see *Agenda 21*, the final document of the major UN Conference on Environment and Development held in Rio de Janeiro in June 1992. In the institutional and political sphere, we see these developments reflected in the creation of separate environmental departments in government apparatus. For a description of the growth and development of the environmental movement, see, for example, J. Cramer 1989.

roots, and this is even truer of the underlying causes.[3] A comparison with other cultures shows that similar problems may also arise under different circumstances, and have indeed done so in the past. The question then arises whether they are not intrinsic to every culture. This makes it no easier to analyse our condition or develop an effective cure, though. To this end, some form of anamnesis and subsequent analysis of the condition's history seems useful, if not in fact indispensable. But how far back must one go to lay bare the roots of the problematique? To the 19th century, or ... the neolithic revolution? In answering these questions, two issues are of relevance.

In the first place, the problematique must be stated in more explicit terms. In environmental science, environmental problems are regarded as social issues.[4] The implications of this position for subsequent elaboration of the environmental problematique forms the principal theme of the present chapter. As a discipline, environmental science endeavours to analyse environmental problems, to explain them and to point to possible solutions. Science can only make a meaningful contribution if the social issue at hand is also 'translated' into a scientifically stimulating and researchable topic. Although the values and standards of the translator play a key role in this proces of translation this may not always be explicit.[5] This act of translation directs the research, the choice of method and the selection of relevant facts.[6] Any quest to identify the roots of the environmental problematique must therefore commence by defining the 'problem' more explicitly, in both social and scientific terms. This as not as self-evident as it sounds. Historical analyses invariably launch a treatise on 'the contemporary environmental crisis' or 'our ecological

---

[3] There is today a very extensive literature on environmental history including new Journals like *Environmental History* (since1995). See, for example, W.L. Thomas 1956, J. Hughes 1975, 1994; D. Pepper 1984, 1996; H. van Zon 1986, 1994; P. Brimblecombe 1987; A.W. Crosby 1986/1993; D. Worster 1988, 1993, 1994; B.L. Turner II et al. (eds.) 1991; C. Ponting 1990, 1991/2; I.G. Simmons 1990, 1993; R.H. Grove 1990, 1992, 1995; R. Sallares 1991; J.R. McNeill 1992; W. Beinart & P. Coates 1995; D. Arnold 1996; A.M. Mannion 1997. For the Netherlands from about 1850, see J.L. van der Zanden & S.W. Verstegen 1993.

[4] *Cf.* the description of environmental problems in J.J. Boersema *et al.* (eds.) 1994, 25. See also M. Schwartz & M. Thompson 1990.

[5] For a recent treatment of this topic in environmental science, see P. Leroy 1995.

[6] For a recent discussion of the concepts 'social' and 'scientific' in relation to the environmental problem, see M. Hajer 1997 and J. van Tatenhove et al (eds.) 2001.

crisis', with no need further explanation apparently felt.[7] The term 'crisis' is particularly popular.[8] Further on in such publications it generally transpires that the author does implicitly hold certain views regarding the nature of that crisis.[9] In the following sections we shall therefore briefly outline the problem definition that forms the starting point of our own quest.

The second point is that the historical depth of study required is determined by the aim of the anamnesis and analysis. If the goal of such an exercise is better treatment of the complaint, one need go no further back than is necessary for that purpose. In this sense, surgeons are in fact generally far less 'incisive' than many psychiatrists. However, if the aim is to elaborate not only a solution but also a causal explanation and scientific analysis, the chain of cause and effect can, in principle, be taken back far further. In the case of the environmental problem, there is generally a dual objective: to identify both the solution and the cause. For both, the help of science is called in. Science plays a different role in each case, though, because the efforts required to solve the environmental problem involve other factors beyond science (including funding, political will, etc.). Scientists can investigate the role of these other factors and incorporate them in their analysis, but they have no say in how these are applied in the service of a solution. Some causes may also be irrelevant to the solution, moreover, because they can no longer be influenced, for example. When it comes to analysing causes, science has a virtual monopoly. Against this background, the solution does not therefore follow automatically, as a 'natural extension' of the causal framework. The assumption that we must first identify the roots of a social issue such as the contemporary environmental problem before it can be resolved is therefore more often valid than its converse: that once we have identified a problem's causes we will be in a position to solve it. The linkage between cause and effect is less direct than is frequently assumed and is probably not as strong

[7] Particularly in the early years of the environmental debate. See, for example, L. White 1967, L.W. Moncrief 1970. For an example of more recent usage of the term crisis see J.K. Sheldon 1992

[8] The word 'crisis' dates back to the writings of the Greek physician Hippocrates, who used it to denote a turning point in the course of an ailment, the juncture that decides whether the patient recovers or dies (D.J. Furley 1989, 223).

[9] As is the case in the collection of essays published under the editorship of W. Achterberg & W. Zweers 1984.

in both directions, as indeed with certain medical complaints. In this study we shall focus primarily on the root causes of our environmental problem, discussing the relevance of these for possible solutions only at the outset and conclusion of the volume.

1.2 Structure of the Study

The study is structured as follows:

In Chapter 1 the environmental problematique is first described as a modern social issue (§ 1.3). The social problem is then further analysed and worked up into a number of questions amenable to scientific inquiry. Many environmental problems prove to be bound up with basic human notions about our dealings with the natural world (§ 1.4 and 1.5). These attitudes have ancient roots and it is argued that, in the western cultural context, our quest must therefore start in the cultures of antiquity, and in particular those of ancient Israel and Greece (§ 1.6). The problem definition, the framework within which these problems are examined, the argumentation of the choices made and the working method adopted are treated in § 1.7 and 1.8. The chapter concludes with an informative appendix intended for readers less familiar with the Bible.

In the second chapter the two creation narratives of the Hebrew Bible (HB) are analysed. The first narrative is described as 'order out of chaos' (§ 2.2). What is the significance in this creation story of God instructing humankind to have dominion over nature? (§ 2.3) The narrative culminates in the seventh day, a day worthy of particular consideration (§ 2.4). In the second narrative, the Garden of Eden is presented as an ideal in which human beings live in harmony with nature (§ 2.5). The events in the garden lead to that harmony being disturbed and to our expulsion from Eden (§ 2.6).

Chapter 3 considers the dietary laws laid down in the HB in Deuteronomy 14 and Leviticus 11. The importance of these laws and the framework within which they are analysed are described in § 3.2, 3.3 and 3.4. Actual textual analysis follows in § 3.5. The chapter concludes by positing an answer to the question of the rationale of the dietary laws. In doing so, a connection with the creation narratives is established (§ 3.6).

Chapter 4 discusses some of the ideas of the classical Greek philosophers. First the concepts of nature and cosmos are clarified, as are the main characteristics of the Greek style of philosophy (§ 4.2

and 4.3). The major philosophical schools are treated in § 4.4, with several 'dissident' thinkers being discussed in § 4.5. The principal ideas of the Greek philosophers are summarized briefly in § 4.6. The chapter concludes by presenting three intriguing motifs encountered in considering humanity's relationship with nature (§ 4.7).

The final chapter presents a general discussion. The conceptions of the Hebrew Bible are compared with Greek philosophical thought (§ 5.2). What interpretations and perspectives are to be found in the early Christian church? What ideas were passed on, and who contributed in this respect? (§ 5.3). What can this information tell us about human attitudes towards the natural world, as they have evolved in western culture, about our notions of sustainability and quality. And what is the relevance of all this today? (§ 5.4)

1.3 The Environmental Problematique as a Social Issue

At the international level, the environmental issue is deemed of major importance. Indeed, it has now been on the public and political agenda for over twenty-five years.[10] This is certainly the case in Dutch society, as numerous surveys have shown.[11] At the same time, surveys do not always provide a clear idea of what exactly those interviewed take to be 'the problem'. People are generally asked what they consider to be major social issues, or they are asked to prioritize a number of issues specified in no further detail ('criminality', 'unemployment', 'the environment', etc.). In either case the result is a hierarchical listing of the issues as perceived by society at that particular instant in time.[12] Asked to characterize the environmental issue in greater detail, it makes a world of difference whether people are questioned about matters confronting them in their own everyday environment or about environmental problems 'in gener-

[10] The United Nations conference on the human environment held in Stockholm in 1972 is generally seen as the first political landmark. In the same year the Club of Rome published its first report *The Limits to Growth* (D.H. Meadows *et al.* 1972), which can be taken to mark the start of the international scientific and public debate on the environmental problem.

[11] 'The environment' has been among the top 5 issues of public concern since 1970 SCP, 1994 and J.W. van Deth & J.C.P.M. Vis 1994, 19,25; Cf. R.E. Dunlap 1991.

[12] If government is seen as having a potential role to play in tackling the *social issue*, we may speak of a *political issue* (J.W. van Deth & J.C.P.M. Vis 1994, 21).

al'.[13] Such a discrepancy is to be expected, because not every general environmental problem is experienced as a problem in one's own living environment. Being aware of problems is not the same as experiencing them.

There may also be differences between the 'general problems' cited by the public and the issues discussed in environmental studies and reference works or in government documents.[14] This can be explained in a variety of ways. A given environmental problem may be quite familiar to a broad section of the public, but still only rarely be cited spontaneously in surveys.[15] It may also be the case that a problem is deemed of vital importance as well as topical interest by policy-makers and public alike, while having little appeal as a 'scientific problem', i.e. worthy of research. Yet other problems may not have permeated the public consciousness, even though scientists and policy-makers may already have been wrestling with them for some time. Due caution should be exercised here, however, as there may simply be a lack of public familiarity with the specific terminology employed.[16] Finally, there are also problems that are recognized by some sections of the scientific community but are yet to be acknowledged by other scientists, policy-makers and the general public.[17] Whether such recognition indeed follows depends partly

[13] To assess the effectiveness of public communications efforts, the Dutch environment ministry reviewed a large number of surveys and found that air pollution, water pollution and exhaust fumes score highest when citizens are asked to name the environmental problems that first spring to mind (G.C. Bartels 1994, 10). The dissertation of R. van Poll (1997, 106,107,203) indicates that noise nuisance (and particularly neighbours' noise) is a dominant factor determining the quality of the physical human urban environment, while at the neighbourhood level burglary, vandalism and 'tidiness and decency' are major determinants of quality. Cf. SCP 2000.

[14] National environmental policy themes such as groundwater depletion are mentioned only rarely, and the same holds for global themes such as loss of agricultural topsoil due to erosion.

[15] A case in point is ozone layer depletion; probed further on the subject, most people not only demonstrate an awareness of the problem but also associate it with (substances in) spray cans.

[16] An example is the world wide loss of biological diversity (now usually termed biodiversity). This problem featured prominently on the agenda of the Rio de Janeiro conference on environment and development (*Agenda 21*, 1992). At that time it was unfamiliar to broad sections of the public. Many specialists even considered the very term biodiversity to be unsuitable for communication with the general public (RMB 1996a, 7,18). This proved to be a misjudgement, for today it is common parlance and widely used in the media.

[17] Although the human and environmental hazards of persistent pesticides like

on the 'hardness' of the data brought forward as evidence, and partly on how the problem is picked up by policy-makers and society at large.[18] If such recognition is not forthcoming, then by definition the issue at hand does not constitute an environmental problem, for public perception of the problematical nature of the issue is a *sine qua non* in this respect. The question, though, is this: when is failure of such concern to materialize to be deemed definite and defining? With many environmental problems, in retrospect we frequently see a shifting of public concern with time. The social scientist Downs introduced the compelling idea of an "issue attention cycle", later reflected in the Dutch environment minister Winsemius' "policy cycle" approach in associated with the same phenomenon.[19] He identified a general pattern whereby every environmental problem goes through successive policy phases, each associated with a different degree of attention on the part of the various actors involved. The approach taken by Downs and Winsemius is a good reflection of how a social issue such as the environmental problematique always has both a factual and a perceptual side. These two aspects stand in complex relation to one another. Problems may well 'drop out of the picture', as it were, while still remaining just as topical as real-world phenomena. Facts can also be interpreted differently.[20]

DDT had been familiar to specialists for some time, it was due to the efforts of Rachel Carson that the gravity, scale and coherence of the problem were brought so convincingly to the attention of the public at large (in the form a series of articles in *The New Yorker*, later on in her pathbreaking work *Silent Spring* 1962. From this time on we can therefore speak of a (major) environmental problem. Similarly, the potential threat to the stratospheric ozone layer posed by halogenated hydrocarbons had been a topic of scientific interest for some years before NASA's measurements of the 'hole in the ozone layer', in 1983, put the topic firmly onto the social and political agenda.

[18] A recent example is 'hormonal pollution' by 'endocrine disrupters', described by T. Colborn *et al.* 1996.

[19] Downs (1972, 39), who holds such that such an issue attention cycle also applies to the environmental problematique (however defined), distinguishes five phases: essentially rise, shine and fall. Peters & Hogwood (1985) have examined the validity of the 'cycle hypothesis' by comparing both political and public attention vis-à-vis a large number of problem areas over ten-year periods. Over this length of time for the majority of problems they indeed found a rise-shine-fall pattern. Winsemius (1986) differs from Downs in that he assumes that an issue, once perceived and acknowledged, will also indeed be tackled and thus ultimately move on to the 'maintenance' phase. Neither does Downs make any explicit distinction between the social actors involved.

[20] I consider it essential as well as useful to continue to distinguish between facts and perceptions, even though they cannot always be clearly separated.

Certain issues may draw massive public attention even though 'objectively speaking' there is little reason for such a sharp rise in interest.[21]

Considered over a longer time span, too, there may be major changes in both public and political awareness of the issues at hand. In just a few decades the smokestack underwent a metamorphosis from a symbol of progress and reconstruction to one of environmental pollution, an emission source to be controlled. Václav Havel goes further, and holds the smokestack that "fouls the heavens" to be "a symbol of a civilization that renounces the absolute, denies the natural world and despises her imperatives".[22]

Finally, we also see national and cultural differences in how environmental problems are perceived and described. Such differences may be due to the factualities of natural circumstance, but cultural perception is also generally involved. The sentiment voiced by Havel is not as likely to be heard in contemporary China, say, and even less likely to be voiced by that country's political leaders. This cannot be explained from the facts, for in large regions of contemporary China the density of smokestacks is considerably greater than in the Czech Republic and the heavens just as heavily fouled.

The fact that we are willing to take the social nature of the environmental problematique seriously apparently makes it more difficult to reach agreement, certainly internationally, on defining what exactly we mean by 'environmental problems'. At the same time, the problems confronting the average citizen often deviate from the wider, more general issues addressed in government reports and scientific journals. In addition, the various actors within (a) society may be 'out of phase' in their attention cycle, and to compound matters still further there are temporal and cultural differences in problem perception. Finally, social consensus may evaporate all too rapidly. There is every reason, then, to include the cultural-historical background of environmental problems and their perception within the domain of environmental research.[23]

---

[21] The 'oil crisis' of 1973 is a case in point. At the time there was little if any physical scarcity, and the economic scarcity (resulting from the price rise) was very modest, certainly when compared with later rises (at the end of the 1970s). Public and political perception of the problem were quite different, though.

[22] Václav Havel 1989 (Dutch translation: 1990, 89).

[23] This is done successfully by Jan Eberg 1997 in his study on waste management and by Richard van der Wurff 1997 in his study on climate change politics.

1.4 The Role of Environmental Research

The considerations discussed in the foregoing section lead me to two basic positions. Both relate mainly to academic environmental research and thus also to the research reported on in the present volume.

First, it is my view that environmental scientists should not align themselves too much with short-term perceptions of problems in society, nor with the problem definitions currently in sway in policy circles.[24] The environmental scientist should seek to lay bare the underlying, more fundamental problems and unravel the relationships between them.[25] This quest may (temporarily) lead the researcher away from the current, often whimsical, public debate, but it is essential: not only because it is the only avenue by which to arrive at more readily practicable answers to familiar problems, but above all because it is an innovative approach that holds out greater perspective for identifying as yet unknown problems and solutions. I am no advocate of social problems being defined or reformulated by science. It is not for scientists to take on this task on behalf of society, nor is that indeed possible. The dangers of such 'constructivism' have recently been highlighted, and rightly so.[26] What scientists can do, however, is endeavour to present a clear picture of the exact situation, what positions have been and might be adopted, and where, scientifically speaking, the important and researchable problems lie. This may obviously include pronouncements on the nature, scale and relative gravity of the problems concerned.

This position thus advocates a more narrowly 'scientific' role for environmental science in tackling today's environmental problems.

On the other hand, environmental scientists should perhaps be less pretentious about their ability to resolve social issues. Two arguments can be given for greater modesty in this respect. First, in many fields there is too little scientific knowledge available for us to expect reliable recommendations for solutions from this quarter. Indeed, in some fields this will probably remain the case.[27] Second,

---

[24] The sociologist C. Wright Mills (1972, 248) gave similar advice to sociologists: "Do not allow social issues as officially formulated or difficulties encountered in your private life to determine the problems you study".

[25] *Cf.* H.A. Udo de Haes 1997, 62.

[26] M. Thompson 1997.

[27] This holds not only for many aspects of human behaviour, but also for our

although scientists may well succeed in carefully dissecting the problem at hand as well as all the associated dilemmas, they are rarely able to resolve those dilemmas on purely scientific grounds. Conscientious, or less 'engaged', scientists return the ball of these dilemmas firmly to the court of society, or to the party commissioning the research. Less scrupulous, or more 'engaged', scientists—when challenged or otherwise—may champion their own personal choices as scientific solutions. Although the first approach may leave the researcher dissatisfied, the latter also has its drawbacks. It involves a twofold risk: the solutions pursued may prove 'erroneous', for the scientist's expertise is not generally in the realm where the ultimate decision is to be taken, and democratic transparency may be lost, with those responsible for decisions hiding behind science or scientists. Scientists should clearly indicate when they are speaking as professionals and when they are participating in a debate as engaged citizens.

This position thus advocates a more 'social' debate on how to resolve the environmental problem and a more transparent role for science and scientists in that debate.[28]

1.5 The Environmental Problematique Explored

In the previous section I have argued that the task of environmental scientists should be to lay bare fundamental problems and chart their interrelationships between such problems. Ever since the environmental issue moved onto the public agenda, many attempts have been made to do just this. The results of that work have, in turn, helped keep environmental issues firmly on the agenda. The diversity of theories presented to explain the environmental problematique has been enormous.[29] By the late 1980s a fair degree of consensus appeared to have been reached at a general, fundamental level. Largely as a result of the Brundtland Commission report *Our Common Future*, published in 1987, the concept of sustainability began to serve as paradigmatic marker of the problematical nature of our relationship with the natural environment.[30] In its negative formu-

knowledge of complex systems such as the earth's climate. On this point, see also H. Tennekes *et al.* 1990; M. Huxham & D. Summer 2000.

28 Cf Jocelyn Kaiser 2000.

29 See J.J. Boersema 1985.

30 WCED (World Commission on Environment and Development) 1987. With

lation (unsustainable) the concept posits the problem, in the positive (sustainable development) the direction of a solution. Many governments subsequently adopted the ideas of the Brundtland Report as a cornerstone of national policy.[31] Universities subsequently also incorporated the notion in their mission statements.[32] The concept was successful in mobilizing numerous scientists to explore and operationalize it further.[33] Has this brought us any further, then? In the remainder of this section, I shall outline some of the main elements of the recent debate in this area, with particular reference to the Netherlands.[34]

Besides a near-endless stream of studies elaborating the concept of sustainability, there has also been criticism of the notion as well as growing irritation at its (over-)frequent use. Within the former group of studies, the main focus was initially on how sustainability should be measured. At this stage a key question concerned the indicators to be used for that purpose.[35] In recent years the focus has shifted towards the relationship between economic growth and the associated pressure on nature and the environment. Key notions in this context are: delinkage, eco-efficiency and quality. 'Delinkage' refers to the notion of achieving positive or zero economic growth while reducing the burden on nature and the environment, thus breaking the historical linkage between these two quantities.[36]

---

this concept, the Brundtland Commission aligned itself with ideas that had been current in scientific circles for some time already.

[31] In the Netherlands, most explicitly in the National Environmental Policy Plan (NEPP) of 1989.

[32] In the European framework, may universities adopted the Copernicus programme *The university charter for sustainable development* in 1993.

[33] Witness, for example, the priority programme set up by the Netherlands Organization for Scientific Research, NWO, on 'Sustainability and environmental quality' in 1990/1991. The availability of additional research funds was certainly instrumental, but this does not adequately explain the positive heuristics of the concept.

[34] In comparison with other western countries, in the Netherlands there was enormous scientific and political response to the Brundtland report. The same was true, it may be added, of the first report to the Club of Rome. Nowhere in the world were so many copies of the book sold as in the Netherlands. The Dutch environmental debate is certainly relevant in the international context and has in some respects also been trend-setting.

[35] Important suggestions in this area have been made by: H.E. Daly & J.B. Cobb Jr. 1989/1990; O. Kuik & H. Verbruggen (eds.) 1991; D. Pearce *et al.* 1993; R. Costanza & B.C. Patten 1995; RMNO 1994; B. Moldan et al. (eds.) 1997; H. Bossel 1998, 1999; P. Ekins & S. Simon 2001.

[36] Gross National Product (GNP) or Gross Domestic Product (GDP) are the

Whether such an aim is desirable or feasible, or both, is still the subject of debate.[37] At the least, it would appear to be a necessary precondition of sustainable development that there be a major reduction in the quantity of energy and material resources consumed per unit good or service, or in other words an improvement of 'eco-efficiency'. Some of our needs will have to be serviced in a completely different fashion. It will be necessary to analyse and indeed achieve sustainability at the level of society's basic needs. A change in systems of taxation is probably also required, so that prices 'speak the truth', in the phrase of E. von Weiszäcker.[38]

A crucial point in the concept of delinkage is the question of how to establish whether, on balance, we are successfully reducing our burden on nature and the environment—for we have no uniform, reliable unit with which to measure that pressure. As a result, there is an inborn tendency to focus on parameters on which there is universal agreement as to measurement method.[39] What cannot be counted may well also remain unaccounted for. If certain parame-

---

traditional measures of economic wealth. Whether delinkage is feasible while these yardsticks are still employed is very much an open question. For a recent treatment of this issue, see E.U. von Weizsäcker, A.B. Lovins & L.H. Lovins 1995; J.B. Opschoor 1996, 1997; H. Daly 1996; Th. Aronsson, P.-O. Johansson & K.-G. Löfgren 1997; S.M. de Bruyn 1997, 1999; Folmer & R. Komen 1997; P. Hawken, A.B. Lovins & L.H. Lovins 1999; J.J. Boersema & J. Bertels 2000.

[37] For a review of the arguments, see the contributions of W.A. Hafkamp, R.U. Ayres, A.B. Lovins and L.A. Geelhoed in: RMB 1996b. On the basis of empirical data, some hold that the general relationship between per capita income and environmental burden can be plotted as an 'inverted U' curve, the so-called Environmental Kuznets Curve (EKC). This has led some to adopt the general position that economic growth is good for the environment (W. Beckerman 1992, 481, 1995). Although the validity of this general position has since been disproved (K. Arrow *et al.* 1995, J.B. Opschoor 1995, 1997, R.J. Heintz & H. Verbruggen 1997 and S.M. de Bruyn 1997), it should be noted that today's growth-oriented economies show little sign of any real delinkage (S.M. de Bruyn 1999).

[38] By which is meant that the true costs of goods and services should be reflected in their price. Prices are currently distorted by subsidies, exemption from VAT, use of unpriced environmental goods and displacement of costs (see also H.E. Daly & K.N. Townsend 1993). In the Netherlands, debate on this issue started very early; see, *inter alia* B. Goudzwaard 1970; R. Hueting 1974; P. Nijkamp 1979.

[39] Using the so-called Distance-to-Target (DTT) method, for example, the Dutch environment ministry at one stage wished to limit measurement of the environmental burden to just six readily monitorable parameters. This restricts the scope of application of this method to issues for which only these six are important, however. In situations involving other environmental impacts which are not factored in, the method loses its validity and may lead to impacts simply being shifted to other areas. See RMB 1994.

ters are not factored into assessments, there is a risk of problems merely being shifted. A particular policy may, for instance, involve a shift from (less) CFCs to (more) CO_2, from (less) CO_2 to (more) space requirements, and from benefits in the present to costs in the future. Any tendency to restrict oneself to a handful of parameters that can be reliably measured and aggregated is therefore a risky undertaking. This risk is inherent to all the methodologically consistent techniques in use today for quantifying multiple forms of environmental pressure (DTT, ecological footprint).[40] This also holds for the most applied technique, Life Cycle Assessment (LCA), because it has still not proved feasible to incorporate biodiversity, nor to factor in the time dimension. Today's 'grail' is therefore to combine a reliable methodology with adequate coverage, at the same time satisfactorily resolving the problem of problems being shifted into the future. There are doubts as to whether such a feat is indeed feasible. It is extremely difficult to discount environmental problems over time.[41] As far as the parameters themselves are concerned, the main problem is with the key parameter *space*, with *biodiversity* following a close second. In the cluster of *energy, raw materials and emissions* there has been significant progress. In this cluster, CO_2 emissions appear to be the best overall indicator. Integration of the environmental pressure associated with all three clusters (space, biodiversity, energy/resources/emissions) in a single index is still a long way off. The question is thus not so much whether environmental pressure should be expressed in a single index as how all the relevant factors can be properly factored in. Perhaps we should work towards composite indices for each cluster. With composite indices,

---

[40] For the 'ecological footprint', see M. Wackernagel & W. Rees 1996, 9: "Ecological footprint analysis is an accounting tool that enables us to estimate the resource consumption and waste assimilation requirements of a defined human population or economy in terms of a corresponding productive land area". They calculate that the Dutch as a whole utilize 15 times as much land as that available within their national borders. The method provides a powerful image of the unsustainability of current modes of consumption. Scientifically, though, there are a number of objections to using surface area as a unit of measurement (as the authors themselves concede: p.17).

[41] On this point, see L. Hendrickx, A. van den Berg & Ch. Vlek 1993 and J.B. Opschoor 1994. In an unpublished study on natural resources and indicators, W. Biesiot (1995) makes a distinction between 'respite time' and 'response time'. The first is the time available, given stocks and technological state of the art, to switch to renewable resources, while the second is the time it takes a society to make such a switch. For a reference to Biesiot's ideas see H. Bossel 1999, 1.

trade-offs can be made more explicit, problem-shifting brought to light and handles created for positive, compensatory measures.[42]

In the debate on sustainable development, the notion of *quality* refers to values felt to be under threat given our current treatment of nature and the environment. Quality is a relative and relational concept, but it is rooted in objective quantities. In humanity's quest to develop a sustainable relationship with the natural environment quality is indeed an extremely relevant concept.[43] We shall be returning to this topic in the following section.

Critics of the concept of sustainability are to be found both at home and abroad.[44] Two of my countrymen are particularly outspoken in their criticism. The political philosopher De Geus was recently moved to state: "Sustainability is the most contaminated term of the nineties". In his eyes, most interpretations of the concept are flawed and misleading.[45] The social philosopher Korthals is also scathing in his criticism. He sees a tension between democratic values and certain notions regarding the role of government in achieving sustainability. According to Korthals, the environmental problematique is entirely novel and none of the old responses are adequate. He advocates processes of open communication and democratic deliberation, not only as a means but also as an end in themselves. Only

---

[42] For a Dutch review of the proposed indices and a discussion of their background, see the following articles in the journal *Milieu*: I. van Dijk *et al.* 1992; P. van Egmond *et al.* 1992; G. Huppes & R. Heijungs 1993; L.J.R. Scholten 1993; H.C. van Latesteijn 1994.

[43] See J.B. Opschoor & S.W.F. van der Ploeg 1990, J.J. Boersema 1993 and *Choosing a Sustainable Future* 1993, 2.

[44] In the international arena there have been a number of publications questioning not only the notion of sustainability but also the gravity of the environmental problem as such, at the same time holding the data presented by environmental scientists to be biased and exaggerated. See, for example, A. Bramwell 1994; D.L. Ray 1994; R. Bailey (ed.) 1995; A. Wildavsky 1995 and M. Zimmerman 1997. For a re-evaluation of the risk of toxic substances: W.H. Baarschers 1996.

[45] De Geus was interviewed for Leiden University's newspaper *MARE* following publication of his book on 'ecological utopias' (Utrecht, 1996). With reference to attempts at interpretation by others, and particularly by the government, he says: "People talk about sustainability, but nobody knows what it is". Luckily, De Geus himself does know, for further on in the interview we read: "Ultimately, only a steady-state economy can bring the sustainability required from a political, economic and social point of view" (*MARE*, 13 February 1997). This is by no means a new idea. It can be traced back to 19th century economists like J.S. Mill and has been defended in the environmental debate for over twenty years by people like the U.S. economist H. Daly and Dutch environmental specialists such as L. Reijnders.

thus can some form of sustainability be attained.[46] Others, too, have expressed concern about whether the changes deemed necessary from the environmental perspective will in fact be accepted by society. It has also been queried on what grounds these changes are to be defended. In a controversial report, the Netherlands' Scientific Council for Government Policy (WRR) argued that sustainability is an ambiguous concept that cannot be scientifically objectified.[47] According to this government advisory body, sustainability is more the outcome of a public debate concerning the risks society is willing to take. In this debate, the various risks must be conscientiously weighed up, and subsequently traded off against one another. The report posits a two-dimensional model with the environment and society as its two axes. The degree to which a person or society is prepared to accept 'ecological' and 'social' risks is then indicated by different positions within the four quadrants of the graph. Justification for this approach is sought in the theories of Schwarz and Thompson.[48] These authors distinguish four *Myths of Nature* as determining how people adopt a position when forced to balance social interests against the interests of nature and the environment. The conclusion of the WRR is that there are several positions that can be termed 'sustainable'.[49]

---

[46] M. Korthals 1994. Environmental problems are far less novel than Korthals assumes, and the ultimate impact of the proposed 'deliberations on nature' is very much debatable. I briefly consider the publication here because it reflects the 'relativizing' climate that has sprung up around the notion of sustainability in recent years.

[47] WRR 1994. The Council is particularly critical of the notion of 'environmental utilization space' (EUS, coined by H. Siebert in 1982), or 'ecospace' as it is also known, because of the suggestion that the limits of our interactions with nature and the environment can be scientifically and objectively established (33-41).

[48] M. Schwartz & M. Thompson 1990. Both authors build on the ideas of the anthropologist Mary Douglas (1975, 1978, 1982, 1985, 1986) and in particular the cultural theory she developed in which a person's cultural orientation ('group' versus 'grid') is tied in with ideas from risk theory (M. Douglas & A. Wildavsky 1982, U. Beck 1992). Such orientations are purely theoretical constructs, it should be added, and it is to be queried whether they are encountered as such in practice (see Å. Boholm 1996).

[49] With the exception, perhaps, of continuation of the present situation. Surprisingly, the Council does assume "that sustainability implies that current environmental risks are considered unacceptably high and that there is a willingness to effectuate social change" (WRR 1994, 45). What limits we are currently exceeding and who is to establish the unacceptability thereof, on what grounds, and why the Council adopts this position, as a *scientific* council, are not specified.

To understand these positions, several factors should be taken into consideration. In the minds of Council members, the idea has become entrenched that when the country's first National Environmental Policy Plan (NEPP) was drawn up in 1989, its objectives were developed solely on the basis of ecological criteria. "There appears to have been absolutely no weighing ... against social objectives," the Council states (33). And elsewhere: "These objectives were based exclusively on ecological desirability" (154). In the NEPP and ensuing discussions, use was made of the concept of 'environmental utilization space' (EUS), to give tangible expression to the maximum admissible damage to the environment. Explicit primacy was thus given to the environment: "Society must conform to the utilization capacity defined by the environment" (33). According to the Council, 'EUS' cannot be established in an objective and unequivocal manner; the notion suggests (erroneously) that it is in principle possible to gain 'ultimate' knowledge in this respect, that is, sufficient knowledge to determine the limits and constraints of human behaviour (36). In the late 1980s and early 1990s it was this 'NEPP/EUS perspective' that predominated in the Dutch environmental debate. The authors of the Council report were irritated about the role that science and (certain) scientists had come to play in the debate, and particularly about use of the term Environmental Utilization Space (EUS).[50] The principal aim of the WRR study, then, was to launch a counteroffensive, as it were, to put the debate on sustainability 'back on the track'. In pursuing this worthy aim, however, the Council has to my mind somewhat overshot the mark.

In the first place, the Council provides no solid evidence that ecological considerations were not weighed against social considerations when the NEPP(-related) objectives were formulated. It has been argued convincingly by a variety of authors that there was a bal-

[50] To cite a magazine interview with two of the report's authors ("Stripping sustainability of its sanctity", *Misset's milieu magazine*, July/August 1994, 17-18), for example: "Science got too big for its boots, moreover" (R. Rabbinge); "By adopting the concept of environmental utilization space, what you're in fact doing is letting science dictate the 'space' for politics and policy" (I.J. Schoonenboom). Personally, I can imagine the irritation that has arisen about the term environmental utilization space, but on different grounds. To my mind, the term puts too much emphasis on human usage; I associate it with an image of the physical environment as a 'larder' for humankind. If the concept is to be seriously challenged, however, one must proceed from the intentions and descriptions of the authors who have defined and used it in the literature.

anced consideration of interests.[51] That doubts about the feasibility of the aims of the NEPP were later to arise, in the years following its publication, detracts in no way from this basic truth. This misunderstanding regarding the alleged 'primacy' of ecology can probably be traced back to earlier, more categorical statements on the issue. In the early 1990s, it may be added, such statements were not usually attributable directly to the scientific community. The national Social Economic Council (SER), for example, had stated explicitly in its recommendations concerning the Brundtland Report that the requirement of ecological compatibility should prevail over considerations of economic growth and full employment, among other issues. In a special report on the theme of environment and economy, the SER refers to the environmental objective as being of a "higher order".[52] In adopting such a stance, the SER was succinctly expressing a vision that had gradually permeated society since the Club of Rome's first report: the natural environment sets limits on human activities and by living as we do we have already partly exceeded those limits and threaten to do so further each day. By the end of the 1970s, then, the idea that "human activity should be compatible with ecological constraints" had already been adopted as an official Dutch policy position.[53] Given this factual background, nobody can hold that such a standpoint does not adequately embody social considerations or 'the public interest', for at this general level that interest is, by definition, not contradictory to the ecological interest.

My second comment relates to the actual substance of the Council's criticism of the EUS concept. This criticism is only partly valid.[54] It is furthermore unclear to whom this criticism is addressed.

---

[51] For example, by W. Alblas & H. Sips 1995.

[52] SER 1989. "In line with the SER's stated position, in this report the environmental objective is considered to be an objective of a higher order" (SER 1991). The Social-Economic Council, representing trades unions and employees, advises the government on matters of socio-economic policy. It is a key institutional player in the corporatist politics of the Dutch 'polder model'.

[53] "Ecological compatibility of human activity became one of the main aims of environmental policy" (E. Tellegen & D.K. Tommel 1989, 67). Tellingen & Tommel also cite then-prime minister J. Den Uyl, who in 1975 described the policy field of 'health and pollution control' as: "The protection of the quality of the physical environment for man, flora and fauna, against the backdrop of the compatibility of human activity with a system of equilibrium between man and the natural environment".

[54] The criticism is valid, to the extent that it challenges the position that EUS

The report lacks reference to any literature in which the disputed interpretation is adopted. The Council substantiates its own (?) objections with reference to studies by scientists who have introduced the concept or used it approvingly.[55] All in all, the Council thus contests an interpretation and usage of the concept that is not to be encountered among the principal scientists using it.[56] As a consequence, these scientists are wrongly associated with the disputed interpretation. Perhaps the Council felt irritated by what it construed as misuse of the term or was at any rate concerned about such misuse. Whatever the case, though, one would expect of a scientific council of such standing that its 'message' is at least correctly addressed.

My last remark concerns the clarity of the course being ploughed by the Council itself. One the one hand, it holds that the approach adopted does not necessarily lead to any particular position and that the road towards sustainability is not an *a priori* given. On the other hand, the Council qualifies current environmental risks as unacceptably high and, in case studies, arrives at a scathing judgement when

---

is a parameter that can be scientifically and unambiguously established in every area and at every scale level. It is invalid, though, to the extent that it suggests that the scientific basis for setting environmentally related limits is principally incorrect. The history of ocean fisheries shows that the natural environment sets limits on what may be termed 'sustainable' usage of these resources. Such limits can be established by fishery biologists fairly reliably on a scientific basis. If, in this case, we weigh the ecological risks against the social risks as perceived by the fishermen we may reach a useful political compromise, but it is unlikely to be a sustainable solution. Evidently, the fact that a sustainable solution cannot count on (adequate) support from the target group in question does not necessarily mean that solutions with an adequate support base are thus 'sustainable'. This does not alter the fact that a lack of support base may certainly constitute a major obstacle on the road towards sustainability.

[55] As an illustration, the Council's report (WRR 1994, 36, note 20) refers to the 1991 study *In Search of Indicators of Sustainable Development*, edited by O. Kuik & H. Verbruggen, and elsewhere (40, note 24) to an article therein by J.B. Opschoor & L. Reijnders. The latter note is appended to the sentence: "Opschoor and Reijnders thus point out rightly that the problems associated with establishing such indicators of sustainable development lie in both the scientific and the ethical/normative realm".

[56] Opschoor, for example, one of the advocates of EUS, in an article written with Van der Ploeg explains it as a dynamic concept that (already) allows for the kind of arguments cited by the Council in its report, all the way through to the role of the public support base (J.B. Opschoor & S.W.F. Van der Ploeg 1990, 84-86). See also H. Siebert 1982, J.B. Opschoor & G.A.J. Klaassen 1991 and, following publication of the Council's report, J.B. Opschoor 1995.

it comes to several topical environmental problems.[57] In the case of groundwater policy, this leads to considerations that virtually exclude certain positions in the Council's own model if sustainability is the aim (188-191). These considerations are at odds with the position that sustainability is ultimately a subjective concept. The facts and the analysis apparently 'necessarily' lead more to certain positions than to others. To my mind, when the Council embarks on a concrete discussion of the issues involved in the four case studies it undermines its own position.[58]

In the public debate that has followed in the wake of the Council's report, however, it is above all the latter position (the allegedly subjective nature of sustainability) that has remained behind as 'the Council's message'. This debate is still in full swing over the last couple of years in scientific journals like *Ecological Economics* and, very outspokenly in *Economisch Statistische Berichten* (ESB). [59] In these fora Huetings and Reijnders, elaborating on the work of H. Daly, R. Goodland and R. Costanza, have defended the objective nature of the concept of sustainability. Verbruggen and Dellink, for their part, have made a useful distinction between sustainability and sustainable usage, on the one hand, and sustainable development, on the other. It is above all the former concepts that are rooted in objectivity and hard science, while sustainable development leaves far more scope for interpretation, the term denoting avenues by which sustainability can be achieved. Hueting and Reijnders clarify their position by distinguishing between sustainability, which can be objectively established, and preference for such sustainability. Preferences

---

[57] The preferential choices made by the WRR in the four cases examined are based generally on a factual and highly scientific analysis and assessment of the material. Following this analysis and assessment, the WRR presents an estimate of the risk posed to society.

[58] This discrepancy has been noted by other scientists, too. In a journal review of the report, P. Glasbergen (1995, 254) puts it as follows: "The remarkable thing is that when these considerations are confronted with the data of the scenario analyses, contrary to what the main message suggests, an appeal for tougher environmental policy is the outcome".

[59] R. Hueting & L. Reijnders 1998. In the *ESB*-debate, studies by the Central Planning Agency (CPB 1996) and the Institute for Environmental Studies of the Free University of Amsterdam have also played a part. See, *inter alia*, E.S. Pelle *ESB*, 21 February 1996, 153; R. Hueting & L. Reijnders *ESB*, 8 May 1996, 425-427; and the discussion ensuing from the latter article in *ESB* of 12 June 1996, 531-539: M. Hofkes (531-532); H. Verbruggen & R. Dellink (532-533); H. Stolwijk (534-536) and a postscript by Hueting and Reijnders (537-539).

are established subjectively and cannot be derived unambiguously from scientific data.

Several members of the Council have also engaged in this debate, with Rabbinge recently distancing himself from the subjective interpretation of the notion of sustainability.[60]

In summary, my conclusions are the following:

1. Humanity's present dealings with the natural environment are generally characterized as unsustainable.[61]
2. In the quest for a sustainable relationship with our natural environment, it clarifies matters to distinguish between sustainability and sustainable development.
3. Sustainability and sustainable usage are notions that can, in principle, be objectively defined and scientifically established.[62] This does not imply that these concepts are to be viewed statically, nor that we shall be able to determine them quantitatively in every given field.
4. The term sustainable development is confusing if it is taken to mean a 'desired state'. It can be better, and indeed more logically, used to describe how sustainability can be attained. In this context, many but not all roads lead to Rome.[63] The decision to take such a (preferable) road, and the tempo adopted on that road are matters to be decided on socially and politically.
5. The acceptance of sustainability as a "guiding framework" (Verbruggen & Dellink) and the will to work towards sustainable development are normative choices made by society. Whether people make such choices, individually and collectively, and the degree to

[60] Rabbinge referred, *inter alia*, to his 'own' (WRR) case study on energy. He laid the blame for the in his mind erroneous 'subjectivity' reading (complete subjectivity, with no options excluded) with the Central Planning Agency (*ROM-Magazine*, 4 April 1997, 18-21). His response made no mention of the fact that it was the Council's report that had done so much to kindle this (now-rejected) interpretation.

[61] In different wording, the same conclusion is drawn by the Netherlands' Central Planning Agency (CPB 1996, 22): "In other words: the current mode of production and consumption is generally perceived to be unsustainable".

[62] In this context I see no objection to the description given by Hueting & Reijnders (1996, 537): "Sustainability is defined as a state in which the potential uses or the functions of our physical environment, *viz.* water, soil, air, natural resources and plant and animal species (also referred to as environmental capital) remain available *ad infinitum*". Note, however, that this definition does not address the issue of distribution; on this point, see H. Mulder 1995.

[63] W. Biesiot & A.J.M. Schoot Uiterkamp (eds.) 1993.

which they express preferences on this point, are matters governed largely by their attitude towards 'the quality of life'. Such attitudes are intimately bound up with values and standards that have a very long history and are rooted deep in western culture – in our present context more specifically the values and standards concerning the way we treat the natural environment.[64]

It is the last of these conclusions, particularly, that forms the basic point of departure of the present study. Essentially, this volume considers the origins as well as substance of our attitudes towards and relationship with nature as these have evolved in western culture.[65] In the following section I shall elaborate this point of departure in more detail.

1.6 Humanity's Relationship with Nature as a Cultural-Anthropological Issue

1.6.1 *A closer look at sustainability and quality*

Human beings live in an intimate and intensive relationship with their natural environment. This relationship, described so aptly by

---

[64] Compare the CPB study, which comes to an identical conclusion with respect to sustainability (1996, 23).

[65] Until now I have used the terms *environment* and *nature* more or less synonymously. For a study concerned mainly with an earlier era, this is perfectly acceptable. However, a closer examination of the last two centuries in particular shows the emergence of a clear distinction betwee the terms 'nature' and 'environment' which have different connotations and are subject of interest of writers and scientists from different traditions. See on this point, *inter alia*, R.P. Sieferle 1988; D. Pepper 1984/1990, P.C. Mayer-Tasch 1991; H.J. van der Windt 1991, 1995. In ideological terms, a more human-oriented '*animal rights*' current (P. Singer 1975, 1976, 1981, 1991, 1994; T. Regan 1982, 1983; T. Regan & P. Singer 1976) can also be distinguished alongside a more ecologically imbued 'conservation' position (Holmes Rolston III 1975,1988; N. Myers 1983; R.F. Nash 1989). In recent years the interconnectedness of the 'grey' environmental and 'green' conservation strand has been emphasized and advocated (J.J. Boersema 1989). From a policy angle, besides nature and environment, water ('blue') can also be distinguished as a separate strand. In the coming years these three strands must be integrated into a comprehensive single policy on the physical environment.

In this study I employ the term *culture* in a broad sense, almost synonymously with the terms 'society' or 'civilization' (see S.P. Huntington 1998 for a similar stance). In my analysis I have based myself principally on written expressions of the cultures in question and only sporadically on other cultural and behavioural artefacts.

Marx as metabolism, forms the basis of every human society. From time immemorial, humans have made use of their natural surroundings to satisfy their basic needs for food, clothing, shelter, warmth, security and transport. In this respect, human beings are in principle no different from other creatures. In addition, though, humanity makes claims to the natural environment to satisfy what Maslov terms social needs and the need for self-actualization. This process of metabolism has had a severe impact; human beings have in fact radically changed the 'face of the earth'.[66] At an early stage of development human societies were already confronted with the consequences of their actions, in the form of soil exhaustion, salinization, erosion and desertification.[67] The extinction of plant and animal species also has a long history, extending back to the Pleistocene (> 10,000 BC) according to some scholars.[68] Of later date, but still centuries-old, is local pollution of the soil, water and air with toxic substances and substances in toxic quantities. The fact that we can nevertheless speak of a modern problem, despite its ultimately long history, is due in the first place to the enormous scale human environmental impact has assumed in our era and our likewise vastly expanded knowledge of the nature of that impact. But that is not all. Just as important, if not more so, is that this is now also increasingly regarded as a problem, with all the variation in problem formulation that implies. This somewhat remarkable reversal of attitude cannot be explained by the fact that people were previously blind to the negative impacts of human action; for the most evident forms of environmental pollution, protest is almost as old as the pollution itself.[69] This change in thinking is due mainly to the fact that today, far more so than in former times, we have become aware of the interrelatedness of the whole variety of environmental problems, including degradation of nature as such, no longer categoriz-

[66] For excellent reviews see W.L. Thomas 1956; A. Goudie 1981; I.G. Simmons 1990, 1993; J. Reader 1990 and the literature cited in note 3.

[67] C. Ponting 1990 and the literature cited in note 3.

[68] P.S. Martin & H.E. Wright Jr. (eds.) 1967; L. van Valen 1969; P.S. Martin 1973; V.L. Smith 1975; M. Harris 1977/1978.

[69] See H. van Zon 1994; D. Pepper 1996. Here I use the term environmental pollution, because protests in earlier ages were directed towards forms of pollution that constituted a nuisance or a threat to human health. Prior to the 19th century there was far less protest against degradation of the natural environment in all its forms or against the eradication of animals considered to be pests.

ing them as being avoidable side-effects of our social evolution.[70] In many people's perception, environmental problems are inherent to that evolution and it is therefore that evolution as such that is, or should be, called into question. In that debate, as we have seen in the previous section, human interaction with the natural environment is described in terms of the key concepts of sustainability and (environmental) quality. This pair of concepts can be elaborated in three domains: in terms of time, space and the material realm to which they apply.

With respect to the temporal domain, this means that in assessing our present dealings with the environment due allowance must also be made for the interests of future generations. We may only speak of sustainability if we leave our children a planet providing the same opportunities as we ourselves were bequeathed (cf. the definition in note 62). As things stand today, we attach greater importance to the near than to the distant future, certainly in economic terms. Discounting is the determining of the current value of future costs and benefits. We usually accord less importance to things that are further away. It is thus paradoxical that despite our focus on the future we often do away with (the variable in) the future. The positive discount rate employed by economists means that the material interests of generations in centuries to come are, to all intents and purposes, assigned zero value.[71] How this is to be remedied is a difficult question. We know little with any certainty about the preferences of future generations. As mere mortals we cannot predict the future, moreover. Indeed the very notion of future generations appears problematical. Duly allowing for the interests of future generations therefore means internalizing important problems within one generation (no passing on of major problems), avoiding irreversible effects (such as human driven extinction of plant and animal species) and leaving open options for securing key natural resources. In many cases, it does not mean static conservation of the present.[72]

In the second place it means extending the spatial domain of the above two concepts to the global level, so that 'solutions' here are

---

[70] C. Ponting (1990, 1992) goes one step further. He holds irresponsible human behaviour vis-à-vis natural resources to be the principal reason for the demise of earlier civilizations. His most powerful paradigm is Easter Island, where between about 1200 and 1700 AD a civilization vanished through overexploitation of their natural resources. For a more balanced view: J.A. Tainter 1988.

[71] CPB 1996, 29.

[72] For an elaboration on this topic: J.J. Boersema 2001.

not attained at the expense of sustainability and quality elsewhere. In particular, the reference here is to developing countries. The basis issue of space (land) and space requirements are coming to feature ever more prominently in the environmental debate, certainly in the Netherlands.[73] In my opinion this is justified, for space is perhaps the only thing that is truly scarce, together with time.[74] This is certainly the case if due consideration is given to the quality of that physical space.

Lastly, we see the concepts being applied to a widening circle of entities: respect for future generations and others elsewhere on the planet has been extended to include nature and non-human organisms, which are increasingly seen as independent entities with an intrinsic (i.e. non-instrumental) value.[75]

In the current debate 'quality' is the concept that is least clearly defined. In the context of Dutch government policy the term is used mainly to refer to requirements set on products or on the environmental media soil, water and air. Others employ the term environmental quality to place our relationship with the natural environment in a broader perspective: that of human well-being, or 'liveability', and the intrinsic value of the environment or elements thereof.[76] This broader notion of environmental quality thus brings us closer to the more general notion of 'quality of life'. This conception of quality, to my mind preferable, also begs the intriguing and absolutely crucial question of whether achievement or maintenance of such quality can necessarily be combined with the second objective of sustainability. Even in the context of such seemingly immaterial affairs as democracy and human rights, quality has a material basis for it makes a claim on finite resources. Even peacekeeping operations require fossil fuels and other materials. At the same time, we hold that sustainability cannot be achieved by 'passing the buck'

[73] See M. de Boer 1995.

[74] In practice, plant and animal species are also scarce. Although new species do evolve, the time frame within which this occurs is (far) longer than that of human societies (R.M. May 1988) and we know almost for certain that the new species will be different from the ones lost. On the scarcity of space: J. Bertels *et al.* 1995; on that of time M.P.J. Pulles 1990; G. Clark 1994 and J. Barbour 1999.

[75] On future generations: R. Solow 1986, J.J. Boersema 2001; on intrinsic value: W. Zweers 2000.

[76] J.B. Opschoor & S.W.F. van de Ploeg in CLTM 1990, 101-103; NWO 1991; J.J. Boersema 1993.

to future generations, to those on other continents or to nature. One need not be schooled in the matter to appreciate that current economic practice is still largely characterized by such processes of transferral. Not until we have thoroughly (re)appraised our notion of quality can we ever hope to achieve sustainability. What we are in fact looking for is a concept of quality that embodies sustainability. Thus conceived and described, the environmental problematique goes to the very heart of western culture. In many fields, our culture has proved extremely successful in qualitative terms, and it is precisely this success that defines not only the crux of the problem but also its radical nature.

In considering the environmental problematique, then, we shall take as our basic point of departure these fundamental notions of sustainability and quality. This takes us to our central theme of humanity's relationship with nature and the cultural ideology from which it has evolved. Focusing on the influence of ideas and *Ideengeschichte* does not imply any rejection or undervaluation of material factors or of the *Wirkungsgeschichte* of ideas, it should be added. In this context, I am in basic agreement with the historian K. Thomas, who observes that ideas, in a cyclical process of interaction, can result both in and from changing material circumstances.[77] Similar views have also been put forward by the Dutch philosopher Achterhuis.[78] It is not ideas from which the environment suffers, moreover, but from the actions associated with them—or assumed to be associated, for it is still not yet entirely clear how exactly behaviour and ideas are connected, in the environmental realm as elsewhere.

1.6.2 *Cosmologies*

One of the most fundamental mainsprings of human action is the urge to survive. In this respect, human beings are in essence no different from any other organism. Besides merely sustaining their lives, though, humans (and perhaps other organisms) are also concerned in no small part with the nature and quality of their lives.[79] Consid-

[77] K. Thomas 1983, 301.

[78] The fact that physical reality may also give rise to certain (immoral) conduct and views moves Achterhuis to advocate a 'moral technology', by which he means that we should construct a material reality that elicits 'right' conduct and promotes 'right' views (H.J. Achterhuis 1992, 1992a).

[79] For a description of the development of human societies from the perspective of evolutionary biology, see C.D. Darlington 1969; J. Diamond 1998.

erations of 'quality' may even be important enough for people to end their own or someone else's life, and this has been true since time immemorial. The attitude embodied in a statement like "Better red than dead" has a very long history indeed. And whatever holds for the human individual also holds to some measure for the community in which he or she lives.[80] Conceived in this way, the notions of sustainability and quality form an ancient pair, which are not always necessarily compatible.

In order for early human communities to survive it was imperative for them to some extent understand (or 'read') their environment and the countless dangers it entailed. It was imperative that they interpret what they saw and adapt their behaviour accordingly. This in itself obviously required a certain conception of nature and natural forces. In developing a notion of 'quality', moreover, communities had to form a conception of their own role and significance in the physical world in which they found themselves. Throughout the world, this process led humankind to develop a self-image and a conception of the purpose of the surrounding cosmos. These two conceptions are intimately related, mirroring one another as it were. From a biologically coloured description of human development we thus enter the philosophical and anthropological realm, for the term 'cosmology' is to my mind appropriate for this mirror image.[81] Cosmologies are then to be understood as cognitive and mental constructs by means of which individual human beings (and societies) manage, to varying degree, both to survive and to give their lives purpose and significance.[82] These constructs define man's role and purpose in the cosmos in relation to the assumed natural order of that cosmos. It is perfectly feasible for such cosmologies to be recognized and described as such by non-believing 'outsiders'. If they are to take root in a culture, cosmologies must be shared collectively, or at least by a sufficiently large number of people. Once a cosmology has become rooted in a given culture, it may come to more or less subconsciously influence those living in that

[80] Up to a point, for the environmental problem, too, has confronted us with the sometimes fundamental differences between individual and collective behaviour.

[81] I use the term cosmology because of its etymological roots (see Chapter 4). A term like *Weltanschauung* 'worldview' is too 'down-to-earth' for my liking, while the terms religion or metaphysics are too charged.

[82] On this point, compare the function of religion as described by W. Burkert in his *Religion and the Creation of the Sacred* (1996).

culture. In modern society this observation is as relevant as ever, because many people are not consciously aware of their deeper cultural roots and do not experience these as being operative within them.

The term cosmology embraces the totality of a person's conceptions of reality, the forces operative therein and their own status as a human being.[83] Löwith, and with him Wildiers, holds three elements to be essential to a cosmology: god, humanity and the world.[84] These mark the corners of a triangle that forms the interpretative framework of reality.[85] In this context, the Dutch environmental philosopher Zweers speaks of a "triangle of meaning".[86]

An interesting observation by cultural anthropologists in this context is that while these three elements, the 'operators' at the three corners of this triangle, are to be found in almost every culture, there is enormous variation in how they are conceived to operate and how they are related. This variation is found not only among different cultures, but also within individual cultures.[87] Among aboriginal peoples and in animistic societies, the sides of the triangle are short and the 'operators' closely interrelated, if not intertwined. The whole of physical reality and the world of experience is infused with gods, mythical forces and sacred places. There is indeed no sharp dividing line between human beings and the world they live in, at most a kind of fuzzy distinction. Likewise, at some point the divine and the human coalesce: in the ancestors, for example. It is held by most scholars that while such cultures have a profound conception of the natural order and the meaning of life, they accept that there are limits to what can be known about that order, and that life's meaning is certainly not to be fulfilled in a linear fashion. While today we conceive of time as moving progressively forward (the "arrow of time"),

---

[83] See, *inter alia*, N.S. Hetherington (ed.) 1993; R.W. Lovin & F.E. Reynolds (eds.) 1985.

[84] K. Löwith 1967; M. Wildiers 1988. The term 'god' should be taken here in its broadest sense, as encompassing everything associated with the 'superhuman' realm.

[85] A. van de Beek (1996, 48-52, 321,322) makes a distinction between creation theology and cosmology. Cosmology becomes creation theology when consideration is given to the relationship with the Creator. However, he proceeds from a modern, narrow meaning of the notion of (a) cosmology. In Chapter 4 we shall see that this is not a useful distinction.

[86] W. Zweers 1994, 70.

[87] See C.J. Glacken 1967.

in aboriginal cultures time is generally a cyclic concept.[88] No lofty aspirations are generally held regarding the capacity of human beings to understand and thus influence reality. From day to day, uncertainty and dependency tend to predominate, and the human community's relationship with nature is thus a blend of respect and interconnectedness, on the one hand, and fear and trembling, on the other.

At this point it is important to note that while cultures with such cosmologies are able to evolve a sustainable (in the sense of lasting, in principle for ever) relationship with the natural environment, their quality of life seems to be fairly static. In cultural and social terms, there is little evolution over time. From the western perspective, we tend to describe this kind of quality in negative terms, as 'underdeveloped' or 'primitive', or at least insufficiently dynamic. What, then, are the cosmologies that have left their mark on our own, western culture? These arose largely in the cultures of the ancient Near East and classical antiquity, and in particular Jewish and Greek culture.[89] These are the principal founts of western culture, although animistic (Celtic and Germanic) elements can also be identified.[90] The prevailing view is that the Jewish heritage, via Christianity, and Greek thought, via Rome, the Renaissance and humanism, have had a deep impact on our culture.[91] In his controversial article *The Historical Roots of Our Ecologic Crisis,* published in 1967, the medievalist Lynn White characterized the western belief in progress as "rooted in, and in-

[88] See J.J. Boersema 2001.

[89] I employ the term 'Jewish' as it is generally used in the modern Anglo-Saxon environmental literature in such notions as 'Jewish thought', 'the Jewish tradition' and so on. With reference to the corpus of ideas embodied in the Old Testament (the Hebrew Bible) the term is less accurate, however. Indeed, its use in reference to the 'Jewish' religious system is also a relatively late, medieval development (see G. Wigoder 1994, 13,14). Here (again) the word 'culture' refers mainly to written expressions of culture, and particularly philosophical and religious writings. A more accurate term is therefore 'the culture of ancient Israel'. See also 'Framework', below, and the Appendix to this chapter.

[90] For the cosmology of the Celts, see G. Herm 1975/1992; E. Valgaerts & L. Machiels 1992; for a magnificently illustrated treatment of animistic elements in western culture: M. Schouten 2001.

[91] It should be borne in mind that these two cultures do not stand alone. Ancient Israel was part of the patchwork of the Middle Eastern cultures of antiquity (Babylonia, Assyria, Egypt), while Greek culture was in close contact with these cultures for extended periods of time. Both indirectly and directly, this cultural *Umwelt* has had an influence on Western culture. On this point, see J. Goody 1995, 1996; W.W. Hallo 1996.

defensible apart from, Judeo-Christian teleology".[92] Christianity cannot simply be aligned with Jewish thought or the Jewish tradition, however. Although Christianity was of Jewish ancestry, it had its cradle in a Hellenistic world. Without wishing to detract from the general thesis, in the following chapters I shall examine whether contemporary ideas regarding our relationship with the natural environment are not more accurately termed 'Graeco-Christian' rather than 'Judeo-Christian' in a number of salient respects. In doing so, I take up the suggestions of Passmore *contra* White.[93] In this context, Passmore made specific reference to the Stoic philosophers.

The main focus of the present study is the Jewish heritage. An important motive for that choice is that this heritage, to the extent that it derives from the principal written source, the Hebrew Bible, today still constitutes a fount of inspiration in our own culture and is authoritative for Christians around the world. By adopting this approach I also hope to make some small contribution to the debate to which Lynn White's seminal article has led. In doing so, with respect to the Jewish tradition I shall restrict myself to two sections of the Hebrew Bible: the creation narratives and the expositions on the dietary laws, both key sections of the Torah. The arguments for this selection are elaborated below, in Section 1.8. From the Greek philosophical tradition, several key themes are reviewed. The aim here is to paint an accurate overall picture while identifying the differences with the Jewish tradition. Greek views on the treatment of animals serve as a useful reflection of Greek conceptions of humanity's standing with respect to nature. In a concluding chapter the two traditions are compared, with the influence of each tradition on early Christianity being very briefly reviewed. In this exercise, I restrict myself to the influence of contrasting views.

These, the focal points of this study, are expressed in the title of the book. In the overall context of the environmental problem, the study

---

[92] In this article, published in the authoritative journal *Science*, White posited a root connection between the contemporary environmental crisis and to his mind typical Judeo-Christian views such as a linear concept of history and the idea of 'dominion' over nature as expressed in the biblical creation narratives. The article generated an avalanche of response. For a clear exposition of the debate around the 'Lynn White thesis', to the extent that it relates to Genesis 1, see J. Barr 1972; R. Attfield 1983, 1999; H. Baranzke & H. Lamberty-Zielinski 1995; H.-J. van Soest 1996, particularly Chapter 1.

[93] J. Passmore 1974, 17; L. White 1967.

focuses on man's relationship with nature. The Torah forms the heart of the Hebrew Bible, and several key sections of the Torah are examined. Set over and against the Torah, the Stoics serve as a *pars pro toto* for Greek philosophy as a whole. Additionally, this choice also reflects the fact that it is above all the Stoic position on man's relationship with nature to which particular importance is attached in our western tradition.

1.7 The Question Addressed by this Study

From the foregoing description of the environmental problem and the chosen approach for analyzing that problem, the following pivotal question emerges:

To what extent can a study of important sources of western culture contribute to the contemporary environmental debate, and what insights can be derived from these sources?

This central question has been broken down into the following subproblems:

1. Why should the environmental problem be studied in its historical and cultural context? (Ch. 1)
2. What sources should be studied? (Ch. 1)
3. What do these sources have to say about humanity's relationship with nature, and what is the significance thereof? (Ch. 2,3,4)
4. tion of our western culture? (Ch. 5)
5. In these sources, can points of reference be found for a relationship with nature that respects the constraints of sustainability and quality? (Ch. 5)

1.8 Method, Framework and Selection

1.8.1 *Historical context*

These writings are based on a literature study.[94] With scientific literature, there is a general consensus about how texts are to be read. With literary and literary-historical sources there are, in principle, two methods of textual analysis. The first proceeds from the text as

[94] In transcribing the Hebrew, the rules of *Vetus Testamentum* have been followed (W. Weinberg *HUCA* 1969/1970).

it has been passed down over the centuries. In interpreting and assigning meaning to the text, no particular importance is attached to the origins of the text or to the original intentions of the author. There is close reading of the text, allowing it to be reinterpreted time and time again; the writings acquire a life of their own, as it were. In its most extreme form and formulation, every reader or interpreter creates his or her 'own' text.

The second method is historical, and proceeds from the premise that we can only understand the classical texts if they are read in their proper setting. Here, 'setting' refers primarily to the circumstances of a text's inception and the reliability of the version that has been handed down to us. What exactly has the author written, and what was his or her intended meaning? What influences were operative? What is the background of the author, or authors? How, if at all, does the text stand in relation to contemporaneous social life?

In the case of the Bible the method of textual analysis adopted is particularly important, because of the very broad range of positions on this point. With the texts of the Greek philosophers the question of how representative they are is important, given the enormous corpus at hand. How do the selected texts relate to the key (philosophical) problems of the Greek age? What was the historical decor of the texts?

In this study I have adopted an historical approach to the texts in question. The Bible and the Greek philosophers are regarded as the 'founding fathers' of our western culture, but they obviously did not simply emerge out of the blue. In considering sections of the Hebrew Bible we shall therefore be pursuing qualities of the author(s) as well as of the contemporary *Umwelt* that can shed additional light on the text. With the Greek sources, too, we shall be considering the relevant historical background, but in rather less depth. To a major extent, we shall draw the contextual background from relevant literature sources.[95] Although the main emphasis is thus on the

---

[95] On the religious history of ancient Israel in relation to the *Umwelt*, glad use has been made of W. Robertson Smith 1889/1894; J.P. Peters 1914; W.F. Albright 1950; G.W. Ahlström 1986; I. Zatelli 1991; R. Albertz 1992; Th.L. Thompson 1992; Sh. Salmon 1995; K. van der Toorn 1987,1996; K. van der Toorn, B. Becking & P.W. van der Horst (eds.) 1995/1998; W.W. Hallo & K. Lawson Younger, Jr. (eds.) 1997, 2000; R.A. Caras 1998/1999 and the provocative work of K.W. Whitelam 1996.

literary texts themselves and the ideas and beliefs to be derived from them, where necessary and useful these will be interpreted in their historical setting. Since the texts selected are concerned with humanity's relationship with nature, in seeking to elucidate the historical context it is only logical that the focus should be on nature, a subject on which there exists a very extensive body of literature. By comparison, I have given relatively little consideration to the general archaeology and geography of the ancient Near East.[96] The main emphasis is thus on the archaeological, historical and literary record as it relates to plant and animal life in the ancient Near East and Graeco-Roman world.[97]

As not all readers will be equally familiar with the Bible or with views regarding its origins and textual genesis, at the end of this chapter I have included a brief Appendix outlining some of the most widely held positions and doctrines on the subject.

As with any other historical and literary source, the Bible can be read in either of the two basic ways described above. The Bible

---

[96] On the subject of biblical archeology and geography, the following principal sources have been consulted: T.M. Harris 1820; J. Duns 1863-1868; G.E. Wright 1957; A. Alon 1969; G. Cornfield 1976; Y. Aharoni 1979; N. Pollard 19-84.

[97] On Egypt: J. Boesneck 1988; D. Kessler 1989; D.J. Stanley & A.G. Warne 1993.

On Mesopotamia: B. Landsberger 1934; A. Salonen 1973; A.L. Oppenheim 1978; F.C. Fensham 1988.

On Greece and Rome: O. Keller 1909; J. Brands 1935; d'Arcy Wentworth Thompson 1936; G. Jennison 1937; W. Foerster 1938; W. Kraak 1940; J. Woodkrutch 1961; I.C. Beavis 1969; J.2M.C. Toynbee 1973/1983; J.D. Hughes 1975, 1975[a], 1980, 1984 J.D. Hughes & J.C. Thirgood 1982; J. Rodman 1976; U. Dierauer 1977; J.R.T. Pollard 1977; F. Capponi 1979; H. Baumann 1982; P. Pellegrin 1982/1986; L. Bodson 1983; I.C. Beavis 1988; I.M.B. Wiman 1990; R. Sallares 1991and R. French 1994.

On Israel and the Bible: S. Bochartius 1663; T.M. Harris 1820; J. Duns 1863-1868; H.B. Tristram 1880; J.G. Wood 1882; A. van Veldhuizen 1928; A. Kinzler 1902; Ch.F. Potter 1933; F.S. Bodenheimer 1935, 1960; F.J. Bruijl 1939, 1948; A.E. Rüthy 1942; G.R. Driver 1955; A. Parmelee 1959; C.F.D. Moule 1964; R. Pinney 1964; A. Alon 1969; United Bible Societies 1972/1980; J. Klaits & B.G. Klaits 1974; H.I. de Smit 1974; D. Hakker-Orion 1975, 1984; A. Shulov 1975; M. Jacobs 1977; J.S. Morton 1979; N. Hareuveni 1980; G. Stein 1980; J. Feliks 1981; M. Zohary 1982; P. France 1986; U. Paz 1986; D. Smit 1990; E. Firmage 1992; A. Schouten van der Velden 1992 and P. Riede 1993. Most of this literature on the Bible is of a descriptive biological nature. Important studies that also consider the underlying world of ideas include J.L. McKenzie 1952; C. Houtman 1982; B. Janowski *et al.* 1993 (an excellent work with a very extensive coverage and bibliography) and Th. Hiebert 1996.

cannot be understood via theology. The only meaningful Bible commentary, then, is a historical study of scripture.[98] The religious nature of the text introduces a new element, however: the personal position of the reader and interpreter, which may infuse the interpretation with an additional, subjective aspect. In popular parlance this has led to such sayings as "The Devil can cite Scripture for his own ends".[99] In more scientific terms, we speak of interpretative traditions and interpretative frameworks. There is a far broader scientific consensus emerging in this area than popular wisdom holds, however, even though there are still differences within and among the various traditions. If these differences also align in some way or other with the interpreter's own personal conviction, communication on the topic may become rather clouded. A further introductory explication may then be useful. On this point I find myself in basic agreement with a remark made by E.J. van Wolde.[100] She stresses the importance of 'open-minded' reading. Certainly this is a noble aim and attitude, but it must be still backed up with some form of historical-critical textual analysis, I feel. Van Wolde herself takes more of a literary and semiotic approach.[101] Although reading and analysis must be grounded in the text, then, some contextualization must necessarily follow. In our case the context is formed at least by the other texts of the Hebrew Bible as well as a variety of nonbiblical sources. Due care should be taken that more recent texts and data are not too readily adopted as a frame of reference in analyzing older texts. In this approach, a conscious effort should be made to avoid overlooking certain aspects of the text, reading too much into the text or constraining it unnaturally within one's own (modern) interpretative framework. Van Wolde is also rather too optimistic, I feel, about the extent to which these aims can indeed be realized in practice. It is axiomatic that a text can never be considered in isolation from the particular tradition in which the reader finds him or herself situated and raised, personally as well as sci-

[98] On the origins of these views in early Enlightenment and the pivotal role of B. Spinoza see: J.I. Israel 2001, esp. 197-217, 447-476.

[99] Writing in the 17th century, in his *Tractatus Theologico-Politicus* (*Theological-Political Treatise* transl. S. Shirley, Leiden 1989) B. Spinoza cites the Dutch proverb *Geen ketter sonder letter* ('no heretic without his verse and chapter').

[100] E.J. van Wolde 1995, 18,19.

[101] Besides 'open-minded' reading, she also advocates reading "from the inside, out" (1995, 18).

entifically and socially.[102] Any attempt to project oneself into the world of ideas of the writer is inevitably made from the platform of one's own world. The past always remains a stranger and to pretend that one can read a text entirely 'open-mindedly' is thus to fool oneself as well as others. To state one's good intentions on this point indicates a readiness to do so, however, as well as an openness that may prove fruitful.

1.8.2 *Framework and selection*

The topic of this study is humanity's relationship with nature. How do we see ourselves, as part of the natural environment as well as set against nature? What kind of rules govern that relationship? As conceived in antiquity, such questions belonged to the 'philosophy of nature' and were a central element of cosmology.[103] In our age, we tend to use terms like worldview or *Weltanschauung* rather than cosmology. Having a conception of the world implies having a self-image. The Israelites of the Hebrew Bible (HB) saw themselves and the world as being divinely created, and we can thus speak of a 'creation faith' based on 'views of creation'.[104] Nowhere in the HB do we encounter the notion of 'cosmos'; nor is there any general word for 'nature' in the Hebrew language. This is no coincidence. The Israelites did not write—nor speak, it may be assumed—about reality in abstract terms: they spoke of the canopy of the heavens, of the earth around them, and of their flocks. Philosophical reflection on the cosmos and its representation in the form of an image, let alone in some doctrinal form, is alien to the HB. In this respect, there is a world of difference from Greek thought. At the same time, though, there are certainly 'cosmologies' to be discerned in the HB, in the sense of coherent answers to the fundamental questions of life. Indeed, some texts (the last section of the Book of Job, for example) appear to have been written with the explicit aim of addressing such questions. In many cases, such answers were given in the form of stories. What we have in the HB is thus not so much a worldview as a general outlook on the world and on life.

I use the plural 'cosmologies' because the HB is not a monolith-

---

[102] In my own case, historically speaking this is the Protestant world of the northern Netherlands.

[103] See, for example, D.J. Furley 1989.

[104] See, for example, D.A. Knight 1985. He concludes that there are six different "views of creation within Yahwism".

ic, 'one-volume' theological treatise but rather a collection of books that has evolved over the ages, each book with its own particular origins and evolution, its own content and setting. The narratives of the HB speak not only of how the world came into being, but also of how the Israelites emerged as an individual people. Some texts are better understood when seen in relation to the phase of the Israelites' history at the (presumed) time of writing. The plural 'cosmologies' also allows for sufficient differentiation, between the priestly texts and the wisdom literature for example, to permit reference to multiple cosmologies without these being assumed *a priori*. It should be borne in mind, though, that with this kind of concept it is no easy matter to draw sharp dividing lines. The differences involved may be relative, and fairly modest in absolute terms. To avoid any misunderstandings, let me state that my own conviction is that in the HB such differences are not pronounced and certainly do not reflect mutually exclusive cosmologies. What we see, rather, are significant and recognizable differences of emphasis and nuance within an overall framework of interconnecting themes. In all probability, we see here the mark of the priestly redactors. Let us first turn briefly, then, to that thematic framework and its constituent themes.

Framework

The fact that the world is viewed in the Hebrew Bible as God's creation has numerous implications. Here, I wish to consider three that are relevant in the present context, because they provide a framework within which to situate the other parts.

1. The Creation was originally good and untarnished. According to the author of Genesis 3, however, the original harmony of creation was corrupted by the events occurring in the Garden of Eden, referred to in the Christian tradition as the 'fall of man'. The implications of these events are far-reaching, having practical consequences for humanity's relationship with nature from that time onwards. For women, the natural processes accompanying pregnancy and childbirth become far more trying; in the relationship between male and female a new, and negative, dimension is introduced: dominion; and the relationship with nature undergoes a dramatic change. Through humanity's doing the face of the earth is 'cursed', and from how this is depicted one can deduce that this is above all a reference to the earth's 'recalcitrance' when brought under tillage.

Humanity's relationship with nature becomes clouded by hostility; death and chaos manifest themselves in the world. This violation of the creation order is not permanent, however; there is a perspective of the schism being repaired. Ultimately, harmony will be restored in a new creation. This eschatological expectation runs as a golden thread through the history of the people of Israel, although less pronouncedly in some periods than others. What little we read about this 'restoration' relates primarily to the corrupted aspects of inter-human and human-animal relations.

The Hebrew Bible's conceptions of humanity's relationship with nature can best be understood from a creational perspective, I feel, and then imagined as a triptych. In the left panel, we have creation as described at the beginning of Genesis: good and immaculate, as God intended. On the right, we see the prophetic, eschatological images of the final restoration, the restored world of the peaceable or Messianic kingdom. In the middle, in a considerably larger panel, we then have the here-and-now, in the HB the people of Israel, but also the world as we now know it. In the Christian tradition, this triptych mirrors the pivotal notions of the Creation, Fall and Redemption. The latter two can be seen as theological characterizations of the hinges between the three panels. Here, we shall be concerned more with the panels themselves. In this, the starting point must necessarily be the left-hand panel, creation in its 'original' manifestation. Even if one holds the view that the creation narratives are to be regarded as meticulously edited, retrospective projections of a human desire for a harmonious world rather than as true accounts of human origins, in our context the crucial point remains valid: in the eyes of the authors there was a difference between creation as they experienced it at that time (the middle panel) and creation as they considered it intended. This fundamental tension forms part of the framework within which the texts will be interpreted.

2. The God of creation is the God of history, here to be taken as denoting a progressive history rather than one of cyclic repetition. This is not to say that there is no awareness of the natural cycles in the HB, though. In the book of Ecclesiastes they play an important role, particularly in reference to the human condition (see, *inter alia*, Ecclesiastes 3:16-22). Even there, though, the reference is not so much to 'cyclic repetition'. Ecclesiastes is anomalous in several respects, and the degree to which the book is a typical reflection of ancient

Israelite thought may be queried. Ecclesiastes was written at a comparatively late date, around the third century BC, and the possibility of other (Greek?) influences cannot be excluded. The dominant and underlying conception of time in the HB (which is not entirely lacking in Ecclesiastes; see the book's conclusion, 12:14) is one of continuity, with a beginning, a direction and an end. God stands at the beginning of history, and He shall complete it. This conception is readily aligned with the notion of (re-)creation, because it engenders strong expectations of a better world. In later western cultural tradition this linear conception of history became linked with the idea that development and progress are possible and indeed desirable. Lynn White sees this view of history as specifically Judeo-Christian and characteristic of the western world.[105] As the work of Albrektson demonstrates, though, this position is contestable. He argues convincingly that on this point the beliefs of the ancient Israelites bore greater similarity to those of the cultural *Umwelt* than is generally accepted.[106] This is not to deny that differences existed, however, nor that these have left their mark on western culture.[107]

3. The creation narrative implies a radical break with the polytheism and nature worship practised by neighbouring peoples and within Israel itself. Although the world had been created by God, creation itself was not to be considered divine. For the human race, this meant that worship of nature or natural phenomena were to be spurned, and that the world in all its manifestations was to be seen as the result of God's creative power. It also meant that human beings should not stand in divine awe of the forces of nature, nor in mortal fear and terror of the religious powers with which nature is imbued. For the Israelite, therefore, nothing was inherently sacred. Trees, the heavens, the sun and moon, everything had been divested of its divinity: it was therefore not to be worshipped or revered. This said, though, this was presumably an idealized concep-

---

[105] For a thorough elaboration see Th. Boman 1968 128,129.

[106] B. Albrektson 1967. Similar conceptions are also to be found in Greek sources: in Pericles' funeral oration, described in Thucydides' History (II, 36), for example.

[107] The idea of accumulation of knowledge is credited to Bacon (1561-1626) and the idea of looking forward to new knowledge rather than back to recovery of old to Descartes (1596-1650). In this context the answer that Newton is said to have given (ca 1675) to the question as to why he could see so far is revealing: "If I have seen further it is by standing on the shoulders of giants" (~his forefathers). See J.J. Boersema 2001.

tion (on the part of the priestly authors probably) to which the people of Israel did not always conform in their everyday conduct. The HB is full of references to polytheism and the worship of people, places and objects.[108] The priestly authors deemed it necessary to take a stand against such practices. Only through God, and in contact with God, the ultimate personification of holiness, can any degree of holiness be attained by a living being or material object. And, as we shall see, there were very clear rules governing the conditions for such contact.

Selection

No study of the literature of the ancient Israelites that seeks to address humanity's relationship with nature can ignore the creation narratives of Genesis 1 to 3. These form the subject of Chapter 2. In his 1967 article, Lynn White in fact based himself solely on the creation stories in tracing the biblical sanctioning of what he termed "the most anthropocentric religion the world has seen".[109] He evidently considers these texts of fundamental significance for Western culture. Although I certainly agree with the latter position, it is undoubtedly necessary to consider the historical and cultural setting in rather more detail than White in fact did. My own stand here is that the first eleven chapters of Genesis are written from a universalist perspective: they concern the whole of humanity and not the Israelites alone, and should be read accordingly. With Westermann (1974/1983), Houtman (1980) and others, I see the primeval myths of the HB, including the creation stories, as having wider relevance than for the later Israelites alone.

Given the immense amount of literature on these issues, it is obviously essential to set limits to one's endeavours.[110] By referring to

[108] On this point see K. van der Toorn 1996 and *DDD* (*Dictionary of Deities and Demons in the Bible*) K. van der Toorn, B. Becking & P.W. van der Horst (eds.) 1999.

[109] L. White 1967, 1206.

[110] The following commentaries have been consulted:

HB, general: W.K.L. Clarke 1953; G. von Rad 1969 and R. Smend 1978/1989, W. Brueggemann 1997.

Genesis: J. Calvin 1554; Rashi (in the 2000 edition of Pinchas Doron), F. de Liagre Böhl 1923; U. Cassuto 1944/1961/1989; G.Ch. Aalders 1949; G. von Rad 1949; E.A. Speiser 1964; Naḥmanides 1965 edition; C. Westermann 1974/1983; J.P. Fokkelman 1975; Bereishis 1977; P.D. Miller 1978; C.F. Keil & F. Delitzsch 1980; A. van Selms 1984; E.J. van Wolde 1994, 1995 and H. Jagersma 1995.

Genesis 1-3: J. Ridderbos 1925; W.H. Gispen 1966; W.H. Schmidt 1967; E.J. Young 1971; P. Beauchamp 1969; H.J. van Till 1986; K.A. Deurloo 1988, Gen. 2-4 and E.J. van Wolde 1989.

other studies, I shall endeavour to clarify my basic position and any choices made. My central theme will be: order from chaos. In discussing the first creation narrative, I shall consider two motifs in greater depth: the 'image of God', with reference to God's command to humankind to have 'dominion' over nature; and the status of the seventh day, the later sabbath, and the role of this day in the creation narrative and the Torah. The theme here is: the sabbath as reflecting the rhythm of life. In the second creation narrative the emphasis is on the changed relationship between humankind and nature as a result of events in the Garden.

The second topic I shall be considering in greater detail, in Chapter 3, are the dietary laws. In the 1960s, scholarly interest in the dietary laws was revitalized from a somewhat unexpected angle, through the work of the anthropologist Mary Douglas.[111] Below, I shall discuss the results of her work at some length, and in particular her proposed systematics for the dietary laws. At this juncture, though, it is her basic premise that is of interest, her motives for assuming the existence of such systematics and for deeming their study important. Douglas considers the record of the HB from an anthropological perspective. She observes that the distinction between clean and unclean and the associated purity laws are set forth in relation to life's fundamental processes: sexuality, eating, the gathering and preparation of food, and human interaction with wild animals, livestock and sacrificial animals. All these activities are important elements in the everyday life of a people, from early morning until evening, and on into the night. Douglas takes it as self-evident that such rules mirror the central elements of a people's faith. It is also reasonable to assume, she holds, that this mirroring involves a systematic rationale, by means of which the Israelites were reminded, day in, day out, of the deeper dimensions of their (humanity's) relationship with nature. To my mind, this is certainly an attractive hypothesis. Below, we shall be discussing her arguments and endeavouring to validate her results, with particular reference to the clean and unclean species of animal enumerated in Leviticus 11 and Deu-

---

Leviticus: K. Elliger 1966; B. Maarsingh 1974; G.J. Wenham 1979; J. Milgrom 1991 and E.S. Gerstenberger 1993.

Deuteronomy: S.R. Driver 1896; B. Holwerda 1957/1958; H.D. Preuss 1982 and C.J. Labuschagne 1987/1990.

[111] M. Douglas 1966, 1970, 1972, 1973, 1975, 1978, 1982, 1993, 1993a and 1993b.

teronomy 14. In this context, the theme of the 'creation order' developed by Douglas dovetails neatly with the theme of order and chaos embodied in the creation stories and with the content of these terms.

In considering the Greek philosophers, I have adopted a thematic approach. In Chapter 4 the principal schools of Greek thought are discussed, an exercise which, in the context of the present study, has been kept fairly general in scope. In considering the writings of the leading proponents of the schools treated, I shall be focusing on their views concerning humanity's relationship with nature and, in particular, any arguments proposed for a demarcation between humankind and other living creatures.[112] I shall consider whether, and to what extent, these were important elements of Greek cosmology. On the subject of our relationship with nature, a number of Greek thinkers held anomalous positions, and these 'dissidents' will be considered in greater detail.

There are two principal motives for considering the Greek philosophers. In the first place, the influence of these thinkers on western culture can hardly be overstated. Secondly, though, it is my hypothesis that in a number of important respects (early) Christianity was influenced to a major extent by Greek thought, a matter already touched upon in Section 1.6. In what ways these ideas have infused western culture is a question that will be considered briefly in Chapter 5. This will be combined with a discussion of the main conclusions of the study. The questions considered in this concluding section are: can we find any points of reference in our own western tradition for recalibrating our relationship with nature and the environment, and what impulses may justifiably be anticipated from this area in the contemporary debate on sustainable development and quality? In conclusion, let me stress once more that my concern here is merely to identify any such points of reference; it is not the intention of the present study to elaborate these in any detail.

[112] The general literature on the Greek philosophers is given in Chapter 4. For the texts of the philosophers themselves my sources were: Loeb Classical Library (unless otherwise cited), C.J. de Vogel 1960/1964 and SVF: *Stoicorum Veterum Fragmenta* Arnim 1903.

Appendix: The Bible

When we speak of 'the Bible', what exactly do we mean? The word bible is derived from the Greek *biblion*. The plural *ta biblia* gave rise to the medieval Latin *biblia*, which became the accepted name for the Holy Scriptures. The term reflects the fact that the Bible is not a single book but rather a collection of books. But which collection? In such phrases as 'The Bible says' or 'The Bible and the environmental problem', the reference is generally to the collection of books considered by the early Christian and Jewish traditions to be authoritative and recognized as the Holy Scriptures. For Jews the authoritative corpus is smaller than for Christians. The Jewish or Hebrew Bible (HB) is referred to by Christians as the Old Testament (OT), as distinct from the New Testament (NT). For Christians, it is the OT and NT that together form the canon: the corpus of books deemed canonical.[113] Apart from a few sections, the books of the OT are written in Hebrew, those of the NT in Greek.

Besides the canonical writings, a number of so-called apocryphal texts are associated with both the OT and the NT. There have always been differences of opinion with regard to the extent of the OT canon, that is, where the line should be drawn between canonical and apocryphal books. The Roman Catholic church recognizes the canon of the Septuagint, the oldest Greek translation of the OT, which dates from the second and third centuries BC. The Protestant churches came to base their worship on the traditional Hebrew text that had been handed down: the so-called Masoretic text (MT). As a consequence, the Protestant OT is congruent with the Hebrew Bible, in terms of the books it contains, while the Catholic OT canon also includes a number of so-called deuterocanonical books such as the Wisdom of Solomon and Jesus Sirach. The Anglican canon is intermediate. In modern ecumenical translations the apocryphal writings are generally included between the OT and NT. While the apocryphal texts hold canonical authority for the Roman Catholic church, this does not hold for other Jewish texts written around the beginning of our era, and (disregarding the Dead Sea Scrolls) referred to by this church as pseudoepigrapha. I mention

---

113 The word canon is derived from the Greek καvών : measuring rod. It was first used with reference to the biblical texts in the fourth or perhaps third century AD (E. Earle Ellis 1991, 3).

these writings here because they contain several texts that will be referred to in discussing our relationship with nature.

Until now, I have used the term Hebrew Bible, in line with usage in an extensive body of (Christian) literature. Although its meaning is generally clear, it does justice neither to the history of Judaism nor to Jewish religious views. The terms most generally used for the Bible by Jewish believers are Mikra (reading), Holy Scriptures and Tanakh, an acronym for *Tôrāh*, *N^ebhî'îm* (Prophets) and *Ketûbhîm* (Writings). Although the term *Tôrāh* (law, teaching; Torah in English) is sometimes used for the Hebrew Bible as a whole, it usually refers only to the Pentateuch, the Five Books of Moses: the first five books of the OT. Besides this written Torah, Judaism also has an oral Torah. The latter is very important, for in the Jewish faith the distinction between Oral Law and Written Law is crucial. The written Torah, the core of the Tanakh, elaborates the prescribed rules of everyday conduct. It is for this reason that the term 'Torah' forms part of the title of the present book. In the oral Torah, 'the Law' is continually interpreted in the light of changing circumstance. This oral Torah has an extremely ancient history which, according to Jewish tradition, parallels that of the written texts. Judaism holds that the prophet Moses received both the written and the oral Torah on Mount Sinai. Following the destruction of the Temple in 70 AD, the Oral Law was gradually committed to writing (in the Mishnah, and developed further in the Talmud, Tosafot and Halakhic Midrashim), but it continued to be called the oral Torah.[114] Torah thus encompasses not only what is written in the scriptures, but also what, according to a century-old tradition, this implies for everyday life with its ever-changing circumstances. In

[114] *Mishnah:* codification of the Oral Law, completed by around 200 AD.

Talmud: compendium of several centuries of discussions on the Mishnah by rabbinical authorities in Palestine and Babylonia. The Palestinian or Jerusalem Talmud (Talmud Yerushalmi) was edited around 400 AD, the Babylonian Talmud (Talmud Bavli) about 500 AD.

Tosefta (Aramaic: supplement): a collection of older traditions similar to the Mishna, rabbinical writings originating 2nd/3rd century.

Tosafot (Hebrew, same meaning as *Tosefta*) additional comments on the Talmud commentary of Rashi (1040-1105), written by a school of rabbis in medieval France.

Aggadah: homiletic narrative and legendary portions of the Talmud and Midrash.

Halakhah: Jewish religious law concerning everyday conduct.

Midrash: rabbinic commentary, mainly on the books of the bible, designed to clarify legal issues or transmit teachings by means of stories and allegories.

B.W. Holtz 1984, 137-139; G. Wigoder 1992 and 1994, 417-420.

interpretation (the Midrash) reference may made to any part of the Hebrew Bible, but the principal focus is on the Five Books of Moses. In the Jewish tradition, contradictory interpretations are not immediately regarded as inconsistent; rather, they are seen as representing complementary aspects. There is no fixed interpretation of the scriptures, but a rich tradition of interpretation. In this tradition, two main schools can be discerned: one concerned more with right conduct in everyday affairs (the Halakhah), the other with the world of ideas and experience (the Aggadah). This tradition found its culmination from the Second Temple Period until the completion of the Talmuds. The writings from this period are referred to as the 'Rabbinic literature'. As Christians, by definition, see the NT as being closely associated with the OT, and indeed often read the OT from the perspective of the NT, so too do Jews perceive the Tanakh as being intimately entwined with its explication in the further Judaic tradition. The term 'Jew' inevitably carries connotations of the tradition briefly described above and the literature associated with that tradition. It was not until relatively late in the history of the people of Israel, however, that the terms 'Jewish' and 'Jewry' gained their current religious and ethnic meaning.[115] For this reason use of the term 'Jewish Bible' to denote only the books of the OT, or 'Jewish thought' to refer only to the written texts of the OT, is not really accurate. Today the term 'Tanakh' is very popular, but it is typical of the (later) Jewish tradition and involves more than mere denotation of the Holy Scriptures. 'Hebrew Bible' is better, but this collocation is still disputable, because the word 'bible' is rooted in the Christian tradition. Some of the writings are in the Aramaic language, moreover. In the context of the present work, though, I still give preference to the term 'Hebrew Bible', because, all things considered, it has the fewest undesirable connotations. In this study, we are concerned with the common written source for Christians and Jews alike, and within this body of writings with the written Torah. In the same vein, I have elected to use the term 'Israelite' to emphasize that the main focus of the present analysis is on the views and beliefs of the ancient Israelites as they have come down to us in contemporary writing, and not on those of later traditions, whether Christian or Jewish. Of the literature of the ancient Israelites, all that remains today, aside from a few fragmentary scrolls and

[115] The term 'Jewish' around 400 BC and the term 'Jewry' only in medieval times, according to G. Wigoder 1994, 13.

inscribed potsherds (ostraca), is the Tanakh, the Hebrew Bible.

Many modern biblical scholars assume that the stories of creation acquired their definitive form during the period of Babylonian exile (6th century BC). Proceeding from the extant texts, priestly scribes and ultimately perhaps even a single final redactor are held to have transcribed major sections of the Torah, including the creation stories. By analyzing their linguistic form and style, use of the names of God, duplications, repetitions and contradictions and, finally, certain 'viewpoints', texts can be assigned to a certain author or redactor (group).[116] In the scientific exegetic tradition this has led to the following terms being introduced to denote the inferred redactor(s) and/or author(s) of individual texts: the Priestly writers, the Elohist, the Yahwist and the Deuteronomist.[117] In the specialized literature, texts originating from the same inferred source are generally referred to by the initial letter of the presumed author(s) or redactor(s). Genesis 1—2:3 is ascribed to P and the second creation story, starting in Genesis 2:4[b], to J. Biblical scholars differ in their opinion regarding the authorship of verse 4[a]. Although a wide variety of views are held concerning the exact oeuvre, origin, dating and nature of the various sources, in the case of the creation narratives there is a broad consensus on the relative dating of J and P: the second creation story (J) is considered to be significantly older than the first (P). Hard evidence is unfortunately lacking, however, and different conclusions can be drawn from the same materials.[118] With reference to the dietary laws

---

[116] The work of Jean Astruc (1753) is considered to mark the beginning of historical criticism in this realm. In the second half of last century this was further elaborated by J. Wellhausen, in his *Prolegomena* 1878. On the development of these ideas in early Enlightenment see: J.I. Israel 2001.

[117] According to E.J. van Wolde (1995, 15) the majority of scientific exegetes proceed, or have proceeded, from the hypothesis that four strands of editing can be traced in the Torah and in the Book of Genesis. She characterizes these as follows: the Yahwist (J, after the German, 960-930 BC), an anonymous company of authors who use the title 'YHWH' to refer to God, completed their writings in Judah during the reign of King Solomon. The Elohist (E, *c.* 850 BC), an anonymous group of writers who use the term 'Elohim' for God and are situated in the northern kingdom, following division of the original united monarchy (*c.* 930 BC). The Deuteronomist (D, *c.* 620 BC), an anonymous scribe whose redactorship and/or authorship has been dated to the reform of Josiah, and the Priestly writers (P, 550-450 BC), the anonymous group of scribes who compiled and arranged the earlier writings into a single work, handed down to us in this form.

[118] C. Houtman 1980 provides an extensive and thorough review of research into the origins of the written Torah. Many of the views held, including some of those most commonly espoused, he deems untenable. He holds that the definitive

of Leviticus 11 and Deuteronomy 14, there is a reasonable consensus that these were compiled and edited into their present form by the priestly scribes referred to as P. In Chapter 3 this general picture will be refined in more detail.

The value of pursuing an extensive differentiation of textual sources for understanding the Torah is a subject on which opinions differ widely. In recent years there has been growing interest in considering 'texts as a whole'. This would seem to be a generally fruitful approach, certainly when it comes to exegesis. In the case of the Torah, it is at any rate highly probable that these texts have exerted an influence for over 2500 years, as sacred scrolls, and as a corpus of law. The Hebrew canon was constituted in the centuries prior to our era and there are indications that a uniform or at least predominant text emerged in the first century AD.[119] As was customary, it was a bare consonantal text, vocalized during reading sessions. It was on these and other unvocalized Hebrew texts that the masoretes based themselves in their work, from the seventh to ninth centuries, to produce a definitive vocalized text.[120] The result of that labour can be seen in the *Codex Leningradensis*, dating from the early eleventh century, which forms the basis of contemporary Bible translations. For this reason the contemporary text-critical edition of the Hebrew Bible is referred to as the Masoretic text (MT). The reconstruction of the origins of the text and the canon will not be treated any further here; these are beyond the scope of the present study.[121] Where the results of such endeavours shed light on the sections of the HB with which we are here concerned, glad use will be made of the available literature.

---

version of the Pentateuch comes to us from the author(s) of the books Joshua through to 2 Kings. This opus magnum is to be dated to the mid-6th century BC, and may have been written in either Palestine or Babylonia. In his opinion, there are good reasons for situating the materials generally ascribed to P in the exilic or post-exilic era (248).

[119] See A.S van der Woude 1992.

[120] The scholars who devoted their efforts to establishing the spelling and pronunciation of the text are known as 'masoretes'. The word 'masorete' is derived from a Hebrew verb that can mean both 'to bind' and 'to survive'. Important masorete schools included those of Ben Naftali and Ben Asjer in Tiberias.

[121] On these issues, see, for example, M.J. Mulder (ed.) 1988; E. Earle Ellis 1991; E. Tov 1992 ; A.S. van der Woude 1992 and the excellent study of J.C. Trebolle Barrera 1998.

CHAPTER TWO

CREATION AND SABBATH

2.1 Two Creation Stories

In the following we shall refer to Genesis 1-2:3 as the first creation story and Genesis 2:4 to 3:24 as the second. The first is held to be a priestly account (P source), while the second is attributed to the Yahwist (J source). Opinions differ regarding the transition point between the two.[1] We shall follow the first account more or less word for word and discuss the most relevant passages of the second.

Unless otherwise specified, English-language Bible citations are from the Revised Standard Version (RSV).

2.2 The First Creation Story: Order from Chaos

The first creation story provides a tersely written account of how heaven and earth came into being. If the term 'cosmology' is anywhere appropriate, it is surely to this narrative. It opens quintessentially: "In the beginning God created the heavens and the earth".[2] The translation of the first word *b^e rēšît* and the meaning of the subsequent verb *bārā'*, to create, beg two fundamental questions.[3] The

---

[1] Several commentators have the first creation story end at verse 2:3, with verse 4a taken to be the beginning of, or 'title' to, the second account (U. Cassuto 1944/1989; A. van Selms 1984). The majority of scholars take verse 4a as belonging to the first account, however (W.H. Schmidt 1967; C. Westermann 1983; E. van Wolde 1994, 1995; H. Jagersma 1995; Th. Hiebert 1996). This part of the verse is ambiguous, and may possibly be a later addition.

[2] According to Rashi (1040-1105) verses 1 and 2 should be seen as a single verse. A possible translation is then: "At the beginning of God's creation of heaven and earth, when the earth was without form and void....". (Pinchas Doron *Rashi's Torah Commentary: Religious, Philosophical, Ethical and Educational Insights.* Jason Aronson, Northvale NJ, 2000, 4)

[3] This is due, *inter alia*, to the fact that the word *b^e rēšît* as a status constructus in Hebrew has no definite article, thus allowing for the translation 'in a beginning'. For Van Wolde this is the only correct translation, for: "The beginning is still indeterminate; it was not preceded by any other beginnings". Nor was that indeed possible, in her view, because time did not yet exist. "Only with the creation does time come into existence" and "Creation does not occur at a moment

first concerns the relationship between time and the creation, the second the question whether the creation is to be considered as a *creatio ex nihilo*, a creation of 'something out of nothing'.

The two questions are related, but the first is more general and philosophical in scope. Behind such questions as: did God also create time or did God create in time? lie questions regarding the nature of the phenomena of time and God. Does time itself have a beginning? Does God have a beginning? Is eternity timeless? Is the Eternal timeless? Debates on these questions already have an extremely long tradition that runs almost uninterrupted from antiquity through to the modern physics of the present age. These questions will not be considered further here.[4]

In the Hebrew Bible the verb *bārā'*, to create, always has God as its subject. It is a uniquely divine activity. The second question thus concerns both the material beginnings of heaven and earth, as conceived in the Bible, and the nature of creation as a divine activity. At the same time, in Genesis 1 the verb *bārā'* is characteristic but not exclusive: it is alternated with and used alongside *'āśāh* (to make) and apparently has the same meaning here. In the first creation story we find *bārā'* six times and *'āśāh* eight.[5] From this it can be concluded that no exclusive meaning is to be attached to *bārā'*. The acts of creation related in this chapter can also be denoted by *'āśāh*.[6] This does not detract from the fact that *bārā'* may indeed have a very specific meaning in the HB. It is therefore instructive to briefly analyse the usage of this word elsewhere in the HB. While God is always the subject of *bārā'*, the object is very varied. Winds, waters, heaven and earth, animals and humans are all cited as objects, but also "darkness and woe" (Isa. 45:7), the "the smith who blows the fire of

in time, but time is created *together with* the world" (E.J. van Wolde 1995, 25,26).

[4] For a good review of how these questions were conceived and discussed in antiquity (500 BC—550 AD), see R. Sorabji 1983. For a more general but less in-depth review, see G. Clark 1992, particularly Chapters 3 and 5. For a perspective on the phenomenon of time in recent creation theology, see W.B. Drees 1991.

[5] E.J. van Wolde 1995 arrives at seven times *bārā'*, because she assigns verse 4a to the second creation story and seven times *'āśāh*, because she miscounts. From this she draws the doubly correct conclusion: "the narrator shows a weakness for the number seven" (35). Correct, because the narrator of Genesis 1 must indeed have had a weakness for the number seven, and also correct because in the original Dutch the word 'verteller' ('narrator') can also be construed as meaning 'someone who miscounts': Van Wolde who herself counts 'twice seven'.

[6] That *'āśāh* can be used to denote an act of creation is reflected elsewhere in the HB, a clear example being Ps 33:6.

coals" (Isa 54:16) and the Ammonites (Ezek 21:30). In all the passages the accusative of substance is lacking. Nowhere is it related from what substance God created. The texts emphasize the fact that God stands at the inception of the cited 'objects' and that it is He who brought them into being—despite the fact that these objects in many cases already existed. Many texts, particularly in Deutero-Isaiah, stress God's exclusiveness ("Besides me there is no God": Isa 45:5,6) and omnipotence.[7] The creation is the proof of that omnipotence. God even created darkness and evil. The manner in which He creates is not always described in the texts, but the 'method' most frequently cited, certainly in Genesis 1, is through an act of speech, by speaking the universe into existence, as it were.[8]

Where God himself comes from is not an issue in the HB. Unlike the Egyptian, Babylonian and Greek creation stories, a theogony is lacking.[9] Matters go back to God, but never beyond. Although creation does not in all texts therefore have the meaning of 'making something out of nothing', many passages in the HB support the idea of God standing at the beginning of the world. Genesis 1:1 may consequently describe a *creatio ex nihilo* in the sense that the creation of heaven and earth also signified the beginning of the 'material universe' as we know it.[10] Whether, in the eyes of the authors, there was 'nothing', or an 'unordered chaos', prior to that beginning cannot be derived from the available sources. It is perfectly feasible that the Bible's authors did not consider the difference between 'nothing' and 'unordered chaos' relevant.[11] Whatever the case, 'in the beginning' chaos still reigned on earth, and the remainder of Genesis 1 concerns the ordering of that chaos.

In verse 2 the earth is said to be initially "without form and void". The Hebrew *tôhû*, here translated as 'without form', evokes a picture of desolate and eery emptiness. The Hebrew *bhôhû* (a standing combination with *tôhû*) reinforces the meaning of the previous word and expresses the same emptiness, void (in the sense of unorderedness) and chaos. The basic meaning of this combination, which we

---

[7] This image is also reflected in the question posed in Job 38:4: "Where were you when I laid the foundation of the earth?", although this passage does not include the word 'create'.

[8] Cf. Psalm 33:9.

[9] See B.C. Sproul 1979 for examples from Egypt (80-86), Mesopotamia (116-117) and Greece (157-166).

[10] See R. Sorabji 1983, 194.

[11] Cf. Isa 45:18, where *tôhû* is translated as 'chaos' in the RSV.

also encounter in Jer 4:23, appears to be: a place that is unsuitable for life and habitation, because chaos (still or again) rules. This is made even clearer in the rest of the verse: "and darkness was upon the face of the deep". The 'deep' (*t^e^hôm*) stands for the chaos (or chaos-causing forces) and the abyss of the watery deep.[12]

"And the Spirit of God was moving over the face of the waters", so the verse ends. It is an enigmatic statement that cannot be satisfactorily interpreted if the RSV translation is followed. It is possible to translate *rûaḥ* as 'wind' and interpret *'^e^lôhîm* adjectivally as 'great, tremendous'. The Hebrew allows this translation and interpretation. The translation then becomes: 'a tremendous wind blew (or raged) over the face of the waters'.[13] Although the sentence may be clear enough, what is its possible significance at this juncture of the creation story? Although *rûaḥ* can be translated as 'wind', in many texts 'breath' or 'breath of life' is more usual. It is the spiritual force, often God-given, which moves prophets to act, but which is also to be found in animals (Gen 7:15,22). However the association of 'spirit' with an independent immaterial entity should be avoided.[14] In combination with *'^e^lôhîm* it is therefore more plausible to see it as the breath of God. Support for this can also be found in Ps 33:6, where we read: "By the word of the Lord the heavens were made, and all their host by the breath of his mouth". It is His vital, spiritual power: the breath/wind with which, in Job 26:13, He makes the heavens and, in Gen 8:1, assuages the floodwaters. It is certainly to be deemed a mighty instrument.[15] In these last two passages there is also a connection with the forces of chaos: in Job 26 with Rahab and the serpent and in Gen 8:1, referring back to 7:11, with the 'great deep' (*t^e^hôm*). In our text, too, it is reasonable to assume a link with

[12] Cf. Gen 7:11, 8:2, Job 28:14, Ps 148:7 and Jonah 2:5. In a number of passages the creation story is echoed (in the hymn about the passage through the 'Reed sea': Exod 15:8, in Ps 148:7, and possibly also in Jonah 2:5). The (vanquished) watery depths may also be a blessing, however (Gen 49:25 and Deut 33:13). The word is etymologically related to the Akkadian *Tiāmat*, the name of the personified 'god of the primordial oceans', who is vanquished in Enūma Eliš by Marduk. On *Tiāmat* and the similarity to the *t^e^hôm* of the creation story, see B. Alster entry Tiamat in DDD 1999, 867-869.

[13] For these interpretations, see, *inter alia*, J.M.P. Smith (1928/1929) and C. Westermann 1974, 148,149.

[14] For an elaboration on the conception of *rûaḥ* in the the Hebrew Scriptures see: H. Hiebert 1999.

[15] Cf. Isa 11:4, where the 'breath' of God's lips is called 'the rod of his mouth', with which He passes a devastating verdict.

what precedes (the void and the chaos) and what follows (the ordering in the seven days of creation). At any rate, an activity associated with the 'Spirit of God'/'breath of God' appears more logical than a description of a 'mere' wind. The question is then: what activity?

Elsewhere, the Hebrew verb that is translated in the RSV as 'to move over', *rāḥaph*, occurs in the piel form only in Deut 32:11, where it is used to describe a mother eagle or vulture (*nešer*) caring for her brood: "Like an eagle that stirs up its nest, that flutters over (*yᵉraḥēph*) its young...". The image is one of a parent bird flapping its wings at the edge of its nest, an action probably partly intended to encourage the young to learn to fly. From the remainder of the passage in Deuteronomy it is not entirely clear what kind of behaviour the author is referring to. In the second half of verse 11 he has the mother bird carry her young on her wings, a behaviour which, as far as is known, has never been observed in either eagles or vultures.[16] Intense flapping and fluttering of the wings on the edge of the nest, on the other hand, has been described for many species of birds. In the qal rather than piel form we find *rāḥaph* once more, in Jer 23:9, where the RSV translates it as 'shake' (of bones). In both passages *rāḥaph* has a connotation of vigorous, physical movement. For this reason, in our passage the RSV translation of 'move over', or the occasionally suggested 'brood', does not seem particularly plausible. If we opt for an 'active' root meaning in Genesis, then the text reads: "the spirit/breath of God moved vigorously over/on the waters". But to what purpose? What is the significance of this in this first creation narrative? Taking together all the elements reviewed, to my mind the most plausible answer is: to combat the forces of chaos residing in the watery deep and render the earth fit for habitation.[17]

---

[16] See H.G.L. Peels (1994), who, in line with the doubts raised by myself (J.J. Boersema 1986) about the standard biological reading of this passage, has arrived at a new interpretation on linguistic grounds, too.

[17] H. Jagersma (1995, 23) formulates it as follows: "The third sentence already heralds the fact that these 'primordial depths' are transformed into 'waters' by the *rûaḥ ᵉˡlôhîm*, 'the breath of God'". Westermann (1983, 145,146) rightly points out that (the) *tᵉhôm* is not personified in the HB. Westermann relativizes the mythological character of Isa 51:9,10, where there is also reference to the waters being assuaged. He holds that with P there is no (longer any) mythological undertone to the word, it signifying simply "Flut" or "Wasserflut". It may be queried whether Westermann is not 'demythologizing' too much here. To my mind the passage in Isa 51 and the other texts in which *tᵉhôm* is linked with the passage through the 'Reed sea' (Exod 15:5,8, Isa 63:12,13, Ps. 106:9) refer to the forces

Prior to, and in a certain sense as an initial step of the work of creation, the *t^ehôm* must be repelled, an activity that requires battle and is anything but quiet.[18] This passage may echo old Mesopotamian creation myths involving 'battle cosmology': the storm god versus the waters.[19]

Now God's work of creation can unfold. In the text, creation comprises three activities:

* division
* making/bringing into being
* naming/appointing.

By means of this tripartite work of creation, reality is ordered. The creation of light allows a division to be made between light and (already extant) darkness. The darkness is driven back (further) into the night and the day becomes an important unit in the work of creation. That light here, in verse 3, appears to be created separately from the 'lights in the firmament' (1:14) may appear strange to us modern readers. We associate light with the sun and wonder how there can be a day and light without that light having a source, the sun, which was created only on the fourth day. These questions arise only if the story is constrained within a modern scheme of cause and effect, however.[20] The ordering that characterizes this narrative is

---

of chaos in the waters of the deep being overcome (by God): a victory perceived as liberation (and creation) by the Israelites. It is perfectly plausible that the echoes of this mythological role of *t^ehôm* still resound in the text of P (B. Alster in DDD 1999 entry Tiamat).

[18] For an early interpretation along these lines, see J.P. Peters (1911[a] and 1914[a]). Because in Genesis 1 creation is effected through the word, the 'breath of God' in verse 2 forms a good transition to God's speech in verse 3 (E. Noort 1997 personal communication).

[19] There is an extensive body of literature concerning the parallels between the accounts of Genesis 1-11 and the narratives of the Mesopotamian tradition. The classical Genesis commentators treat this matter in great depth: U. Cassuto 1944/1989, G. von Rad 1949, W.H. Gispen 1966, W.H. Schmidt 1967, C. Westermann 1974/1983 and A. van Selms 1984. Among other scientific literature the reader is also referred to a classic and original article by W.G. Lambert 1965, the publications of A. Heidel (1971,1974) and, more recently, a lucid review by R.J. Clifford 1994. For the storm god versus the waters in *Enūma Eliš*, see: W.G. Lambert 1994, S. Dalley 1991, 249 *et seq.* and DDD 1999.

[20] A similar 'problem' concerning cause and effect arises if one queries whether the plants, which are created on day three, could have survived without the sun (day four) for their photosynthesis. For one day perhaps, but if one extends the lengths of the days to extremely long periods in order to tally with the assumed immense age of the earth, serious difficulties arise. These problems are avoid

of a different nature: the chaos is driven back by means of hierarchically ordered, successive acts of creation, and the elements of the order created in each of these divisional steps are gradually elaborated. Light and darkness are not congruent with day and night, but form a fundamental duality with a significance that extends beyond their later restriction and elaboration. The division of light and darkness, of waters above and water below, of the Earth and the Seas, gives rise to the temporal and spatial units within which life can unfold. Today we would speak of the three environmental 'segments' soil, water and air. These major ordering, divisional acts of creation are concluded in 1:4 and 1:10 with: 'And God saw that it was good'. Here, 'good' (*ṭôbh*) may certainly be taken to mean: functionally good, suitable for the purpose, meeting the requirements. We encounter this judgement a total of five times following individual acts of creation and a sixth time as a final judgement ("very good") in 1:31. Light is the first thing to be termed 'good', in verse 4. Halfway through the third day, in verse 10, this qualification is then encountered for the second time. Perhaps we may conclude from this that the previous acts of creation and division belonged together in functional terms, and did not yet deserve the epithet 'good'.

Now that the fundamental a-biotic ordering has been completed, there is scope for the first forms of life: vegetation, seed-bearing plants and fruit-yielding trees. With this plant life there is, for the first time, mention of something having a certain quality of its own. This life is created 'according to its kind' (*lᵉmînô*, from: *mîn*). Besides the plants, in the first creation story the animals are also created 'according to their kind'. This characteristic phrase—which occurs ten times in Genesis 1—recurs in almost identical form in Gen 6:20 and 7:14, in the story of the Deluge. We encounter it once again in the dietary laws of Leviticus 11 (nine times) and Deuteronomy 14 (four times) and, finally, in Ezekiel 47:10. With the exception of Genesis 1, the phrase is used only in association with animals and in Genesis 1 with plants as well as animals. The majority of translators assign to *mîn* the meaning 'species'.[21] This modern-sounding translation does proper justice to the systematic character of the passages in question. In the eyes of the author of Genesis 1, plants and animals were

ed if the story is read as it in my opinion should be: as the Israelites' religious answer to the pivotal questions of existence and, in terms of the story of creation, as answers (pl.) with a timbre and significance all of their own. See this chapter.

[21] K-B (1951, 519) gives: *Art (naturwissenschaftliche Spezies).*

created as individual species and in systematic relation to one another (species by species).[22] This taxonomy has a functional undertone, too, though. The description of the plants in verses 11 and 12 is a prelude to their use by animals and humans in verses 29 and 30. The same kind of functional colouring of the word *mîn* is also reflected in the further text, and the following passage reports on the relevance of this for fruit-trees. They do what they are intended to do: bear fruit, fruit with seed from which new fruit-trees can germinate. In the eyes of the authors of Genesis 1, the ordering of creation is highly functional in nature: all elements have a part to play in the whole.

The division between day and night is elaborated further on the fourth day with the creation of the 'lights in the firmament'. This act enables not only night and day to be distinguished, but also fixed periods, monthly and annual cycles. By being named, or appointed, everything gains a purpose, a function in the whole. In verse 17 the heavenly lights are 'set', or appointed to 'rule': the greater light over the day, and the lesser light, together with the stars, over the night. There is no doubt that the reference here is to the sun and the moon. In Ps 136:8,9 we find the same reference and here the sun and moon are named explicitly. The description of these heavenly lights in the first creation story has undeniable apologetic and probably also demythologizing undertones. The apologetic flavour discernible in these verses is understandable, given the relatively late dating of the first creation story and its priestly authorship. It should not be exaggerated, however, nor should this be seen as the exclusive vision of the Israelites, over and against that of the surrounding cultures. The apologetic is directed both internally and externally. Etymologically, the Hebrew names for the sun (*šemeš*) and the moon (usually *yārēaḥ*) echo the names of gods of other ancient Middle Eastern cultures.[23] In other passages of the HB we find traces of cult worship of the heavenly bodies as well as continual opposition to such worship.[24] Although in this account of creation the heaven-

[22] A similar conclusion is drawn by P. Beauchamp (1969, 244), who concludes his study of the etymology and usage of *mîn* with the sentence: "Dieu créa chaque catégorie animale espèce par espèce".

[23] With regard to the sun, the god Shamash in Akkadian and the goddess Shapshu in Ugaritic. With regard to the moon, Yarikh, a lunar god from Ugaritic texts (*DDD* Introduction, XVI).

[24] Traces are to be found, for example, in Jer 8:2. See, further, *DDD*, under the entries *Shemesh* (E. Lipinski 764-768) and *Moon* (B.B. Smidt 585-593), and also I. Zatelli 1991.

ly lights themselves are stripped of divine status and appointed anonymously, matter-of-factly and functionally as rulers of the day and the night, as lights in the firmament, they are also appointed to be 'signs' (*'ôtôt)* of the times. In the HB, signs (frequently in combination with wonders) have a denotative and supernal significance.[25] There is therefore no reason to assume more apologetics and demythologization on the part of the authors of Genesis 1 than the text allows.[26]

In the HB we encounter the Hebrew verb used with these heavenly bodies for 'to rule over', *māšal*, with a very broad range of subjects and objects. As subjects we find:

God (Job 25:2, Ps 59:13, Ps 89:10),
kings (Jos 12:2,5, 1 Kgs 4:21, 2 Kgs 7:18, Is 19:4, Zech 9:10),
judges, such as Gideon (Judg 8:22),
peoples, such as Israel (Deut 15:6) and the Philistines (Judg 14:4),
slaves (Prov 17:2),
'your husband' (Gen 3:16),
women (Isa 3:12),
the rich (Prov 22:7), and
the wicked (Prov 29:2).

Impersonal entities can also 'rule over', for example:

'sins' (Ps 19:13),
'the hand of the diligent' (Prov 12:24),
'babes'[27] (Isa 3:4).

---

[25] In particular, the phrase 'signs and wonders' is employed to denote the events in the land of Egypt, including the ten plagues (*inter alia* in Deut 4:34, 6:22, Neh 9:10).

[26] To my mind this tendency is apparent in the line taken by N. ter Linden (1996, 16), who, in his treatment of the creation story, assigns no role to the signs and, with his playful explanation that the heavenly bodies "may have no name", suggests more than is actually written. This line follows the 'Amsterdam school' (F. Breukelman, K. Deurloo and others, all strongly influenced by Martin Buber). The exegetes of this school situate the first creation story entirely against the backdrop of the Babylonian exile and view it as a hope-inspiring, highly apologetic story of liberation for Israel and all those socially oppressed (for a contrary standpoint, see A.S. van Woude 1969).

[27] The Dutch NBG translation of the Hebrew *ta'ᵃlûlîm*, a plurale tantum, as 'moedwil' (malice) is probably incorrect. It is more likely that the word is derived from the stem *'ll*, which in the hitpael, according to Koehler-Baumgartner, means 'to act recklessly'. The derivative then means 'arbitrariness' or 'bad treatment'. The English RVS and JPS translations have 'babes', which, used in a hendiadys

With regard to the object, a similar variety of entities is encountered: peoples, kingdoms, Jacob, Israel, a son, womankind, the poor, the mighty waves of the sea and geographical features such as villages and mountains. The texts give a clear indication of who or what rules in a given situation and who or what holds decisive influence. That influence may be either negative or positive, depending on the entity ruling. No concrete, negative pronouncements are made on the nature of the rule in question. Nowhere do we read of 'oppression' or 'violence'. The subjects mentioned do not always even allow for such in relation to the given object. It is difficult for the moon to oppress the night. As such, *māšal*, 'to rule over', has no negative undertone, and there are even neutral and positive connotations to be found (e.g. Judg 8:22, 2 Sam 23:3). The context is all-determining. Two other words used in Genesis 1 for 'to rule over' will be discussed below, in treating verses 26-30.

In the HB, a name may be pregnant with meaning. Far more so than in our society, names may have a significance that is known and recognized by the bearer and/or users of that name. In this connection, the Dane Pedersen has defended the view that knowledge of a given name implies insight into the nature of the bearer of that name. The name (and the giving of that name) then determines the course of the life and the role of the named (he or she who receives that name) therein.[28] In this perspective, naming someone or something is an expression of authority over them. This sense is also reflected in the English usage of 'to name' as 'to appoint'. Not everyone is in a position to appoint; it is characteristic of rulers. God is consequently the namer or appointer *par excellence.* The view that assigning a name is an expression and proof of ruling has particular relevance for the interpretation of Gen 2:19,20, concerning man's naming of the animals. In Genesis 1 the naming process is encountered four times. In verse 5, where the light is called 'Day' and the darkness 'Night'; in verse 8, where the firmament is called

(a pair of words with a similar meaning), is a better match for the 'boys' from the first half of the verse. This translation is based on the root *'lm*, and the associated pair *'elem* (young woman) and *'ālem* (young man), or on the root *'wl*, which can mean 'baby'. In Isa 66:4, the only other place in the HB we find the word under discussion, it is translated in the NBG as 'misfortune'. 'Malice' would there be inappropriate. In this passage, English translations opt for 'affliction' (RSV) or for 'to mock' (JPS).

[28] J. Pedersen 1926. Reiterated many times subsequently, often with explicit reference to his work (*inter alia*, D.T. Asselin 1954, 289, note 25).

'Heaven'; and in verse 10, where the dry land is called 'Earth' and the waters 'Seas'. As is apparent even from these texts, Pedersen's position requires considerable nuancing, for neither these names nor the name-giving as such appear to hold any particular significance. The name-giving, the names and the non-usage of the names are in line with the 'neutral', functional undertone of the story. When considering Genesis 2, we shall return to this name-giving and the presumed connection with knowledge, ruling and the exercise of power.

On the fifth day the animals are created, and they appear in the same sequence as the environmental segments to which they belong. First we see the aquatic animals, then the animals of the air and finally, on the sixth day, the land animals. In the description of the aquatic animals, it is noteworthy that a distinction is made between 'swarms of living creatures', on the one hand, and 'great sea monsters' (*tannîm/tannîn*, pl. *tannînîm*), on the other. In all likelihood, the latter do not (only) denote large marine animals such as whales and sharks. It is indeed very questionable whether the Israelites had themselves ever observed such creatures; they may well have been familiar with them only from stories (Jonah) and songs (Ps 148). The taxonomy of the HB regarding the aquatic animals is relatively simple. The word 'fish', *dāgh*, is found only as a generic term, with names for individual species lacking. It is reasonable to assume that this is tied up with the customs and eating habits of the Israelites, who lived as town or village dwellers, as cultivators and herders, but rarely if ever as (sea) fishermen.[29] The question, then, is whether this constitutes an adequate explanation for a society in which it was permitted to eat fish, as clean animals, and in which they were certainly also 'on the market'. With the knowledge at our disposal this question cannot be answered, however.[30] Neither does it have any di-

[29] The HB contains no description of Israelites fishing at sea. Freshwater fishing is sufficiently familiar from the New Testament (on the Sea of Galilee, for example). Information on fishing (and hunting) is extremely scarce in the HB. Due caution is required, however, in employing the 'socio-geographic' characterization of the Israelites' lifestyle as an explanation for the absence of a differentiated nomenclature for fish species: on the one hand, because of the paucity of available knowledge on the Israelites and their means of subsistence during the various periods of their history and, on the other, because of the uncertainties surrounding the authors of the texts and the knowledge they may be held to have had.

[30] In Ezek 47:10 there is mention of fishermen and nets used for fishing in

rect bearing on the present text. It can reasonably be assumed that the 'great sea monsters' allude not only to the *tannîn* but also to the *liwyātān*, which we encounter in several other passages of the HB, where (in the books of Job and Isaiah, for example) they (may) fulfil a more symbolic role. The significance of the separate mention in this verse now becomes clear: it serves to emphasize that these primordial animals, still strongly associated with the deep waters and the chaos, are also part of the work of creation.

As with the heavenly bodies, with these primordial animals, too, there is a 'down-to-earth' conception of reality. Nature and the forces of nature should not be identified with their maker; in this narrative, nature is not conceived of as being divine. However wonderful, imposing and frightening certain elements and forces of nature may be, like the grass and the cattle they were created by God and there is no reason for special veneration, nor for fear. In the first creation story they are described in a sober fashion and, in the case of the heavenly bodies, functionally, too.

That the terms 'move' and 'swarm' are used so emphatically in describing the other aquatic creatures may be due to the observation that smaller fish are seldom found alone, but generally in shoals. There are, moreover, other species of aquatic organism (apart from fish) for which these terms are appropriate.[31] The mode of locomotion is one of the characteristics of the order introduced and seen in creation. With the creatures of the air, too, it is noteworthy that in connection with the term 'fowl' reference is first made to 'flight' and soon after to 'winged', as if to further emphasize their specific mode of locomotion.

At the end of the fifth day, for the first time God blesses what He has created. To bless (stem: *brk*) is one of the key terms in the HB. In the qal form, *brk* occurs only as the passive participle *bārûk* in its meaning as 'to bless'.[32] This form is found frequently, in the psalms

freshwaters. These fish are cited as being "*of very many kinds*" (*l*[e]*mînāh*; cf. what has been said of this term earlier in this chapter) and, in this, "*like the fish of the Great Sea*". This knowledge of species richness did not yield any nomenclatorial differentiation, however.

[31] This characteristic of aquatic life is further emphasized in the Hebrew original of verse 21, which translates to '...all swarming living creatures with which the waters teemed". In modern parlance, too, there appears to be a connection between use of such terms as 'teem' and 'swarm' and the inability (or unwillingness) to give expression to species richness (as in "It's teeming with insects here").

[32] On three occasions in the HB, forms of *brk* are found that are translated as

for example, to establish (or 'voice', in the context of a religious service) a certain state of affairs (blessed is the land.., the man.., the people.., the work; blessed is he..), to which an explanatory condition can then be attached (the Nation whose God is Lord..., he who considers the poor). Besides its meaning of 'to bless', the most common form of the verb, the piel, can also be translated as 'to praise', 'to prize', 'to worship' and 'to greet'.[33] The translation 'to curse' forms a special case.[34] God is not the only subject that blesses something or somebody. People may bless one another and on certain occasions specially ordained persons may grant a special blessing. People of lower social standing may bless those of higher standing (kings), that is, wish them blessed. The (standard) phrasing employed often leaves unstated who is to accomplish the blessing and what the blessing is to actually entail. It may be concerned with material wealth, or with a long life and such immaterial qualities as wisdom. It is implicitly understood that the ultimate blessing is by divine agents.

In the first creation story, blessing is accompanied by, or consists in, the order to be fruitful and multiply. The link between (abundant) offspring and blessing recurs frequently in the HB. Abundant offspring thus also constitute proof that one is blessed. This holds for many other cultures from this era, as well as later, it may be added. It is not only male offspring that are concerned, although these are very important.[35]

In our text, the blessing concerns the animals created on the fourth

'to kneel'. In Ps 95:6 and 2 Chr 6:13 we find the qal (impf.), to be translated as 'to kneel down' and in Gen 24:11 a hiphil (impf.), to be translated as "to make kneel down" (of camels). The three consonants are seen once more in the word for knee: *berek*. The words are from the same semantic field; they are homonyms. In the meaning of 'to kneel', there may be an allusion to a posture commonly adopted when being blessed. Kneeling was also the customary attitude of prayer in Israelite worship (Ps 95:6, 1 Kgs 8:54, Isa 45:23 and Dan 6:11).

[33] As a general description of blessing, Koehler-Baumgarter (1951, 153-154) gives: "heilvoller Kraft"/"fortunate power". Blessing by God then becomes "mit heilvoller Kraft begaben" and the praising (blessing) of God: "Gott als den Ursprung heilvoller Kraft bezeichnen". In the context of people blessing one another, the meaning is: "als mit heilvoller Kraft bezeichnen", or "heilvolle Kraft anwünschen".

[34] It is then always the piel (perf.) form *bārak*. Koehler-Baumgartner indicates that this can be used 'euphemistically' instead of the more commonly used words for 'to curse': *'ārar* and the piel of *qālal*: *qillēl* (K-B 1951, 154). In its meaning of 'to curse', it is found in Job 1:5,11 and 2:5,9, for example.

[35] The link between blessing and (male) offspring is particularly strong in the Patriarchal narrative (Gen 17:16,20, 28:3, 35:10,11, 48:3,4), but blessing may also be concerned with other matters (Deut 7:13, 1 Sam 2:20).

day. Animals generally play a 'passive' role in blessings, as an element of the blessing. Acquiring plenty of livestock (traditional farm animals such as sheep, cattle and goats, as well as draft animals like asses and camels) is thus considered to constitute a sign of blessing, and possession of such animals proof that the owner is blessed.[36] In Genesis it is the animals themselves which are blessed. It is also notable that here, in verse 22, the blessing is directed towards fowls and fish. Elsewhere, these animals are never associated with blessing, not even passively. It is also characteristic of the biblical creation narrative that animals are blessed specifically, separately from humans.

On the sixth day, the land animals are created, denoted here by the generic term 'living creatures' (*nepheš ḥayyāh*).[37] There is a further, tripartite subdivision: 'cattle', 'creeping things' and 'beasts of the earth'. The word for cattle, *b^ehēmāh*, is the generic term used virtually throughout the HB for farm livestock. Given the other two categories, this is certainly its intended meaning here.[38] In the eyes of the author, the tame, domesticated animals were apparently created as such. That they are tame is part of the creation order. This association was also made for fruit-trees, in Gen 1:11, but it is

[36] Compare Job, who regained twice as much livestock as he had lost (Job 1:3 and 42:12).

[37] The noun *nepheš*, which occurs frequently in the HB, is generally translated as 'soul' in the RSV. It is bound up closely with the verb *nāphaš*, which can be translated as 'to be refreshed' (Exod 23:12 and 31:17) and possibly also as 'to refresh oneself' (1 Sam 16:14). *Nepheš* denotes the life spirit or breath of life (cf. Gen 1:30) which, together with the heart ("with heart and soul": Deut 11:13, 2 Kgs 34:31) or alone ("my very soul"), is described as the characteristic element of humanity (or the animals, cf. Gen 9:4) and, *pars pro toto*, is taken to represent a person as a whole ("O my soul" in Psalms 42 and 103). A number of texts consequently reflect the idea that it is the soul that disappears upon death to the realm of the dead (*š^e'ôl*), or that is saved from death and the realm of the dead (Job 33:18,22,28,30, Ps 49:15, 86:13, 89:48). The HB also refers to the pair "body and soul" (Isa 10:18), but the strict demarcation between body and soul of Greek philosophy is unknown in the HB. The view that animals can also have a *nepheš* was widespread in Antiquity and is also to be found in other (ancient) cultures (for the Far East, see: H.P. Blavatsky 1886, 1887, for Egypt: J. Boessneck 1988 and for Greece: U. Dierauer 1977 and J.N. Bremmer 1983).

[38] The category of livestock is congruent with the category of domesticated animals, since nowhere in the HB is mention made of any (other) animals of the 'pet' kind. Cats, which are mentioned and depicted regularly in Egyptian sources, are entirely absent in biblical texts. Dogs are mentioned only in a negative context. They had a poor reputation (1 Sam 17:43) and were associated with the devouring of flesh and blood (1 Kgs 14:11, 22:38). They were certainly not kept as pets.

less clear there, because there are also 'wild' fruit-trees. This fact supports the idea that in this first creation story the author projects back and/or evokes an idealized picture of the creation, as intended by the Creator Himself. And in that idealized picture the tame animals fit perfectly. It is thus not humankind but the Creator himself who creates the domesticated animals according to their kind. 'Their kind' is part of the creation order. We shall return to this issue below, in considering verse 30.

2.3 Humankind as Ruler

2.3.1 *The image of God*

At the end of the sixth day there follows the account of the creation of man and woman. Here we encounter a key passage frequently cited in the contemporary environmental debate: verse 28, with its instruction to "have dominion" over the earth and over living nature.[39] In the tradition, the doctrine of *dominium terrae* has evolved on the grounds of verses 26-28 and related passages in Gen 2:15, 9:2,3 and Ps 8:6-9.[40] In the contemporary debate this term is taken to denote the (unbounded) dominion of humanity over the earth. Here, we shall discuss verses 26 to 30 as a textual unit.

In explaining the expression "Let us make man" (1:26) considerable attention has been paid to the significance of the plural 'we'. This has been deemed important in relation to the issue of latent polytheism in the (original) creation account. The creation of human beings as a decision of the gods is also encountered in the Babylonian creation stories. These accounts go on to provide a motive for this heavenly decision: people are to take over the heavy work from the gods.[41] In Genesis no motive is given for this decision. We leave this point for what it is. Here, our interest lies in the relationship between the three elements of the verse, viz.: *man, God's image*

[39] The imperative character is clear in verse 28. In verse 26 'dominion' can still be interpreted as a description of humanity's potential. For a thorough treatment of Gen 1:26-28 in a Century of Old Testament research see Gunnlaugur A. Jónsson 1988.

[40] U. Rüterswörden (1993, 81-83) and H.-J. van Soest (1996, 11,12) point out that the expression *dominium terrae* does not occur in Latin translations of the Bible, but is first found in medieval commentaries on Genesis.

[41] *Atraḫasis* tablet I, iv (S. Dalley 1991, 15) and *Enūma Eliš* tablet VI (S. Dalley 1991, 261).

and likeness and the *dominion* over the animals and the earth. This relationship has been a focus of study for virtually all commentators. In the tradition, the expression 'image of God' has become most familiar in the Latin translation *imago Dei.* In his commentary on Genesis, Westermann points out that virtually all exegetes assume that the text has something to say about humans, so that the key question is then: what does the *imago Dei,* or, in his own words, the "Gottesebenbildlichkeit der Menschen", comprise? From among the many answers given to this question, we shall mention two in particular because of their relevance to our subject and then go on to consider Westermann's own reading of the text.

First of all, there is the interpretation of humanity as god's representative on earth. Analogous to the image of the Near Eastern king as the 'deputy' of the Godhead on earth, in Genesis 1 humankind on earth is then held to represent the mighty King of heaven.[42] And because in the creation story it is not one person but humankind that is described as image-bearer, such an analogy only makes sense if 'humankind' acts a representative with respect to God's other, non-human creatures. Humankind may thus rule over the animal kingdom as a kind of viceroy. In Egypt and Mesopotamia this kind of regal ideology was fairly commonplace.[43] In the HB, support for this reading is sought in the use of the words for 'in our image' and 'dominion'. In this context, particular attention is given to Psalm 8, the only other place where reference is made to the 'near-divine' (*m*e*ʿaṭ mē*ʾ e*lôhîm*) nature of humankind in relation to its dominion over creation.[44] The fact that in this story it is not a single person, but all humankind that is to act as a representative is deemed revolu-

[42] F.M.Th. de Liagre Böhl (1923, 62) gives very clear expression to this stance when, in his commentary on Genesis, he writes: "And thus, here on earth, man is, in miniature, what the great Godhead himself is in heaven".

[43] One of the many commentators to follow this line is U. Rüterswörden 1993. On the basis of iconographic material, he sees correspondences with Persian regal ideology, in particular.

[44] *m*e*ʿaṭ* (from the stem *mʿṭ*, to decrease, grow less) occurs over 100 times in the HB. The general meaning is: little, a little bit. The RSV translation of Ps 8:5 is: "...little less than God". This line is also adopted by P. Beauchamp (1969, 359), who proposes the translation: "tu lui as fait manquer de peu la condition divine". He considers the expression employed by the Psalmist to be older than the text of Gen 1 and on this basis posits that it is more justifiable to explain Gen 1 with reference to the Psalm than vice versa. In the Jewish tradition "god(s)" ("condition divine") also refers to the realm of angels (JPS, 1985 note to Ps 8:6 and Bereishis 1977, quotation from Ps 8:6 on p.71).

tionary by some commentators.[45] In this reading, the message of Genesis 1 and Psalm 8 is uniquely 'democratic'.[46]

Objections to this interpretation have been raised from many quarters. On closer inspection, the iconographic material used by Rüterswörden to support his thesis proves to relate mainly to dominion over people and far less to dominion over animals.[47] On theological grounds, Westermann deems the regal ideology implausible. In his words: "Daß der Mensch der Repräsentant Gottes auf Erde sei, ist für P kein möglicher Gedanke" (212). To the extent that this rejects the implication that, with the appearance of the human race on earth, (the 'deputy' of) God also appears, I would agree with this position. But whether this puts an end once and for all to the regal analogy remains to be seen.

As to the question of what is embodied in the 'likeness' between God and humankind, the faculty of reason has also been cited as a trait by which humans are distinguished from animals and linked to God. In this explanation, it is the rational mind that allows the human species to rule over the animal kingdom. Reason is the human instrument of rule *par excellence*. However, it is highly debatable whether a dualism between animal 'irrationality' and human 'rationality' is congruent with the conceptual world of the ancient Israelites. Nowhere in the HB do we see it explicitly reflected.[48] This is a way of thinking and terminology that emerged only later, under Greek influence, in the centuries prior to our own era. Hebrew has no real equivalent for the Greek philosophical notion of reason (*λόγος*).[49] In the philosophy of the Greeks, *λόγος* is intimately linked

---

45 See, *inter alia*, F.M.Th. de Liagre Böhl 1923.

46 See K. Koch (1983, 30), and, for Psalm 8, J.L. Koole 1966, 7,8.

47 Of the 32 drawings included by R. Rüterswörden in his book (1993, 178-193), nine clearly show people being trampled on by the ruler. The images showing animals are far less explicit and are not as clear in reflecting dominion over animals.

48 The only text that can be held to give any support at all to this thesis is Ps 73:22. The RSV translation of this verse reads: "I was stupid and ignorant, I was like a beast toward thee". The second half of the verse runs literally *bᵉhēmôt hāyîtî ʿimmak*, 'I was cattle with (or with respect to) You'. The JPS translates this as "I was brutish toward You", but I would suggest the translation 'I acted like a cow before You', with 'cow' in the sense of 'an inept person' (as also in the Dutch language). From this lament by the psalmist regarding his own conduct, which he considers more fitting for cattle, we can scarcely derive a measured conclusion concerning the question of whether animals are held to possess 'reason' in the HB.

49 The verb *yākaḥ* (to reason, to argue; in Greek: *λογίζομαι*) comes close, but occurs in this meaning in the HB only twice (Lev 19:17 and Job 23:7). In con

with the divine, and in some schools even with the Godhead in person. Animals are held to have no λόγος and the term ἄλογοι is therefore used in referring to them. In the New Testament, too, we find traces of this heritage.[50] As we shall see further on, in Chapter 5, early Christianity is widely permeated by this 'Greek' way of thinking.

This said, though, even if human beings are gifted with 'reason' and animals are not, this is not to say that this quality is at the core of the *imago Dei*. On the contrary, in the HB it is to be queried whether reason is even seen as a divine quality at all, let alone as most characteristic quality. Throughout the HB numerous human qualities and emotions are attributed to God, but nowhere does reason emerge as the dominant trait. God is pictured as loving, wise, wrathful, vengeful, merciful, sorrowful and forgiving, but never as the personification of reason. In references to humanity, too, we nowhere find an explicit association between reason and divinity. The wisdom which, according to Prov 3:19 and 8:22, was also involved in the creation of the world has a far wider connotation in the HB than our term 'reason'. People and animals can be 'wise', too (Prov 30:24-28); and according to Job 38:36, wisdom is to be found even in clouds. Possession or lack of reason cannot therefore be taken as an argument or justification for humanity's ruling over or subduing the animals. There is here a world of difference from the thinking of most Greek philosophers.

Both the exegeses here discussed posit a clear link between 'ruling over' and being created 'in God's image', the former being held to be directive and therefore explanatory for the latter. According to Westermann, this is demonstrably false, since the "Beherrschen der übrigen Kreatur nicht Explikation, sondern Folge des Erschaffens nach dem Bild Gottes is". His own exegesis starts with the remark that the text is not concerned with a characteristic of *humanity*, but with an activity, an act of *God*. In his view, the phrase "Let us make man in our image, after our likeness" is itself of an explanatory nature and indicates what God has in mind for all people. Westermann agrees with the standpoint of Karl Barth, who described humans as

---

trast, in the Greek apocryphal text 4 Mc the term λογισμός, which occurs dozens of times and is generally translated as 'reason', can be termed very characteristic.

[50] In 2 Pet 2:12 and in Jude 10, where there is reference to "irrational animals" (ἄλογοι).

being created as "Gottes Gegenüber", implying that something "zwischen dem Schöpfer und diesem Geschöpf geschehen kann". Conversely, this implies that "die Eigentlichkeit des Menschen in dem Gegenüber zu Gott gesehen wird" (218), which also gives a religious-anthropological slant to this highly theological exegesis.

One weak point in Westermann's approach appears to be that his exegesis actually severs the link with the subsequent text, concerning 'dominion'. After all, there is no reason why an event enacted between God and humanity should or even can logically lead on to human dominion over animals. At the same time, with all the obscurity surrounding the exact character of the *Gottesebenbildlichkeit*, both in Genesis 1 and in Psalm 8, it is precisely 'dominion' which immediately follows and with which it appears to be inextricably bound up. In Gen 9:1-7, too, where we encounter the phrase "in his own image" in verse 6, this link is emphatically made. The borderline between the concept in question and that which follows, or follows from it, becomes rather theoretical if no independent sources are available with which to explain the former, the only information reported being in terms of what follows, or follows from it. In many expositions, 'dominion' is a consequence of humanity's being "little less than God" (Psalm 8). While this may well hold true in terms of cause and effect, it would seem more fruitful to take the effect as our starting point in endeavouring to understand the passage.

In explaining the 'image and likeness', it is defensible as well as more logical to think immediately of 'dominion'. The 'likeness' is embodied in the dominion itself. Human beings do not rule over others of their kind, nor as men over women, but over the earth and the animals. This is quite remarkable, in that era and in fact in any era. Although implicit, in this first creation narrative there is already discernible reference to the fundamental equality of the sexes. We shall return to this equality in discussing the second creation story in Section 2.5. It is by 'having dominion', and in that dominion, that human beings bear the image of God. On this point, K.A. Deurloo arrives at a similar conclusion: "It is possible to distinguish between but not to separate what someone is and what he does. Indeed, the essence of God and humankind lies precisely in their 'doing'".[51] It is not the case that human beings are pictured as a 'God in miniature', or as a being with a divine core, but rather that by their

---

[51] K.A. Deurloo 1986, 26.

dominion they shows a likeness to their Maker. This capacity for dominion is something given to humankind in God's Creation. In 'lording over' the living world, humanity shows a likeness to the Lord God. This is also the picture found in Psalm 8: compared with the majesty of God, human beings may indeed be insignificant, but they were still created near divine and 'crowned with glory and honour'. Verse 6 of the psalm then states: "Thou hast given him dominion over the works of thy hands". The verb used for 'ruling' in this verse is *māšal.*

Besides the unmistakable connection with dominion, the text also appears to establish a second link: between 'the image of God' and humans being created as 'man and woman', or 'male and female'. In this reading the second part of verse 27 would then constitute an explanation for the first. The image of God indicates a relational aspect of being human (cf. Westermann's "Gottes Gegenüber") and this finds analogy in 'male and female'.[52] This is the view taken by Karl Barth, but I find it highly disputable. Van Soest rightly points out that there is little support for this position in the context of the history of religion.[53] Sexual differentiation is not specific to human beings, moreover, but is also the norm in the plant and animal kingdom. If the image of God were associated with male and female aspects of the human species (and of God), it is unclear why such an image of God should hold only for humankind.

The expression at the end of verse 27 is found again in Gen 5:1-2. Gen 5:3 states that Adam creates (a son) in his own likeness. It is not reported what this likeness entails. It may be a similarity of appearance, but equally well a similarity in behaviour, nature or position. With respect to Genesis 1, this passage from Genesis 5 provides too few clues to choose between the two interpretations suggested. Following F. Breukelman, Deurloo sees 'the generations' (a translation of *tôlᵉdôt* in 5:1) as a central theme of the book of Genesis, giving rise as they do to the people of Israel, with Abram and Sarah as the main procreational link. It is God's creation "in his own likeness, after his image" that provides the basic figure of human procreation, "in his own likeness, after his image". Arguing further along these lines, according to Deurloo: "it first becomes

---

[52] Verse 27 first has a singular (*'ôtô*), and then a plural (*'ôtām*) form: "in the image he created him; male and female he created them."

[53] K. Barth 1970, 204-233; H.-J. van Soest 1996, 177,178.

manifest *in Israel* that man has been created as God's counterpart".[54] It is to be queried whether this explanation does not embed the narrative of Genesis 1 too much in the (sacred) history of the people of Israel, rather than treating it as a universal creation narrative. This 'sacred history' interpretation of Genesis 1-11 by Deurloo and Breukelman stands diametrically opposed to the universalist exegesis of Westermann.

The Jewish tradition ties together a number of the interpretations treated above. A recent translation of Genesis, with annotations from the Mishnah, the Talmud and the rabbinic literature, gives the following interpretation of 'image and likeness': "Taken in sum total then, the two parallel terms *ṣelem* and *d^{e}mût* describe humans in their spiritual resemblance to their Creator: their endowment with intellectual perception that gives him pre-eminence over the animals, that guides them consciously in the exercise of his free-choice, their moral sense of right and wrong, and finally that gives humankind its fundamental distinction of approximating some spiritual resemblance to his Creator."[55]

2.3.2 *Dominion and subjugation*

How, then, are we to interpret the concept of dominion in verses 26 to 30? This certainly seems an important issue, given the problems that that 'dominant behaviour' have given rise to in numerous realms, including that of the environment. Interpretation pivots on the meaning of the two Hebrew verbs *rādāh*, used in 1:26 and 1:28 in relation to the animals and translated consistently as 'to have dominion over', and *kābaš*, used in 1:28 in relation to the earth and translated as 'to subdue'. Several studies proceed from the assumption that these verses are not prescriptive but descriptive. History teaches us that this description (dominion and subjugation) of human behaviour towards nature has proven fairly reliable and indeed pertinent throughout the ages. But even though this characterization may be correct for our western civilization, the reading is disputable. Verse 26 allows for a descriptive interpretation, given the verbal form employed (impf. of the qal), but in verse 28 we encounter the imperative form of *rādāh* and *kâbaš*. In this verse we must therefore interpret dominion and subjugation prescriptively, as a com-

[54] K.A. Deurloo 1986, 36.
[55] Bereishis 1977, 71.

mand or assignment. It is an assignment that is intimately connected with the blessing given by God in the beginning of verse 28.[56] For this reason, an exclusively descriptive interpretation of verse 26, too, is extremely implausible.

To our modern mind, the traditional translations of the verbs *rādāh* and *kābaš*, 'to have dominion over' and 'to subdue', sound harsh. To what extent do we find support for these translations, and their harshness, elsewhere in the HB? Are there nuances to be found that may have a bearing on our reading of Genesis? Both verbs are indeed used in a veriety of contexts and the relevant passages have been subject of several recent studies.[57] The results can be summarized as follows:

Rādāh is found twenty times in the HB and once in the Apocrypha (Sir 44:3). Only in Gen 1:26,28 do we find *rādāh* with animals as its object; in the other texts it always refers to interhuman relations. On nine occasions (Lev 25:43,46,53, 26:17, Num 24:19, Isa 14:2,6, Ezek 34:4, Ps 110:2) the dominion clearly involves violence or harshness. Two passages (Ezek 29:15, Neh 9:28) describe dominion negatively, in combination with the power of the enemy, and in Ps 49:14 dominion by 'the upright' also has a negative undertone.[58] In three texts (1 Kgs 5:30 MT, 1 Kgs 9:23, 2 Kgs 8:10) 'ruling over' has the meaning of providing leadership, the activity of a commander or supervisor.

We find an apparently positive meaning in 1 Kgs 5:4 HB and in Ps 72:8. Both these passages concern the rule of a benign king: Solomon. Here, 'dominion' is placed in a geographical context: "on this side of the river" and "from sea to sea, and from the River to the ends of the Earth". Analysis of these two passages shows that here, too, there is reference to dependent subjects (including servants/slaves) who, according to Rüterswörden, were obliged to pay taxes to the ruler under the law of nations.[59]

In this context, Joel 3:13 contains a controversial passage. In this

---

[56] An imperative may also take the direct form of a wish (or blessing), as in the expression 'Go in peace'.

[57] Most extensively in: U. Rüterswörden 1993, 81-126. For an up-to-date and comprehensive review as well as a clear-headed evaluation of viewpoints, see: J.-H. van Soest 1996, 181-211.

[58] The text is possibly corrupt. Many exegetes doubt whether the verb *rādāh* was in fact employed in the original. For a review of this debate, see: P. Casetti 1982, 132-136.

[59] U. Rüterswörden 1993, 95,96.

verse a metaphor from the grape harvest is used to call the soldiers to battle. The passage reads: "Go in, tread, for the wine press is full. The vats overflow". In the Hebrew text we find the imperative form *r^e^dû*, which also occurs in Gen 1:28. With the translation 'tread' the translators of the RSV take *r^e^dû* to be an imperative of *rādāh.* The 'ruling over' the grapes is here therefore synonymous with treading, because the grapes are pressed by treading on them. However, *r^e^dû* may also be an imperative of *yārād*, to descend, yielding the translation: 'descend'.[60] Archeological data render this translation less probable. The wine presses excavated to date lie higher than their surroundings, and are generally only about a hand's width deep. One can hardly speak of 'descending', and the translation 'tread' is to be preferred.[61] Another controversial classification is to be found in Lam 1:13, although most scholars here opt for *rādāh* and translate the derivative as "overcome" (Driver) or "zertreten" (Kraus).[62]

Etymologically, the verb has been associated with the Akkadian *redû*, which denotes the 'leading' or 'driving' of animals.[63] This derivation has been critically analysed by Rüterswörden, who proceeds from the similarities between the verbs *rādāh*, *yārād* (to dismount, to descend) and *rādād* and, following others, arrives at a common (Egyptian) root *rd*, foot. The verbs denote actions undertaken with the feet. From this root meaning the Hebrew and Akkadian meanings developed in parallel. For *rādāh* the meaning would then be 'to tread', 'to bring underfoot'.[64] It is against this background that 'ruling' should be seen. The ruler assumes authority over the land by 'treading' (on its soil) and those over whom dominion is exercised, the population, are subjugated to him as *'^a^bhādîm* (servants/slaves); they have been 'brought underfoot'.

Summarizing, the conclusion is thus that there is a harsh undertone in the meaning and usage of *rādāh* in the HB. In most translations this harsh undertone is unavoidable. There is always some inequality of power.

The verb *kābaš*, translated in Gen 1:28 as 'to subdue', is found thirteen times in the HB. The object of the verb is more varied than

---

[60] The Dutch Statenvertaling ("daalt af", descend) and the Vulgate ("descendite") follow this reading.

[61] See: J.-H. van Soest 1996, 183, note 135.

[62] U. Rüterswörden 1993, 100, note 59.

[63] K-B 1951, 874; AHW 965; N. Lohfink (1988, 22) and K. Koch (1983, 25) here assume an influence of Akkadian on the Hebrew.

[64] M. Görg 1986, 144; U. Rüterswörden 1993, 101,102.

in the case of *rādāh.*: land, people, the kings, slaves, nations and iniquities are all mentioned. In all cases there is subjugation by violent means. Num 32:22,29, Josh 18:1 and 1 Kgs 22:18 describe the subjugation of "the land" and 2 Sam 8:11 that of "nations". A number of passages make reference to people being "brought into subjection as slaves": Neh 5:5, Jer 34:11,16 and 2 Kgs 28:10. In Esth 7:8, where the object is the queen, the verb is generally translated as 'to assault'. The descriptions in Zech 9:15 and Mic 7:19 evoke an image of (iniquities) being 'brought' or 'trodden' underfoot. There are clear correspondences between *kābaš* and *rādāh*; in the LXX the verbs are translated identically on a number of occasions. With *kābaš*, too, there is a distinct connection with the servitude of servants/slaves, but there are also differences. The harsh undertone of *rādāh* has almost become the dominant tone in *kābaš*. More so than in the case of *rādāh*, *kābaš* denotes the actual substance of dominion and subjugation.

Modern commentators have endeavoured to refute accusations that the seeds of the environmental problem are bound up in these words. It has, for example, been argued that the usual translations with their harsh undertone are inaccurate.[65] In this context it has been suggested to translate *rādāh* not as 'to rule over' but as 'to manage', with the implied meaning of to watch over and protect (German: 'hegen'). The main argument for opting for such a translation is the fact that *rādāh* is frequently used to characterize the activities embodied in a king's rulership. And, just as a benign king leads and protects his people with wisdom and in this sense rules over his subjects, so too should *rādāh* be understood as a term used to express the 'regality' of humankind over the animals. In relationship to the earth and the land, *kābaš* is then held to mean above all 'to work', 'to develop', 'to open up' or 'to shape', something that in the ancient Near East obviously required considerable effort, since the soil in this region is generally difficult to work. In the context of an agricultural society this is a natural line of thought.

[65] J.-H. van Soest (1996) has clearly demonstrated that this 'reversal' in interpretation and translation is connected with the initial expressions of concern about the environmental problematique. Examples of 'environment- and animal-friendly' exegesis begin to appear in the early 1970s. C. Westermann (1974, 219,220) can be seen as the first representative. Subsequently, and more outspokenly: K. Koch 1983, 23,24; C. Labuschagne 1990, 11; J.C.M. Holman (1992, 83) and a range of Anglo-Saxon literature on environmental ethics (*inter alia*, L.H. Steffen 1992).

As we have seen above, however, these arguments prove to be weakly grounded. The word *rādāh* is not characteristic of the rule of a king, for example, even though the verb may often have a king as its subject. In a 'regal' context, the verbs *mlk* and *špṭ* are far more frequently used for 'to rule'. Another argument for a 'kinder' interpretation is derived from the earlier passage from Joel 3, in which the word is used to denote the trampling of grapes to make wine. Although the grapes are stepped on with force and '(op)pressed', in this reading that 'treading' does not serve to rule over the grapes or treat them cruelly, but rather to free the costly juices. Although this is undoubtedly an original interpretation of this 'grape text', it is erroneous, as I have argued above. The question remains, moreover, whether this is thus the most eligible general meaning of *rādāh*. *Rādāh* has a slightly wider range of meanings than *kābaš*, and a number of these certainly have a less negative and possibly even a positive undertone: in relation to the rule of Solomon, for example. In many passages, on the other hand, it is difficult to remain oblivious to the harsh undertone. Just as with *kābaš*, there is a danger of texts being read somewhat too selectively or too interpretatively. Behind these readings there also lies an overidealized picture of the kings and rulers of the ancient East.[66] A similar idealized picture is also reflected in the descriptions 'open up' and 'shape' for agricultural tillage of the earth.

In attempting to imagine what the author of the first creation narrative might have meant by 'subduing' the earth and having 'dominion' over the animals, we must make due allowance for the fact that in Gen 1 (and in the beginning of the second narrative, which we shall be discussing below) creation is described as it was intended by the Creator in the eyes of the writer. Again and again, the acts of creation are termed 'good'. In interpreting verses 26 and 28, the exegetes discussed make too little allowance for this fact. There is thus no reason to assume that these two verbs possess explicitly unfavourable meanings and to translate them as, say: to trample and tread underfoot. Nor need the terms 'dominion' and 'subdue' in the present translation be interpreted as denoting some mission to ruth-

[66] On many occasions the HB sketches a picture of ruthless, oppressive kings, with Solomon forming an exception and an ideal rather than a representative example. This oppression extended beyond human subjects and enemies. C. Houtman (1982, 17) points out that a benign king (or shepherd) was also able to protect his subjects (or herd) against 'human-unfriendly' nature.

lessly exploit nature.[67] Neither is there any good reason to interpret the other passages of the HB in a 'kinder' light than intended, retroactively, in order to arrive at a 'kinder' reading of Genesis. There is no reason to assume that the writers of the creation story had somehow become entirely divorced from the realm of experience of the agricultural society with which they were familiar. Within the hierarchy and the order described in Genesis 1, humans rule not over other human beings but over the animals and the earth. While the latter are to serve humankind, this is not their sole vocation, though, for they are held to be good and blessed in their own right. The hierarchical relationship between humankind and nature is different from that holding today, however, lacking elements that are characteristic of the world as we know it.

To clarify this we must at any rate establish a connection between verses 26 and 28 and verses 29 and 30.[68] In the latter verses, we find explicit rules about what humans and the animals are to eat. The latter are given green plants to serve as food, while humankind is provided with seed-bearing plants and fruit from the trees. In the creation order of Genesis 1 humans and animals are thus vegetarian. Death is absent, at least the death that attends the acquisition of food. The prophet Isaiah sketches an evocative picture of this ideal of a 'peaceable kingdom' (chapters 11 and 65). It is well possible that the author of Genesis shared or was familiar with this prophetic, eschatological image and projected it back onto the creation order.

This said, though, an image of this kind may still entail 'subjugation' of the earth, in the sense of 'opening up' land for the purpose of cultivation. From his own experience, the author was familiar with a world in which 'wild' nature formed a threat to the security and survival of the (agricultural) community.[69] In this context, arable farming, viticulture and fruit-growing are not only permitted, but

[67] If we follow P. Beauchamp (1969, 359) in his view that Psalm 8, on the basis of its earlier dating, can serve more appropriately to explain Gen 1 than vice versa, then use of the verb *māšal* (for 'dominion') in Psalm 8 may also be taken as an argument for a 'non-despotic' interpretation of the ruling of Gen 1.

[68] J.-H. van Soest (1996, 193, 194) considers the efforts to bring Gen 1:26-28 into harmony with Gen 1:30 as stemming from a desire on the part of the exegete, it being "perfectly feasible that the dividing line between the *real* world and the *ideal* world is not drawn that sharply in Gen 1". In my view there is a tension between the world in which the writer (P source) lived and the ideals he had. It is conceivable that Gen 1:26-28 is formulated more in terms of the *real* world than Gen 1:30, which clearly reflects an *ideal* world.

[69] See C. Houtman 1982, 17.

also constitute a mission and ideal, so that the imperative remains in force and gains new significance. Humanity's 'dominion' over the animals can be seen as embracing their use as pack and draught animals, as well as the milking of cattle and the shearing of sheep. However one envisages this situation, at its heart lies the fact that there is no place for death in human relations with animals, nor in relations among the animals themselves. Human beings have fruits and crops to eat and in Gen 1:30 the animals are given green plants to feed upon. On the basis of these passages, it seems arguable that, in the idealized picture sketched by the author, there existed (can exist) a relationship with nature whereby the 'dominion' was (is) still as the Creator intended.[70] This stands in contrast to dominion that, in the world of the author (as well as our own), has a predominantly harsh undertone: harsh, because it is accompanied by death, degradation and exploitation when it comes to the animal kingdom and the earth and harsh, also, because it has come to extend to interhuman relations, too, where it originally has no place in the eyes of the author of Genesis 1.

It is undeniable that Genesis 1 at the same time fundamentally establishes the human status relative to the rest of creation. Humans are unique creatures, situated separate from and in a certain sense above the others. The notion that the human beings represent merely one more organism in a long evolutionary lineage is reflected nowhere in the creation narratives, nor in the rest of the HB. More generally, in addition to the recognizable theme of the order that gradually emerges from the chaos, the image of the first creation narrative appears to be one of hierarchical order. And in that order humankind stands above the other creatures, as the apparent 'crown' of creation.[71]

There is an alternative interpretation that states that, if there is indeed a hierarchy, humans should be placed at the bottom, because they appears last, thus 'closing the ranks'. To my mind, this reading is implausible. The chosen sequence from a-biotic, via vegeta-

[70] These elements recur (or were already present) in the image of the ideal king. The benign king protects his subjects against the threats posed by wild beasts. Cf. what David states to Saul to indicate that he is equipped to fight Goliath (I Sam 17). I shall return to this 'regal motive' in Chapters 4 and 5.

[71] This view predominates in the Jewish tradition, too. In a recent publication Geoffrey Wigoder, chief editor of the authoritative *Encyclopedia Judaica*, formulates it thus: "Man is the apex of creation, because he is created in God's image" (G. Wigoder 1994, 27).

tion, fish, birds and land animals through to humankind, the increasing use of the expression "it was good", the fact of a blessing not being given until the animals were created and, finally, the creation of man and woman in God's image and likeness: all this goes to support the impression of a hierarchical creation order.[72] The Talmud indeed already warns about drawing erroneous conclusions from this order of creation.[73]

The metaphor of the human race as the 'crown of creation' evokes considerable protest in the context of the contemporary environmental debate, in which humankind is seen as the curse rather than the crown of creation. Given the fact that here, too, we encounter a degree of tension between the ideal and the real, it is worth elaborating the former more precisely on this point. This holds even truer because the boundary between human beings and animals in the *scala naturae* has always been a major point of debate in Western culture and in that debate the creation narrative has played a significant role.[74] In this context the difference between man and beast and the unique status of the former have always been heavily emphasized, with certain other aspects to some extent neglected—aspects that are most certainly of relevance for human relations with the animal kingdom. In the first place, it is noteworthy that the human race, although created as the final living creature, appears on the same day as the land animals. This leads one to conjecture that, in the order of things, there is affinity alongside difference. At the least, humans and beasts belong to the same domain. Furthermore, the idea that the main purpose of the creation should be to prepare the earth for (the arrival of) humankind strikes me as un-

[72] P. Beauchamp 1969 adopts a different position. He sees a concentric structure in the narrative, with the creation of the heavenly bodies on the fourth day constituting the climax. A concentric reading has been defended for other parts of the HB, too, and exegetically it often leads to interesting accents. Its application to the creation story has attracted few adherents, however, and, partly because of what is written about the seventh day being holy, this reading also appears less plausible.

[73] This exercise in putting things in their true perspective is elegantly phrased in the title of a book by C.D. Stone: *The Gnat is Older than Man* 1993. Stone was inspired by the Talmud, which states that the world was created for man, but also cautions against pride and calls for an attitude of humility, for "the gnat is older than he" (Stone vi).

[74] See, *inter alia*, A.O. Lovejoy's classic work *The great Chain of Being* 1936 and also C.J. Glacken's masterly *Traces on the Rhodian Stone* 1967, 35-79.

tenable.[75] It is certainly the case that the author, unfettered as he was by the presumed negative implications of an anthropocentric perspective, modelled the order of creation to a large extent on the ideal pastoral and agricultural society, in which humankind plays a pivotal role. But in that order all things have their true place, function and value, even divorced from the significance they (may) have for humanity. They are termed 'good' long before humans makes their appearance. There is, consequently, no linear hierarchy, with the 'lower' existing solely for the benefit of the 'higher'. The idea, which we shall encounter in Chapter 4 in connection with the Greek philosophers, that the value of an organism is derived exclusively from its value to (or usefulness for) an organism 'higher' in the hierarchy is not found in the creation narrative. The 'intrinsic value' of non-human life is also frequently encountered elsewhere in the HB. The concept of intrinsic value is thereby given substance by direct, i.e. non-human-mediated, ties being assumed between the Creator and non-human nature.[76] The hierarchy thereby assumes the form of a scalene triangle (see Figure 5.1 in Chapter 5).

In the creation narrative itself, finally, the metaphor of humankind as the 'crown of creation' is brought very firmly into perspective. This is at the end of the first creation story, which concludes not with the creation of humankind, but with the seventh day, a day that elsewhere in the HB (in Exod 16:29, for example) is associated with the sabbath.[77] Perhaps, as I hope to argue convincingly in the following section, this is the conclusion not only of the narrative, but also of creation itself.

---

[75] The Dutch *Statenvertaling* (completed in 1637 and, just like the King James, for several ages very influential) goes a step further in its comments on the text of Isa 45:18: "He did not create them in order that they might be empty, [but] formed them in order that man might dwell therein". The translators hold the view that human habitation, as the purpose of creation, applied specifically to the Church: "And, especially, partly to that end, in order that He might gather together from among men on earth a church where He might be known and affirmed".

[76] It is very characteristic in this connection that God, according to the Bible, establishes a covenant with the animals (Gen 9:8-17 and Hos 2:17), that animals can be redeemed (Ps 36:7b) and that creation can sing praise to the Lord (e.g. in Ps 19).

[77] The probable authorship of P makes it very plausible that a foundation for the seventh day as the Sabbath can already be read into the first creation narrative.

2.4 Creation Completed

2.4.1 *The seventh day as the sabbath*

Chapter 2 of Genesis starts by looking back: "Thus the heavens and the earth were finished, and all the host of them". In their wording, the first three verses belong unmistakably to the first chapter. At the start of Genesis 2, P is still the scribe. It is possible that P did not intend these verses to serve as the conclusion of Chapter 1, but as a 'title' to the following creation story. Although the text reads as a logical conclusion, what follows in 2:2 sounds strange at first: "And on the seventh day God ended his work...". Strange, because here one would expect the sixth rather than the seventh day. Everything had already been completed, so what was there left to end on the seventh day? The enigmatic nature of this passage presents us with a problem of exegesis. A number of interpreters as well as several modern translations resolve this problem by reading 'sixth' instead of 'seventh' here, in verse 2^a. This reading has a very long tradition and is also supported by other, older versions of the text, such as the LXX, the Samaritan Pentateuch and the Peshitta. The same holds for the rendering of this passage in the book of Jubilees (2:2).[78]

Understandably, this problem was deemed important from a very early age, for what Genesis states here about the seventh day—even in the absence of the actual word 'sabbath'—can be considered a foundation for the sabbath and for observance of the sabbath in the order of creation. Because of the central and unique status of the sabbath and its observance in the religious tradition and life of the people of Israel, an accurate reading of the text is of the greatest importance, particularly in the later Jewish tradition.[79] If there was

[78] The book of Jubilees played an important role in the Qumrân community. Fifteen fragmentary copies of the lost Hebrew original of Jubilees have been found in the various caves of Qumrân. Fragment 1, column V of 4Q Jubilees contains the passage from Genesis with the 'sixth day' reading (F. García Martínez & A.S. van de Woude 1994, part 2, 322).

[79] Early (8th century BC) indications for the existence and the observance (or otherwise) of the sabbath in the Northern Kingdom are to be found in the books of the prophets Amos (8:5) and Hosea (2:10) and in the Southern Kingdom in Isaiah (1:13). Non-biblical sources appear to give some credence to the idea that the Israelites were already observing/celebrating the sabbath in the pre-exilian era, because Israel was attacked not once, but twice precisely on that day (G.F. Hasel 1992, 853). Reference is also made to the possibly allied Akkadian word *šapattu*, but the mere existence of this word says nothing about observance of the sabbath in pre-exilian Israel. Certainly, though, the sabbath and the 6+1 rhythm

anything still to be completed by God on the seventh day, then creative work would have been performed on that day. Particularly in the religious climate in which the texts with the 'sixth day' reading were written, this was a difficult notion to accept. As the New Testament, the apocryphal books Judith and I & II Maccabees, the Dead Sea Scrolls and a multitude of non-religious, non-biblical writings attest, the Jews had clear views about respecting the seventh day, the sabbath.[80] Indications about actual celebration of the sabbath (in the sense of sabbath worship) by people other than the priests are sparse in the HB, however.[81] There is a virtual absence of any clear description of religious activities (by lay people) in connection with the weekly sabbath. Nor is there any extensive description of what positive meaning sabbath celebration actually had for them. In the HB, observance of the sabbath appears to consist solely in stopping normal work. In Lev 23:25, in the description of the Day of Atonement, we encounter the commandment to refrain from *servile* work. The texts provide more detailed information about what was forbidden on the sabbath than on what was allowed. Certain domestic activities such as baking and cooking were forbidden, for example.[82] The pressing of grapes for wine-making and trading

---

of the week must be taken to be a typical Hebrew construct (see on this point the extensive study of the origins of the sabbath by J.H. Meesters (1966). Meesters' findings have recently been confirmed by W.W. Hallo (1996, 127-135). For a treatment of the question to what extent ordinary Jews truly engaged in active, religious celebration of the sabbath in the era of the HB and later, see: H.A. McKay 1994. McKay considers the difference between "sabbath observance" and "sabbath worship" to be of the greatest importance (23) and states that this distinction is not, or insufficiently recognized in conventional studies on the sabbath. As a result, the impression is given (in her eyes, wrongly) that active religious celebration of the sabbath was already common practice among ordinary Israelites in pre-exilic times.

[80] The importance of the sabbath in Jewish religious life is also made abundantly clear in the later rabbinic literature. Interesting proof that there have also always been doubts about its actual celebration is presented in the following words, ascribed to rabbi Simeon ben Jochai: "If Israel were to observe the sabbath twice strictly according to the rules, it would be redeemed at once" (*Babylonian Talmud, Shabbat 118*).

[81] Following a discussion of the relevant texts, Heather Mckay (1994, 18) comes to the following conclusion: "So, apart from the allusions to future sabbath worship supplied by these two intriguing texts, there are no texts about regular sabbath worship for the ordinary worshipper in the Hebrew Bible". In her view, the two texts she considers 'intriguing', Isa 66:23 and Ezek 46, are not concerned with actual everyday situations, but "are couched in the language of the prophetic vision".

[82] Cf. Exod 16:23.

activities also appear to conflict with the sabbatical rest.[83] An account in Num 15 relates that the 'gathering of sticks' on the sabbath led to the stick-gatherer being stoned to death.[84] These clues all point towards an interruption of normal everyday activities. It is to be assumed, however, that it was not until the post-exilic era that the law was further developed and intensified to yield an elaborate canon of rules to which worshippers were to strictly adhere. In the centuries leading up to our era this development can be explicitly traced, for example from the documents from the Geniza of Cairo, copies of which have been found among the Dead Sea scrolls.[85] Particularly among the Pharisees and Essenes, the religious leaders of the Jewish people immediately prior to our era, the strict character of observance of the sabbatical rest became a veritable shibboleth. On the seventh, sabbath day, every person was not only to refrain from his or her normal everyday work. What one was still permitted to do also became part and parcel of the 'path of the devout'. Only the most essential of activities were to be carried out, and these were minutely described. Most were routine activities, and only a minimum of these were to be undertaken. If there was one thing that was prohibited, it was creation, an act generating something entirely new.

Most exegetes and commentators, including the medieval Jewish commentators Rashi and Maimonides, agree, however, that the Masoretic Text (MT) is to be deemed original in this context. In the

[83] Neh 10:31, 13:15, Amos 8:5.

[84] Num 15:32-36. Meesters 1966 hypothesizes that stick-gathering was punished so severely because the purpose was to build a fire. In the later Jewish tradition fire-making on the sabbath was held to be fundamentally opposed to sabbath observance. In the HB there is no support for this position, apart from this (indirect) reference (?) in Numbers.

[85] The documents from the Geniza of Cairo were found at the end of last century in the lumber room of the Ezra Synagogue of Cairo. They included two copies of a previously unknown work. These are known as CD (A and B). Following the discovery of the Dead Sea scrolls, it was suspected that CD belonged to the same collection of sectarian writings as the Qumrân scroll (1QS). That it was an important scripture for the Qumrân community became clear when it was discovered that no less than eight copies of the work had been found in cave 4. Columns X and XI of CD (A) contain rules pertaining to the observance of the sabbath. These rules are considerably stricter than those in the NT. "On the sabbath, nobody may help an animal give birth to its young... If it falls in a well or pit, one may not pull it out on the sabbath" (F. García Martínez & A.S. van der Woude 1994, vol. 1, 220, 241). Cf. Matt 12:11, where a rescue of this nature is cited by Jesus as precisely something that one would do on the sabbath.

'sixth' versions it seems likely that the text has been adapted, a change probably intended to avoid problems of interpretation.[86] In his commentary on Genesis, Cassuto reports that talmudic sources cite this (*i.e.* 'sixth') reading as being one of the textual changes introduced in the Greek translation of the Torah made for Ptolemy Philadelphus (II).[87] Against the described historical background, an 'adaptation' of this nature is understandable.

If, like most commentators, we proceed from the 'seventh' reading, the interpretation problem shifts to the following word, the verb 'to end'. In this respect, either an answer must be given to the question of what was ended on the seventh day, or the active (translation) of 'end' must be revised. Rashi opted for the first approach and his elegant answer to the question of what was still to be created was: rest. On the seventh day God created rest, because that was all that creation lacked.[88] Although Rashi's answer can be termed a good 'find', other Jewish commentators have provided a clever but far less convincing explanation. They have the creation end in a split second on the boundary between the sixth and the seventh day.

In his commentary on Genesis, the modern Jewish exegete Cassuto also treats these verses extensively.[89] He focuses on the translation of 'to end' in 2:2. In his view the problems surrounding this verse are due not to the literal text itself, but to its interpretation. He points to the coherence of verses 2 and 3, which are highly structured through the triple repetition of both "on the seventh day" and "his work". On these grounds alone, he argues, the first 'seventh', in verse 2[a], cannot possibly have been intended as 'sixth', nor should it be read as such. On the other hand, Cassuto is convinced that all the work (of creation) had been completed by the end of the sixth day. On the seventh day God performed no work. He rejects the translation of *wayyišbôt* by 'rest(ed)'. Comparison with such texts as Exod 20:9-10 and 23:12 makes clear that there is an antithesis here between working six days, and doing all your work, and not working on the seventh day, and not doing any work. The verb *šābhat*

---

[86] W.H. Schmidt (1967, 156 note 2), following Schaller, concludes: "Dieser Eingriff in den biblischen Text ist vermutlich der älteste Beleg einer alten rigoristischen Sabbattheologie, deren Spuren sich später bei den Pharisäern und vor allem bei den Essenern finden".

[87] Unfortunately, Cassuto does not name the talmudic (being the midrasj *Genesis Rabba* 10, 9) sources (U. Cassuto 1989, 61).

[88] *Rashi's Torah Commentary* Pinchas Doron 2000, 11.

[89] U. Cassuto 1989, 61-64.

has a negative connotation: something is not done; work is stopped.[90] The rest associated with the sabbath in later texts is a result of that 'not-working'. In his dissertation, J.H. Meester comes to the same conclusion. The root meaning, in his view, is 'to cease', from which 'to cease work' and, finally, 'to celebrate the sabbath', but nowhere: 'to rest'.[91]

Cassuto translates the form of 'to end' in 2[a] not actively, but as a past participle: "And since God *was finished* on...'. On the basis of other, parallel occurrences, in particular Exod 40:33 (where Moses 'finished' the tent of the congregation, even though activities there had already been completed), he arrives at the following reading of the verse: "Since God was on the seventh day in the position of one who had already finished His work, consequently He abstained from work on the seventh day".[92] In his interpretation, Gispen follows Cassuto. Although he does read the passage as: 'And on the seventh day God ended his work...', he does hold that, analogously to Exod 40:33, an alternative reading would be: "And on the seventh day God had finished his work".[93] Westermann, and with him many others, follows the same exegetic line. Westermann approvingly cites Delitzsch, for example, who writes: "Der Sinn ist nicht, daß Gott am siebenten Tag die noch unfertige Arbeit zu Ende führte". Westermann also agrees with A. Heidel, who sees a declarative element in this verbal form: "And on the seventh day God declared his work finished" and observes finally, following Cassuto: "Das Verb 'er (sie) beendete' bedeutet in jedem Fall ein Konstatieren des Abschlusses, nicht aber ein Zuendebringen".[94]

Although there is thus scarcely any difference of opinion among these exegetes, none of them follow classical or several modern Jewish commentators in opting for a translation consistent with their views. Alongside the already cited translations from the NBG and Gispen, Westermann, for example, gives: "Und God vollendete am siebten Tag das Werk...". Even several recent Dutch translations that presume to follow the Hebrew as closely as possible are no different on

---

[90] See, *inter alia*, M. Buber (1964, 95) and A.B. Ehrlich (1968, 7). In his dissertation *Van ophouden weten*, C.B. Posthumus Meyjes (1993, 15) follows Buber. G.F. Hasel (1992, 849) also gives the verb the meanings 'to stop' and 'to cease'. In his view the precise relationship with the word for the day, *šabbāt*, is still open to debate.

[91] J.H. Meesters 1996, 14.

[92] U. Cassuto 1989, 62.

[93] W.H. Gispen 1966, 90.

[94] A. Heidel 1974, C. Westermann 1974, 233.

this point, reading as: "God completed his work on the seventh day..."[95]; "Elohim finished on the seventh day..."[96].

Neither, indeed does the original Hebrew text provide any good reason to opt for any other translation. What is written is *way^{e}khal*, an imperfectum consecutivum piel of the stem *klh*, and in a narrative text this form of the verb is generally translated as a preterite (perfectum).[97]

2.4.2 *Hallowing and blessing as completion*

Upholding this translation means that we are still faced with the problems and the associated exegetic questions outlined above. The main problem appears to be that something is completed that has already been finished. How are we to conceive of this? All the non-Jewish exegetes cited leave this question undiscussed. They suffice with the explanation that there can be no question of work (of creation) being carried out on the seventh day, but leave the translation in the preterite essentially as it stands. The Jewish commentators are aware that this question remains to be resolved, but none of the proposed solutions is entirely satisfactory. A more acceptable explanation can perhaps be found by reading the verses as a coherent whole and once again taking note of the statement regarding the seventh day:

> He finished his work which he had done.
> He rested on the seventh day from all his work which he had done.
> He blessed the seventh day and hallowed it: because on it God rested from all his work which he had done in creation.

The second parts of these three sentences relate, each time in slightly different wording, that the work had been done. In these verses the word for 'to do', *'āśāh*, which elsewhere in the HB is also used for 'normal everyday work', is employed for the work of creation. There is an elegant progression of usage. First we have the work that has been done; one might say: all the products of creation. Then we have the work as a whole. Reisel here translates as: "the task He had set himself".[98] And, finally, the link is established with

---

95 M. Reisel 1966, 20.

96 L. Buijs & M. Storm 1994, 16.

97 See: J.P. Lettinga (1976, 135,168,169). The verbal forms are reviewed by: B. Davidson 1970, 379,380.

98 M. Reisel 1966, 20.

bārā', the characteristic creative work of God.[99]

The repetition in the second halves of the sentences renders it probable that there is repetition involved in the first halves, too. The first halves each relate something about the seventh day. The three statements are mutually explicative. God finishes His creation by not working on the seventh day and he blesses and hallows this seventh day, because He had ceased from His work.[100] In the act of blessing and hallowing, creation is completed.

Blessing and hallowing can be regarded as two mutually related concepts. Blessing refers to the future. As we have already seen, in the HB it is often connected with the fecundity of humans, animals, crops and work. In a more spiritual sense it can be taken to mean: to state that the blessed will serve its purpose. Here, we encounter the term for the third time in the creation narrative. With good reason, Cassuto takes this repetition to signify the importance of blessing. We again see a progression, from physical fecundity in the case of the animals, via the double blessing of man and woman, entailing 'fruitfulness' and 'ruling' (note, therefore, that 'dominion' is also to be considered a blessing), through to the blessing of the seventh day, which is of a spiritual and all-embracing nature.

'To hallow' means 'to make sacred' or 'to declare sacred'. In the HB the term 'to hallow' (root *qdš*) is often taken to mean 'to isolate' or 'to set apart'.[101] Although this translation cannot be deemed incorrect, it is more of a description of the consequences of hallowing than the root meaning of the actual concept. Westermann suggests a different meaning when he posits: "Die Wurzel *qdš* enthält in sich das Moment des Sonders. Wenn Gott den siebten Tag heiligt, so bestätigt er damit die Besonderheit des siebten Tages gegenüber den sechs Tagen der Schöpfungswerke".[102] But what is then "das Moment des Sonders"? Wherein lies this 'special nature' of the seventh day? Cassuto gives as the "real meaning" of "holiness" (*q^edhûšāh*): "elevation and exaltation above the usual level".[103] The seventh day is

---

[99] A.B. Ehrlich (1968, 7) suspects that the text originally had *bādā'* rather than *bārā'* at this point. Instead of the usual unfavourable meaning of *bādā'*, 'to make up something/to lie', he proposes the translation: as God had planned it ("sich ausgedacht hatte").

[100] Cf C.F. Keil & F. Delitzsch 1980, 68.

[101] See W.H. Gispen (1966, 93) and many other commentators. K-B (1951, 825) gives a similar description of 'hallowed': "dem gewöhnlichen Gebrauch entzogen".

[102] C. Westermann 1974, 235.

[103] U. Cassuto 1989, 65.

elevated above all other days. To obtain a full picture of this important concept, however, another meaning of *qdš* should also be mentioned, and that is 'radiant' or 'shining'.[104] That which is hallowed distinguishes itself from all else by radiating and shining. It elevates itself above the rest like a radiant crown. In their usage in the HB, 'hallowdness', or 'holiness', always have a strong association with God. He is 'the Holiness' in person, and this radiates onto His surroundings. Conversely, people and objects must be 'hallowed' for them to be fit to be present in His surroundings and/or fit to serve their purpose. A similar line of thought is to be found in the rabbinic tradition concerning the laws of cleanliness. Here, a potter may not simply offer for use the pottery he has made, even though it is well-made and unblemished. He must first declare it fit, or 'hallow' it, before it is clean and completely fit for use.[105]

The Torah makes it clear that there also degrees of holiness, depending on the proximity to God.[106] This is very clearly reflected in temple services and in the design of the temple, for example. God Himself resides in the most sacred part, the Holy of Holies, and only the most hallowed priest, the high priest, was permitted to enter this space and then only on a special day: the Day of Atonement (Lev 16).[107] In the next chapter we shall see that the concept of holiness also plays a key role in the so-called dietary laws.

The six days enjoy their radiant crowning in the seventh day: a day that concludes the days of creation and thus yields the '6+1'

---

[104] K-B (1951, 825) traces this meaning back to the Akkadian *qadāšu* (glänzen, to shine). For an extensive treatment of the concept of sanctity in the priestly writings, see: P.P. Jenson 1992. Jenson places the concept in opposition to 'unclean' and links it to the Israelite cult. For the meaning 'radiant', see also: J.A. Emerton 1967.

[105] Cf. Mishna tractate Tohorot 7:1 (J. Neusner 1988, 1050). The ritual cleansing of pottery and tools is based on Num 21:22,23. See also: S.Ph. de Vries 1986 172-175.

[106] *Ergo*, the title of Jenson's work *Graded Holiness*.

[107] Although radiant light also plays a part in the divine manifestations described in the HB, the descriptions are ambiguous. Some of these manifestations were so overwhelming (blinding?) that normal mortals were unable to look. Nor were they indeed permitted to look (Exod 19:21, 24:17, and the NT account of Saul on the road to Damascus, in Acts 9). On the other hand, some were not overwhelming at all, as is reflected in the subtle appearance (?) before Elijah as the "still small voice" (1 Kgs 19:12,13). The most hallowed is often linked inversely to darkness. Sacredness and taboo are juxtaposed. Transgression of the sabbath commandment was punishable by death, for example (Exod 31:14, 35:2). This may explain the apparently paradoxical fact that the Holy of Holies was the darkest place in the temple.

rhythm. But it is not only the days that find their culmination in this hallowed day, as Westermann states; the work of creation itself is crowned on that day. The work is done, it was good, very good, but the creation is only truly complete after it has been blessed and sanctified. It is not humankind but the seventh day that forms the radiant crown on the work of creation.[108]

This aspect of sanctification appears to be of crucial relevance to the significance and celebration of the later sabbath. What does it mean to 'keep' the sabbath? Not working for a day? Recovering? To refrain from everyday work obviously remains a central element in 'keeping' this day. This already implies an element of rest, of getting one's breath back, so to speak. Resting and recovering are part of the weekly rhythm.[109] Without recovery, one cannot embark on a new working week. But there is also the sanctification, which introduces an active element into celebration of the sabbath. This active element might in any case consist in a reflection: reflection on the six days of creation that preceded it and on the works performed.[110] Celebration may go beyond reflection, though.[111] In the Christian (Sunday) tradition, little attention has been devoted to this

---

[108] M. Schlitt (1992, 144) follows Westermann in speaking of a "fast liturgisch klingenden Schlußakkord".

[109] Cf. Exod 31:17: "and on the seventh day He rested, and was refreshed", and Exod 23:12: "that your ox and your ass may have rest, and the son of your bondmaid, and the alien, may be refreshed". It is striking that (farm) animals share in this sabbatical rest. This fact accentuates the special position of animals in the Torah and also illustrates the exceptional status of the domesticated land animals. H.A. McKay points out in her study (1994, 13 note 8) that in both versions of the list of those that are to respect the sabbath the 'woman of the house' is absent. This she finds remarkable, but does not draw any conclusions. C.B. Posthumus Meyjes (1993, 24), who makes the same point, suggests that the chosen formulation evokes the wrong associations. Women do not belong in the list of 'possessions' (your...your). In his view, rather, she is "included in man".

[110] In English the combination of 'to stop' and 'to reflect' is familiar from the expressions 'to dwell on' and 'to linger on'. Although the modern term 'reflection' cannot be derived directly from the Hebrew text, the idea seems to be in line with the significance of the sabbath.

[111] In the HB such active interpretation was mainly the affair of the priesthood, although there are certainly references to the activities of others. For example, the expression "to do the sabbath" (from *'āśāh*) in Exod 31:16 HB indicates a more active interpretation than that embodied in the RSV translation 'to keep the sabbath'. In the New Testament we find this reiterated. The active character is reflected very clearly in the acts performed by Jesus on the sabbath: feeding and healing the hungry and sick. These healing activities he considered fully in line with the character of the sabbath ("Wherefore it is lawful to do well on the sabbath days"; Matt 12:12).

active/reflective side of sabbath celebration. This aspect has been elaborated far better in the Jewish tradition, with its celebration of the sabbath at home, on the Friday evening.

Only if justice is done to both elements can we truly speak of a sabbath celebration, for it is only then that such a celebration can have a healing, recuperative effect.[112] Thus can the sabbath serve to generate the future for humankind, nature and society.[113] It is this role of the seventh day in the rhythm of the week (of creation), in particular, that recurs in the sabbath year and the jubilee year.

2.5 Second Creation Story: The Garden as an Ideal

In the second creation story, described in Gen 2:4 to 3:24, three sections can be distinguished. First we have verses 2:4-7, which illuminate the creation of humankind as a living being in a kind of miniature story of creation. Next, in 2:8-25, follows the first part of the Garden of Eden narrative, a description of the situation prior to the 'Fall'.[114] And finally, in Chapter 3, we have the narrative of the snake, the eating from the Tree of Knowledge and subsequent banishment from the Garden. Altogether, this narrative has a very different flavour from the first, in Genesis 1. The Yahwist (J) is assumed to be the author of the second creation story; the assumed date of writing is the ninth century BC.[115] The style of writing is less terse, and the narrative is less systematic and 'cosmological' in character; it is geared rather to relationships and to man's activities in the Garden. It is almost a pastoral idyll, which is brutally disturbed by the snake. The whole is written very much in a narrative style, with the key message being: why things are is no longer as God intended.

2.5.1 *Man: living being and 'earthman'*

Some translations have the first sentence of Verse 4 read "This is the history (*tôl^e^dôt*) of heaven and earth", and in doing so establish

[112] For an elaboration of this idea, see: J. Moltmann 1985 and C.B. Posthumus Meyjes 1993.

[113] See P.W. van Boxel 1987, 18.

[114] D.U Rottzoll 1998.

[115] See Th. Hierbert 1996 for a treatment of the relationship between nature and religion as found in the texts of J.

a characteristic link between Creation and history. If we take this sentence as constituting a conclusion to the preceding text, the term *tôlᵉdôt* can be translated fairly neutrally as 'the origin' or 'the generation' (as a verbal noun). *Tôlᵉdôt* is not history in the absolute sense of what has occurred in the past, however. The word is related to the verb *yālad* (to bear, to beget) and thus refers to history as it unfolds in the lineage of generations. Elsewhere in the book of Genesis *tôlᵉdôt* is consequently translated as 'generations', and then refers to what is to follow, the coming generations.[116] A translation such as 'that is what it came to' also gives a good idea of the term's essence.[117] The creation of heaven and earth, in which we hear an echo of Genesis 1:1, is thus drawn into the narrative. The creation implies history, and history implies a creation. This 'linear' concept of history, with a beginning, a development, and a direction in which it is heading, 'what it comes to', is of the essence in the HB and was later to be adopted in the Christian tradition.[118] The use of the Name of God, *YHWH* ('I AM THAT I AM'), the God who is and ever shall be, is seen as supporting this historicizing interpretation.[119] There are good grounds for making two serious reservations with regard to this view, however. In the first place, the work of Albrektson has amply shown that Israel distinguished itself less from its *Umwelt* on this point than is generally assumed.[120] The second reservation proceeds from the

[116] This is indeed the RSV translation here. The JPS translation reads 'line', "this is the record of Adam's line" (Gen 5:1). In the Dutch language, the NBG has 'history', but a recent translation of H. Jagersma (1995, 11,33) has "verwekkingen" ('generations').

[117] For an extensive treatment of the meaning of *tôlᵉdôt* and the structuring role played by this concept in the book of Genesis, see C. Houtman (1980, 244-248) and H. Jagersma (1995, 11-13). E.A. Speiser (1964, Introduction xxiv) deems the term characteristic for P: "one might call it P's signature".

[118] The general opinion is that we are here concerned with not only an essential but also an exclusive view adopted by Christianity from the ancient Israelites. N.H. Ridderbos (1958, 1), for example, writes: "'Israel's *Umwelt*' knows the gods from nature; Israel knows its God from history, as explained in the prophecies". A similar position is held by modern theologians like H. Berkhof (1968, 12-14), by historians like Lynn White (1967, 1205) and by philosophers like J. Passmore (1980, 78-82). In the opinion of Hiebert 1996, this position is characteristic of the Enlightenment and, through the work of such commentators as G. von Rad, is firmly anchored in theological thought.

[119] Exod 3:14. For an overview of possible origins and meanings of *YHWH* see DDD 1999, 910-919.

[120] B. Albrektson (1967, 11) describes the view that Israel had a "religion of history" and the neighbouring peoples a "religion of nature" as "a commonplace". His study shows that it is difficult to characterize the "history" viewpoint as being

recent work of Hiebert on the role of nature in the religion of the Israelites. In his view there is no reason to delink 'history' and 'nature', let alone oppose them to one another.[121]

In the philological tradition, the occurrence of the Name of God is also a powerful argument for assuming a different source for this section, this 'second creation story'. However different in language, timbre and purpose the two stories may be, though, the connection is established between the two and is fundamental in character.[122]

Verses 5 and 6 provide a fine image of the interdependency of Creation, as perceived by the author: for the land to yield crops, there must be cooperation between humankind and nature. This reflects accurately on life in the agricultural communities of the Middle East, where pasture fertility and harvest yield were wholly determined by careful manipulation of the precious moisture (dew, mist or rain) and by measured tillage of the land. Verse 7 makes clear how tied to this earth 'man' is. There is an elegant play on words between the name of the man Adam (*'ādām*) and the word for field, Adamah (*'^{a}dāmāh*), possibly meaning 'red earth'. It is a phrasing characteristic of J.[123] 'Man' is thus an 'earthman' in the most literal sense. To gain the full measure of human life, though, requires a second 'segment' of the environment: the air, a divine gift recieved in the form of the 'breath of life'. It is God's breath that transforms 'man', essentially 'dust of the earth', into a living being (a *nepheš ḥayyāh*, a 'living soul'). The choice of words draws attention (once more) to the basic likeness between humans and other 'living beings' that dwell on earth, because further on in the narrative the animals are also referred to as *nepheš ḥayyāh*.[124] There is also a difference, though, for the life-giving breath of humanity can and indeed may be seen as deriving directly by God.

---

exclusive to Israel, given the fact that it has been proven to exist in the *Umwelt* (Assyria, Mesopotamia), too. This said, though, he draws the following conclusion: "One thing may however be stated with some confidence: it would seem that the idea of historical events as divine manifestations has marked the Israelite cult in a way that lacks real parallels among Israel's neighbours" (115).

[121] Th. Hiebert 1996, particularly 3-29.

[122] In the approach adopted here, verse 4a can best be characterized as an 'interpolation'. It links the (older) second narrative, by P, to the first, as if it were a sequel. Partly for this reason, I also discuss them in the same sequence as they appear in the HB rather than in historical order.

[123] Th. Hiebert 1996, 66.

[124] In verse 19. The expression is found on many occasions in the HB. See H.I. de Smit 1974, a curious but interesting work full of interesting biological facts.

2.5.2 *Life in the garden*

Verse 2:8 marks the beginning of the story of the garden of Eden, a garden in which man is set (2:8 and 2:15) in order that it may be 'tilled' and 'kept' (2:15). According to the text, the garden is to the east.[125] The word for garden, *gan*, indicates a sheltered space; the word 'Eden' means 'loveliness', 'delight' and perhaps also 'rich in water'. Elsewhere in the HB we find 'Eden' as meaning God's garden as well as a geographical concept. In a world plagued by drought, a place rich in water and adorned with trees certainly evokes pleasant associations. Eden is a safe haven of delights. In the garden God makes all kinds of tree grow that are 'pleasant to the sight' and 'good' (i.e. suitable) to eat. Two trees receive particular mention: the 'tree of life' (2:9) and the 'tree of knowledge of good and evil' (3:3), both of which stand in the middle of the garden. These trees are obviously a crucial element of the story. Here, however, we shall not dwell on the many meanings ascribed to the two trees in the story of the garden and elsewhere.

Eden is the source of a river that waters the garden, flowing further as four separate rivers to regions or countries that cannot now be precisely located. The first region is described with reference to the minerals and products to be found there: gold, *bᵉdôlaḥ* (the gum resin bdellium?) and *šōham* (onyx stone? lapis lazuli?). The second river flows around the whole land of *kûš*, probably Ethiopia, and the third flows to the region 'east of Assyria'. The only thing written about the fourth river is its name: the Euphrates. Despite many sophisticated studies, little can be said for certain about the names of the rivers and regions and their exact geographical location, *inter alia* in relation to the mentioned products. It is well feasible that the author penned this geographical sketch with the intention of indicating that the economic arteries of the contemporary world had their origins in the (God's) garden.

The narrative then turns to man, who is once again said to be 'put in the garden', but now with a concrete task: 'to till and keep it'. The use of *ʿābad*, to till, indicates that the purpose was to serve, by working the garden. This is the most usual meaning of this very

[125] Besides designating place, 'eastward' (*miqedem*) can also designate time; cf. Neh 12:46, Isa 45:21, Isa 46:10, Ps 74:12, Ps 77:6 and Ps 143:5. An alternative translation might thus be 'for the first time' (as in Neh 12:46) or 'from ancient time' (other texts).

common verb: to serve, as the people of Israel were to serve God, but also as a slave serves his master, since the word for a male slave is derived from the stem (*'bd*). In the HB serving implies acting. A number of texts indicate that *'ābad* can also certainly denote agricultural labour such as tilling, ploughing, sowing and so on.[126] To what extent the author had such farmwork in mind as concrete activities for the human species cannot be judged from the passage alone. The text speaks only of 'eating of trees', i.e. their fruit. Even if some kind of farming is indicated, which seems plausible, given the name 'Adam' and projected back from Gen 3:17,18, then it is at any rate farming without 'toil'. There is an unmistakably functional flavour to the passage: there is clearly a 'human purpose' within the garden.

'To keep' (*šāmar*) occurs very frequently in the HB and generally has the meaning of (conscientiously) 'keeping' the commandments and ordinances of God. It is a verb of implicit activity. A second meaning is 'to guard over', with the connotation of being on the look-out for any threats to items of value. The image of a guard on duty springs to mind. 'Keeping' in the passive sense of 'conserving', or 'keeping as it is' is a less plausible translation. Translation as 'to guard over' also poses problems, however, as it is by no means clear what the threats to the garden exactly are that man must 'keep' (from happening). Was humankind, ignorant as it was of good and evil, even aware of danger? Some exegetes make a connection with the events of 3:1. Since the passage in question refers to humanity being given a general mission in the garden, before the eyes of Adam and Eve were opened, this seems an unlikely explanation, however.

In addition to and alongside the task to "be fruitful and multiply" in 1:28, in our (European) culture the 'tilling and keeping' of 2:15 has had an enormous historical impact as a model, as a cultural mission for humankind on earth.[127] In this connection, it is in-

[126] In addition to serving, honouring and "serving as a slave", K-B (1951, 670,671) also gives: to cultivate, "to work the soil" and to till (e.g. in Gen 4:2,12, 2 Sam 9:10 and Ezek 36:9,34).

[127] For historical reviews of the these two passages, see, *inter alia* J. Cohen 1989, C.J. Glacken 1967. In St. Augustine's commentary on Gen 2:15, also followed by Calvin, we find the foundations of the Protestant ('Calvinist') work ethic (St. Augustine *De Generi ad Literam* 8, 8,15; J. Calvin 1554/1900, 68,69). For a Dutch example, see: *Bouwen en bewaren*, published by De Groen van Prinsterer Stichting (by that time the scientific office of the Calvinist-oriented political party GPV) on the so-called "cultural assignment" of humankind (Amersfoort 1974). In the party

teresting to note that in Eastern Europe early Christian traditions produced images of paradise in which human figures were portrayed as resting and meditating. In the Western tradition a more active role predominates, with Adam depicted as a farmer and gardener.[128]

Verses 16 and 17 contain the command not to eat of the tree of knowledge of good and evil, under penalty of certain death. Here, too, the account begs numerous questions that are difficult to answer, such as: while still in the garden, did humankind know what death entailed? And if so, how? And if not, what is then the sense and effectiveness of a conditional command of this nature? The author of the narrative clearly sees death and dying as being connected with eating from the tree, and in the tradition death even constitutes the most fundamental consequence of human disobedience.[129] Whether this is indeed true will be considered below, when we discuss Gen 3:19. Here, the question is also relevant: what death? Life in the garden is characterized as a life without death, or at any rate a life in which death does not predominate. This ties up with (or rather: conforms with) the vegetarian diet described in Gen 1:29,30. Biologically, life presupposes death. This mutual relationship was also familiar to biblical scribes in their day, as is reflected, for example, in passages in which the word-pair 'sow and reap' (or similar) is employed.[130] These everyday, biological implications did not deter the author of this passage of Genesis from writing these lines, however. Here, he was not interested in death as part of the biological cycle of life and regeneration. In this passage, humans are the key player. The reference to 'certain death' may reflect the idea that the author saw humans-in-the-garden as immortal beings. The text allows for an alternative explanation, however. The author may

programmes of all the Christian-inspired political parties in the Netherlands, 'working/tilling' and 'keeping' serve as a fundament for humanity's stewardship (see also Chapter 5).

[128] On the differences between the Greek East and the Latin West see: P.M. Gregorios (1987, 1992) and Ph. Sherrard 1992.

[129] The motive of human mortality versus divine immortality is found in many creation narratives. The Atraḫasis myth provides an etiology of death. Every culture seeks answers to the enigma of human mortality versus the desire for eternal life. In some narratives, such as the Gilgamesh epic, human immortality is lost. In this respect, the second biblical creation story may (also) serve as the Israelites' answer to these fundamental questions.

[130] In Gen 8:22, Deut 14:32 and Isa 17:11, for example, and very explicitly in the NT, in John 12:24 and 15:2. For general ideas about mortality in nature, see Isa 40:6,7.

not have been concerned with the (natural) termination of life, but with death as the 'adversary' of life, not only human life but also animal life. In this reading, we are concerned with death marked by the flow of blood, death whereby living beings are deprived of their 'lifeblood'. This would encompass manslaughter and misadventure, the slaughter and consumption of animals by humans and the killing of animals and humans by predators. Formulated more generally, the reference is then to the premature ending of life, with all its associated unresolved questions and pain.

There is also a third possible interpretation, in which the reference is not so much to the fact of death (in whatever form) as to humanity's awareness of death. We shall return to this point in Section 2.6.

2.5.3 *The other half*

"It is not good that the man should be alone; I will make him a helper fit for him." This divine pronouncement, in verse 18, marks the start of an intriguing passage concerning relations between humankind and animals and between man and woman. The first surprise is how the text continues. Following an introductory verse of this scope, one would expect the narrative to continue with a woman, or at least another human being, coming to help. There follows an intermezzo, however. First, all the beasts of the field and birds of the air are formed from the same substance as the human species: the soil of the earth, and these are brought to Adam to see what he will call them. Verse 19 then states: "and whatever the man called every living creature, that was its name". Following this work of naming, at the end of verse 20 comes the somewhat sobering and even disappointing statement: "but for the man there was not found a helper fit for him", as it reads in the RSV translation. This follows the Masoretic Text, which opts to vocalize the *lāmed* to give the reading 'Adam'. "But for Adam no fitting helper was found" in the JPS translation. Given the run of the overall narrative, however, this is an arguable choice. The Dutch exegete Van Selms, for example, translates this passage as: "but for man He found no...".[131] Given the Hebrew, this is certainly a feasible translation, arrived at by different vocalization of the *wāw* and the *lāmed* (*w^{e}lā* instead of *ûl^{e}*). By explicitly naming God ("He") as the subject of (not) finding, Van Selms also brings

[131] A. van Selms 1984, 59. For *'ādām* in Gen 1-5, see R.S. Hess 1990.

this verse into line with the translation of verses 18, 19 and 21, 22, in which God is the active agent. The RSV translation uses the passive voice here ("there was not found..."), but God is the implied agent. Some translations choose to render verse 20^{b} with man as the active agent of the seeking. The NBG, for example, even uses a reflexive verb and runs: "but he found himself no help fit for him". To me such a rendering appears more plausible and indeed preferable, given the first half of the verse, in which (hu)man(kind) is acting as a 'namer', and read in tandem with verse 23. Cassuto has the translation: "but as for man, he did not find a helper corresponding to him...".[132] Westermann shows a similar inclination, with "für ein Menschliches Wesen, wie er ist..".[133]

Perhaps most importantly, though, it is almost certain that the reference here is not to Adam as a named individual, but rather to some contrast between the animals and the human race. From the perspective of contemporary ideas about the differences between humans and the animals this is a remarkable passage. In the first place, we can only assume that the help was both sought and expected from among the animals, the living creatures that had been formed from the same raw material as man. And, secondly, it is remarkable because of the outcome: the help was sought, but it was not found. Serious consideration must be given to both these elements of the account.

The *seeking* indicates a close affinity between human beings and animals in the overall order of creation.[134] This affinity is reflected,

---

[132] U. Cassuto 1989, 126.

[133] C. Westermann 1974, 249, 312. According to Westermann, Adam is not intended by name here. With the reading cited in the text, he takes the same line as F. Delitzsch and deems that no change in the Masoretic vocalization is required if the passage is read thus.

[134] U. Cassuto (1989, 127-129) rejects this reading with a series of arguments. His main objection is that God is thus described as a hesitant creator who engages in experiments. Cassuto holds the view that the so-called 'seeking' among the animals served to impress on 'man' the condition of his aloneness and evoke in him the desire for "a soul-mate". This interpretation is also unsatisfactory, however, for the passage then again jars with the announcement by God that: "It is not good that the man should be alone; I will make him a helper fit for him." The objections raised by Cassuto are equally pertinent to his own reading. He brings forward yet another argument: that there is no mention of those animals which are perhaps most suited to helping humankind (cattle and other farm animals). The text refers to "every beast of the field" and "every bird of the air". Citing Lev 17:13, Cassuto then takes the stand that we are here concerned with "two classes of creature that man can catch only by hunting". In his eyes, this

among other things, in the mention of the two being formed from the same raw material. It is an affinity that is also fully in line with what has already been observed about the creation of humans and the animals on one and the same day and about their both being denoted as *nepheš ḥayyāh*. And it is an affinity, finally, that should not be underplayed solely because of the negative outcome of the search.

The fact of *not-finding* points to fundamental differences between human and non-human creatures. These differences are at any rate so great that there is not a single animal that can fulfil the role of helper. From the following passage, in particular verse 23, it can be gleaned what the animals lacked. They may have been formed from the same basic material, but they are not of the same 'flesh and blood'. This is in contrast to the woman, or womankind, who, according to the text, is "bone of my bones and flesh of my flesh": a true soul-mate.[135] Only someone of the same, human species can be a help, a true 'other half' or 'opposite number' to man. Animals fall short of the measure; as living beings they may be related, but not closely enough.

Alongside this predominantly biological argument, there is another 'unfitness argument' that may be cited. We must then take a closer look at the Hebrew expression *'ēzer k^enegdô*, translated in the RSV as "a helper fit for him". Van Selms is surprised at the use of the male word for 'helper', given the existence and usage of the female form *'ēzrāh*. From this he concludes that it is not the sexual bond

---

interpretation also does away with the contradiction between the first and second creation story regarding the sequence in which the animals and humans are created. Cattle are mentioned in verse 20. "From this it may be inferred that the cattle were already to be found with man in the garden of Eden". This explanation is artificial and fails, moreover, to resolve the inconsistency identified.

A. van Selms (1984, 58,59) proposes a different reasen for the domesticated animals not being mentioned in verse 19. In his view, all animals were originally wild. God brings the animals to 'man' for him to select those that are to serve as 'cattle'. In this interpretation, cattle 'emerge', as it were, from the confrontation between man and the (wild) animals. This I find far-fetched. On the grounds of the second part of verse 19, where 'the beasts and the birds' are termed *nepheš ḥayyāh*, it seems probable that the reference is to all land animals and all birds. In verse 7 this term was encountered in reference to humankind. The fact that it recurs here indicates that humans and beast have more in common than Cassuto and Van Selms presume.

[135] The translation of *sēlā'* as 'rib' is disputable. There are good arguments for translating it as 'side' (cf. 2 Sam 16:13, Exod 25:12,14, Ezek 41:26). At the same time, this would do justice to the idea of woman being created as 'the other half'.

and opposition that are the main focus of this passage.[136] This cannot be induced from the non-usage of *ʿēzrāh*, however, because this term is generally employed in the abstract, as 'help', rather than to denote a 'female helper'.[137] The sexual identity that certainly does play a role in the rest of the narrative is not being 'muffled away' through usage of the male form here. 'Being a helper' should probably be interpreted broadly, certainly also including the sexual aspect. It may also be remarked that a 'helper' in the HB is often a fully independent person who provides aid and succour.[138] In the text the character of the *ʿēzrāh* is indicated by *kᵉnegdô*, which is derived from *neged*. This latter word is used to denote something visible, something that is before one's eyes because one is *face-to-face* with it. This root meaning is reflected in virtually all the various translations: in the sight of, before, before the eyes of, in the presence of, and so on. From this the character of the helper and the nature of the help can be derived. The helper is not a 'useful assistant' to man, nor somebody there 'along with him', and certainly not somebody hierarchically 'below' him.[139] The helper is 'an opposite number', someone seen face to face and recognized as the missing human partner.

From the perspective of the end of Genesis 1:27, let us now generalize to 'male' and 'female'. This story is not about one man and one woman; it is concerned with a crucial element in the existence of men and women. In the order of creation, male and female complement one another; without one another, they are (equally) incomplete. With the animals there was no such mutuality, and the question then is why? The answer is to be found in the name-giving described in verse 19. As we already saw in discussing Genesis 1, in the Near East the activity of naming was by and large not bereft of significance. It may mark insight into the nature and fate of the other

[136] A. van Selms 1984, 57.

[137] See: K-B 1951, 696.

[138] In Ps 10:14, for example, where God is called the helper of orphans and in Hos 13:9, where God is referred to as the helper of Israel.

[139] There are certain similarities between this narrative and the epic of Gilgamesh, here with respect to the element of equality. The subjects and especially the young daughters complain to Anu about Gilgamesh. Anu passes the complaints on to Aruru, the mother goddess. She asks Aruru to create someone in the image of Gilgamesh, someone who will equal him and compete with him so that Urduk can live in peace and the young daughters are no longer troubled. Aruru does so and creates Enkidu, the warrior hero, from the clay of the earth (Tablet I, ii; S. Dalley 1991, 52).

and thus be evidence of dominion and power. 'May', because there is every reason to be wary of exaggeration on this point.[140] Due caution should thus be exercised in taking the giving of names to be a mark of power.[141] The fact that a human being is charged with this task by God, the name-giver and 'appointer' *par excellence*, is certainly an indication of the special status of humankind. It may also be interpreted as indicating that there is no symmetrical relationship (possible) between the human and other species. Between man and woman a different relationship holds. Although, in verse 23, (the) woman is given a name by (the) man, it does not appear to be a 'personal name'.[142] She is called Wo-man, or, in Dutch, 'mannin'. Name-giving was not at issue here; as things stood, she already had a name, because she was 'out of man', and at the same time proved to be the female 'man' that the male 'man' still lacked. In the Hebrew, the man *'îš* is given a complementary half in the 'woman' *'iššāh*. This passage embodies (far) more equality than is sometimes assumed.[143] At the same time, though, we should beware of projecting modern ideals of gender equality back onto the authors of Genesis 2-3.

All exegetes are in agreement about the phonetic similarity between this pair of names, but it is unclear whether there is more than phonetics involved. The two words may be etymologically related on the basis of a presumed shared root *'nš*.[144] There are literary as well as linguistic arguments on this point. The second creation sto-

[140] Giving names to villages may be an indication of the authority being exerted over these villages (cf. Num 32:41,42, Judg 1:17 and 2 Sam 5:7-9). See also A.S. van der Woude (1973 entry on *šēm* in *TWAT* 2, col. 943). Name-giving certainly does not always indicate the possession or exercise of power, however. In some cases, for instance, a name with a particular significance may be assigned by a narrator after events. Many names are merely functional, moreover, being used to denote that someone is 'the son of...', for example, or 'hailing from...'. Neither is it entirely clear whether the bearer of a name was always aware of the (original) meaning. Finally, there are many names whose meaning is unclear.

[141] On this passage, see G.W. Ramsey 1988.

[142] In Gen. 3:20 (the) man gives his wife a different name: Eve (*ḥawwāh*). It is a name which is connected with women as the source of life and it may be etymologically related to the Aramaic word for snake (see: K-B 1951, 280). This association between Eve and the snake is also familiar from the visual arts. The 'serpent' from the second account of creation is often found depicted with an 'Eve-like head'. Both the new name itself and the giving thereof can be taken as supporting the idea that the 'old' name reflected the former *status quo*.

[143] Among others by W.T. de Groot 1992, 484.

[144] It is not uncommon for the *nûn* to be dropped or assimilated to *š*.

ry contains a number of other 'punlike' word-pairs. Other commentators hold that these two names have different roots.[145] They assume for *'îš* the root *'yš* or *'wš* and for *'iššāh* the root *'nš*.[146] There is, finally, the possibility that these words have no basis in existing verbs, but are primary noun forms.[147] Whatever the case, translators through the ages have always been very aware of the close similarity. The various translations as 'man and woman' (Latin: 'vir et virago', Dutch: 'man en mannin') reflect the perceived parallel between the original Hebrew *'îš* and *'iššāh*. Only this related 'woman', different yet equal, is a helper 'fit for him'.

"Therefore a man leaves his father and his mother ..." the text continues in verse 24. The change in narrative style is immediately apparent. The story is elevated out of the context of the garden and into the everyday reality of the writer and reader. Most commentators see here an interpolated, explanatory gloss (comparable, for example, to Gen 32:32: "therefore to this day the Israelites do not eat the sinew of the hip..."). It is also feasible that the author here interrupts his narrative to comment on his own story.

In the orthodox Protestant-Christian tradition it is assumed that this verse posits marriage as part of the order of creation. The Dutch exegete Gispen formulates this view as follows "Monogamous marriage was introduced by Him and intended by Him (cf. verse 18)".[148] The passage makes no reference to marriage, however, although the author/commentator may well have had this in mind.[149] If he did, it is more likely that he was concerned with marriage as a generic term for a form of human relationship. To assume reference to (modern) marriage as a social institute is, I feel, to go rather beyond the original designs of the text.

Other interpretations discern a pendant to patriarchy in this verse: the man must leave his parent's home; the woman does not take up home with him. Given that the passage appears to be in the nature

---

[145] For example, U. Cassuto 1989, 136.

[146] K-B 1951, 40.

[147] N.P. Bratsiotis treats the entries on *'îš* and *'iššāh* in *TWAT* (vol. 1, 238-252). The author rejects the assumed roots *'wš* and *'nš*, along with their respective meanings ('to be strong' and 'to be weak') as 'unproven'. Bratsiotis considers it likely that with both *'îš* and *'iššāh* we have "Primärnomen ohne zugrunde liegende Verbalwurzel" and attributes any relationship between these two key words of Gen 2:23 to popular etymology.

[148] W.H. Gispen 1966, 160.

[149] See K. van der Toorn 1994.

of a commentary, this appears relatively unlikely, because matriarchy and matrilocality were not the norm in the world of the writer/commentator. Neither does the passage embody any explicit indication of a matriarchal order. It is not stated that the man takes up home with the woman, but that he leaves his parents and enters into union with his wife.[150] There is, in all likelihood, no antithesis between patriarchy and matriarchy intended here.

The text is more general in scope. Because male and female complement one another in the creation order, it is logical for man (a man) to have to reinstate his two-in-oneness with 'wo-man'. He should therefore sever his attachments to those who have raised him and leave his parental home and enter into union with his wife. The verbal forms (perf. cons.) of 'to leave' and 'to cleave' are frequently used not to denote a single event in the future, but an activity of longer duration. The choice of words makes clear why things must necessarily be so between man, or male, and woman, or female. Male is more than a man and female more than a woman. We are here concerned with general human characteristics, and the passage suggests that in this context male and female must complement one another, possibly within one and the same gender, if human beings are to arrive at their full potential.

"To become one flesh" obviously refers back to verse 23. There is no reason to assume either a purely physical or a purely spiritual meaning here. 'Flesh' stands for 'the whole person'. Because 'bone of my bones and flesh of my flesh' often denotes family lineage in the HB and 'flesh' is a standard term for 'relative/family', it is certainly also reasonable to presume a reference to offspring, to setting up a new family.

Verse 25 concludes this passage on the male-female relationship and at the same time establishes a clear link with verses 7, 11 and 21 of the following chapter. "The man and his wife were naked", but they were not ashamed before one another. In the light of Gen 3:7, this verse is often interpreted as signifying that, in the eyes of the author/commentator, there was no cause for shame, because the nakedness was not perceived as such. There was no knowledge of 'good and evil', and the negative connotations of sexuality were

[150] K. van der Toorn 1994 sees in this passage a confirmation of the traditional marriage ceremony in the house of and with the wife's parents. One reason for this was that the parents disposed over proof of their daughter's virginity.

therefore lacking. By extension, the chain of association between shame, guilt and sin (the Fall) was established. This is indeed defensible. Shame about something is born from a link with the 'evil' with which it is associated. In the case of sexuality, that link is indeed so direct that in some translations (the Dutch NBG, for example) the word *'erwāh*, 'nakedness', is often rendered as 'shame', which in many texts is then even synonymous with the male or female sexual organs. The exposure or sight of 'shame' has an ethical undertone and cannot remain without consequence.[151]

It is to be queried, however, whether this direct link between corporality and sexuality is the prime concern of this verse. The verb employed here for 'shame', *bwš*, has no immediate association with sexuality.[152] At the same time, though, the word for 'naked' (*'ārôm*, pl. *'ᵃrûmmîm*) appears to be phonetically related to the Hebrew for shrewdness, or 'subtlety': the *'ārûm* quality of the 'serpent', or snake, in the next verse. Although this play on words is undoubtedly intended as such by the author, its meaning is not immediately clear. There is, in fact, no satisfactory and/or widely supported interpretation at hand. The most obvious reading, to my mind, is to simply assume a connection between knowledge, shrewdness and nakedness, however. The serpent is the 'most naked' of animals and is at the same time the symbol of shrewdness (wisdom) in various ancient cultures. The tree was good to eat from and "to be desired to make one wise" (verse 6). The man and his wife ate of the tree and acquired wisdom, the reason their first 'discovery' after 'eating of the tree' concerns their nakedness. Viewed in this light, the story is also possibly an allusion to the acquiring of maturity and adulthood by humankind.

Westermann makes no attempt at all to explain this passage. He does, however, warn against an overly narrow, sexually oriented interpretation of verse 25 back from verse 37 and also queries whether the traditional Christian view that sexual shame is a consequence of 'the Fall' can justifiably be read into the conclusion of the second chapter. In his view, shame is a relative notion and a lack of shame can indicate both utter innocence and a state of depravity. Westermann is wary of having the passage say more than it does.

---

[151] See, for example, Gen 9:18-27 and Lev 18. In this latter chapter the expression 'to uncover nakedness' is even synonymous with engaging in sexual contact or intercourse.

[152] K-B (1951, 115) gives a range of variants, but none in a sexual context.

2.6 The Expulsion from the Garden

Genesis 3 relates the story of how and why humankind is expelled from the garden. In the Christian tradition, the events leading to the expulsion from the garden are called 'the Fall', a summarizing as well as interpretative term. It is a fascinating account with a sense of drama that has left an indelible mark on the history of Western culture. In the eyes of the biblical author, this is a decisive juncture of history. In theologizing about this turn of events and in their exegesis of this chapter, scholars have devoted considerable attention to what the story has to say on the origins of evil, sin and death in the world. Related questions include: What role is played by the serpent in all this? Is it 'merely' an animal, or is it the personification of evil? If it is no more than an individual of a species (created by God), why did it speak, and whence its knowledge of the consequences of eating of the tree? If the snake is more than this, what are its origins and what is its status in the overall 'goodness' of creation? Why did it begin by posing a misleading question, and why did it want to seduce the woman into eating from the tree, and indeed into 'being like God' (verse 5)? And what is the deeper significance of verse 15, in which 'enmity' is put between the woman and the serpent and between the woman's offspring and that of the serpent? Is death a direct consequence of the events described here?

This is not the place to embark on an extensive discussion of the fundamental issues raised in this passage. With respect to the origins of evil, such a discussion would not only lead us too far away from our central theme, but would also probably bring us little further. Nowhere in the account is the notion of 'sin' mentioned. Anything that might here be stated about the etiology of evil remains implicit, and in interpreting the text too great a weight is often placed on sometimes arguable assumptions. I am thus in agreement here with the general conclusions of Westermann (who follows Zimmerli): "für die Herkunft des Bösen gibt es keine Ätiologie", and: "Die Verführung... wird als Rätsel stehen gelassen".[153]

How terse and enigmatic the account of the origins of evil may be; but on the *consequences* of these actions for the serpent and for humankind it is fairly elaborate and unambiguous. The description of those consequences contains many elements of great significance

[153] C. Westermann 1974, 325.

for humanity's relationship with nature. This juncture of the story marks a clear turning point, the general message being that all relationships previously described are transformed. Even nature itself and its fundamental processes appear to undergo a change. In this, we see Chapter 2 reflected: the aspects there described recur, but now in inverse sequence.

The most immediate consequence of eating the fruit of the tree is that the prediction of the serpent proves true: the eyes of the man and the woman are opened. Although some translations continue by stating that 'they saw that they were naked' (the NBG, for example) the RSV gives a more accurate rendering of the Hebrew *yāda'*, which has a more figurative meaning. 'They knew', 'they became aware', 'they realized; all these reflect better the fact that it is here cognitive rather than visual perception that is involved.[154] This realization is a direct consequence of 'knowing good and evil', as a result of which, according to the serpent, man and woman will "be like God" and, according to God himself, in verse 22, "like one of us". It is remarkable, to say the least, that the first thing the man and his wife realize as a consequence of their deed is their own nakedness. They had expected to acquire Divine knowledge, but what they discover, once they have acquired it, is their own nakedness. The story makes clear that 'naked' is here intimately related to the primary sexual organs, since these are specifically covered, with loincloths or 'aprons' (*h^{a}gôrôt*) of fig leaves. Aware of their own nakedness, the man and woman also become afraid and hide from God (verse 10). The fig-leaf aprons are not enough to still their fear when they hear God arriving in the garden.

With all due respect for Westermann's cautious exegesis of Gen 2:25, here, in 3:5,7,22, it is very difficult to avoid the highly sexual connotation. Knowledge of good and evil is apparently closely connected with sexuality and with 'uncovered nakedness'. The opened eyes and the acquired knowledge lead to an awareness of sexuality. The 'fear' that arises may relate to the keen awareness that, in the human capacity for 'procreation' and sexuality, good and evil stand in close proximity to one another.

In his reading of the text Westermann follows Pedersen, who sees the conflict between God(s) and humankind as the leitmotif of Genesis

---

[154] The verb *yāda'* is sometimes also employed in the HB in a sexual context: "to know a woman". In this meaning, we already find the term in Gen 4:1.

2-3. The fact that human beings are created with a strong urge to live and acquire knowledge sets up a tension. Human life has its limits, but knowledge does not. It is in this respect that man and woman can become 'like God'. In Gen. 2-3 'knowledge' should be read as being synonymous with 'wisdom', as it is in the ancient myths of the Near East and in other passages in the HB such as Job 15:8. Fully in line with his reading of 2:25, here too Westermann distances himself from what, in his view, is a too narrowly sexual interpretation.[155] If these verses are to be properly understood within the overall context of the HB, it may be necessary to draw in other motifs, too. It is tempting to read the passage as a reference to the gaining of self-awareness and self-knowledge, whereby unreflective unity with the world makes way for the individual, 'separate' consciousness. It is also possible to go further, and interpret the phrase as a "das ganze Menschsein umfassende ... Erkentniss" and a "Daseinsmeisterung". However, none of this detracts from the basic fact that, in the text, the first, immediate stated consequence of becoming "like God, knowing good and evil" is an awareness by man and woman of their own, literal and physical, nakedness. This basic fact appears to have vanished somewhat in Westermann's 'umfassende Erkentniss'.

When man is asked who has told him that he is naked and whether he has eaten of the tree he was commanded not to eat of, he points to the woman. "The woman whom thou gavest to be with me", he adds, by way of apology, although there is almost a tinge of accusation. Once again there is an elegant play on words: that which was given has now given the fruit. The woman, for her part, names the serpent as the one that "beguiled" her. This 'passing the buck', as we would say today, may seem unsympathetic when it comes to one's personal responsibility, but this narrative is devoid of any moralism. God appears to continue in the same vein, for He immediately addresses the serpent, without any preliminary questioning, let alone time to give an excuse: "Because you have done this...". These words mark the start of the curses and 'punishments' pro-

---

155 C. Westermann (1974, 337) concludes with the following lines: "An allen Stellen geht es um Erkenntnis (oder Weisheit) in allgemeinem, umfassendem Sinn. Damit ist jede partielle Deutung der 'Erkenntnis von gut und böse' ausgeschlossen. Es kann weder die sittliche noch die sexuelle, noch eine andere Teilerkenntnis gemeint sein, sondern nur das ganze Menschsein umfassende und bestimmende Erkentniss. Wenn sie in den beiden babylonischen Mythen und in Hi 15:8 "Weisheit" genannt ist, so ist Daseinsmeisterung gemeint, genau wie in Gn 3:5 "Erkenntnis des Guten und des Bösen"".

nounced on the serpent, the woman and the man, respectively, in verses 14 to 19.

Westermann subscribes to the view of Schottroff that these curses and punishments were added to the account at a later date. In Westermann's eyes the 'original' punishment was expulsion from the garden or, in other words, from the presence of God. This reading does not detract from the relevance of considering the significance of these curses.[156] As we shall see below, though, their interpretation may be derived from this view concerning their later addition.

The serpent is cursed. In the verse in question there is a reference back to the description of Gen 3:1. Parallel to the snake being distinguished in its subtlety from all other animals in 3:1, now, in 3:14, it is cursed above all cattle and all wild beasts. Here, too, there appears to be a pun intended between 'subtle' and 'cursed' (*'ārûm* and *'ārûr*). Are the two indeed linked? Does the quality that placed the serpent above the other animals now render it cursed above them? The curse is phrased in concrete rather than general terms. From now on, the serpent is to 'go on its belly' and 'eat dust'. There is little point in speculating on the question of how the author conceived of the mode of locomotion and feeding habits of snakes.[157] The reference here is to the changed status of the serpent in the order of creation. In the light of the law-giving of Lev 11:42, 'going on its belly' means that the serpent must (henceforth) be seen not merely as being among the unclean animals, but as a truly abominable creature. In the following chapter of this book we shall be returning to the demarcation between 'clean' and 'unclean' in the dietary laws and to the 'order of creation' in the creation narratives.

The second part of the curse, 'eating dust', should be seen a sign of deep humiliation. We can deduce this from several other passages in the HB, where there is reference to enemies 'licking the dust'.[158] This description obviously has nothing to do with eating habits. Nor should it be interpreted here as an accurate description of (a change) in feeding habits. It is a metaphor for the subjugated status of the serpent. The author as well as later readers were well aware of how

---

[156] All the more so as the *Letztgestalt* holds major significance for us.

[157] Based on present ontogenetic and phylogenetic understanding, it is assumed that snakes have evolved from creatures with legs. This will certainly not have had any influence on the author of Genesis. It is absolutely inconceivable that anything was known (handed down?) about these distant, walking ancestors of the snake at the time these passages were written.

[158] Ps 72:9, Isa 49:23 and Mic 7:17.

snakes can bite and poison their prey; this is also probably the reference at the end of verse 15.

The serpent is to remain subjugated 'all the days of its life'. In Isa 65, where prophetic words are spoken about the creation of a new heaven and a new earth, it is once again predicted that "dust shall be the serpent's food" (verse 25). In this recreated world neither the serpent nor the wolf nor the lion will do evil or sow destruction. In his interpreting this passage, Cassuto considers the serpent more as a symbol of evil than as a 'normal' animal species like the cited lion or wolf. He sees a parallel with Isa 27:1, for example, where punishment is pronounced on "Leviathan, the twisting serpent". In his view the verse in Isaiah 65 has a similar scope: God wants to rid the world of evil.[159] In the HB a very particular role is evidently assigned to 'the serpent', or snake, and in many texts it cannot be considered simply as another 'normal' species of animal.[160] In the context of the second creation narrative, this has led to wholly allegorical interpretations, with scarcely any place for the snake as an actual animal. Westermann distances himself from such an analysis. Still, it must be said that Cassuto provides a more satisfactory explanation of the 'curse' under discussion than that embraced in ignoring the 'symbolic' elements of the text. From a wealth of different cultures, comparative religious studies have presented materials reflecting the special, symbolic significance ascribed to snakes.[161] In this context, reference should certainly be made to the Gilgamesh epic, in which, following the deluge, the snake steals a thorny plant from the naked (!) Gilgamesh. In this way he is deprived of the opportunity to recover the eternal life he has lost.[162] Here, too, in Genesis 3, the snake is more than a mere animal, but at the same time does not entirely vanish behind that 'more'.[163]

Even after these concrete elaborations of God's curse, there is still more in store for the serpent. In verse 15 judgement is pronounced

[159] "God will eradicate the principle of evil from the world" (U. Cassuto 1989, 160).

[160] Earlier in this chapter reference was already made to the association of the serpent with the primordial depths. The serpent has also become a symbol of healing, it may be added, because of the story of the healing properties of the 'bronze serpent' (Num 21:4-9), the story from which the medical tradition derives its serpentine symbol. See also K.R. Joines 1974.

[161] See, *inter alia*, R. Lewinsohn 1954.

[162] Gilgamesh XI, vi (S. Dalley 1991, 118,119). See also R.J. Clifford 1994, 148.

[163] See K.R. Joines 1974.

on its future relationship with the woman. There is to be 'enmity' between the woman and the snake, and between their respective offspring. As to the nature of this enmity and the course it is to take, the second half of the verse states that the woman's offspring will 'bruise the head' of the serpent, while those of the serpent will 'bruise the heel' of the woman. There is a pun between the two words translated in the RSV as 'bruise' (both of which are derived from the root *šûph*). It is not entirely certain whether they are to be translated the same because they are derived from the same root.[164] Cassuto proposes the word-pair 'crush/crave'.[165] This said, though, the translation makes little difference to the behaviour being referred to here. The 'craving' of the serpent for the heel of the woman's offspring most likely denotes the snake's propensity for biting therein. To unravel the profounder significance of this verse remains the province of exegetes. Here, too, an explanation whereby the serpent is merely a 'normal' animal does not satisfy, while other exegeses, for example those assuming a 'protoevangelic' reference, perhaps read more into the passage than is strictly reasonable.[166] We shall leave this point for what it is.

After God has spoken to the serpent, in verse 16 He turns to the woman. She is not cursed, but 'punished' it would appear. She is told that the pain associated with childbearing will be greatly multiplied, and that she shall bring forth children in pain. The translation 'multiplied' indicates a (negative) change in a phenomenon that already exists, but the Hebrew also admits of the translation: 'will multiply thy sorrow *and* thy conception'. In this reading, pregnancy and childbirth themselves as well as the great pain associated with them are both part of the 'punishment' and as such are both 'novel' for womankind. This becomes even clearer if the childbearing is taken to be explicative and the text read as: "your pain, which consists in your childbearing".[167] This interpretation introduces a connec-

---

[164] K-B (1951, 956) assigns two root meanings to *šwph*, I: to trample, to step on, and II: to bite at, to 'snap' at.

[165] In his view the second *šûph* is related to *šā'aph*, which can be translated as "to crave, desire" (U. Cassuto 1989, 161).

[166] The term 'protoevangelic' refers to the reading whereby the 'bruising' of the serpent's head is taken to be a future reference to Jesus' triumphing over death in the NT. For a treatment of this and other exegeses, see, *inter alia*, C. Westermann 1974, 354,355.

[167] See, for example, J. Calvin 1554/1900, M. Naḥmanides 1965 and G. von Rad 1949.

tion between reproduction, the discovery of sexuality, the 'opening of the eyes' and the mortality of the human race. Only as mortals do human beings have an interest in reproduction, because otherwise they will vanish not only as individuals but also as a species. Reproduction enables mortals to conserve life, as it were, and pass it on. One argument against this interpretation is that the command to multiply predates the 'Fall'.[168] In the Atraḫasis we find a contrary motive; there, death is introduced to control overpopulation.

Against the background of the overall account, however, there is an alternative reading which is, moreover, better in line with the translation "greatly multiply" in verse 16.[169] The consequences of eating the fruit are to be characterized as changes, not as entirely new phenomena. This is also indicated by the following element of verse 16: "in pain you shall bring forth children...". The nature of existing reality changes, as we shall see reflected in the coming verses. The text leaves unspoken whether this change occurs relative to a reality that was already manifest before the emergence of woman, or relative to reality as it is generally assumed to operate. In other words: (the) woman does not already have to have experienced childbirth to allow for a 'change'. Pregnancy and childbirth, necessary prerequisites for being 'blessed with children', are to become very painful processes. Although the main reference is to pregnancy and birth itself, it is well conceivable that the author, living in a world in which death in childbirth and infant mortality were commonplace, also had these phenomena in mind.

The second part of God's judgement addresses the woman's relationship with the man: "your desire shall be for your husband, and he shall rule over you". The two halves of this passage can be taken as being separate and independent in meaning, but it is also possible to translate *wᵉhû'* not as 'and he', but as 'although he'. As a description of the new relationship and as a characterization of the enormous tensions embodied therein, the latter reading is to be preferred. The word used here for 'to rule' is *mšl*, encountered in Genesis 1 in the description of the heavenly bodies, the sun and the

---

[168] Although this command is indeed incorporated in the first, but later, creation narrative, in the 'final editing' it has nonetheless remained in conjunction with the second narrative.

[169] The Hebrew reads: *harᵉbāh 'arᵉbeh*. Both (hiphil) forms are derived from the verb *rābhāh*: to be great, or to be many. The hiphil form is translated as 'to make great', 'to make plentiful', 'to multiply'. To my mind, then, "to greatly multiply" remains the most reasonable translation.

moon. Of the three words employed in Genesis for the concept of 'ruling', *mšl* is the 'kindest'. It says more about the mutual position of the two entities, 'one set above the other', than about the nature of their relationship. None of this detracts, however, from the fact that this 'ruling' implies an entirely new kind of relationship than that embodied in the 'complementary opposite number' and the 'becoming one flesh' of the previous chapter.[170] An inequality of power arises; the relationship becomes unbalanced and loses its natural harmony.[171]

After the woman, it is now the man's turn. He is addressed as 'earthman', and his relationship with the earth is indeed so intimate that the very ground is cursed because of him. From now on it is to bring forth thorns and thistles. Its tilling and keeping is to be accompanied by sweat and toil, "all the days of your life", so says God to Adam. To obtain food will mean hard toil on unworkable soil which, besides (and instead of) edible cereals, will yield thorns and thistles. Compared with Gen 1:29, we see the fruit trees (which were also characteristic of the garden) falling away into the background as a source of food, with the new emphasis on the crops of the field. Some commentators place this passage in the context of human cultural evolution. They see the first chapters of Genesis as reflecting that evolution, with the fruit-gathering phase being followed by arable farming and, soon after, by the keeping of animals (for clothing, in Gen 3:21) and by cattle-farming. Although there may be such allusions in certain elements of this chapter, it seems unlikely that the author intended this story to describe such an evolution, let alone that this was the prime motive. The assumed age of the second creation story, which appears to be far older than the first, contradicts such an interpretation. In a number of respects, moreover, the story itself runs counter to the conventional picture of the 'dawn of civilization'. Among other things, this is reflected in the absence of any mention of hunting. If the story was intended to provide an account of the development of human civilization, there should be

[170] In connection with this passage, H. Jagersma (1995, 55) notes that: "From now on, moreover, her *t^esjûqā*, 'desire', will also certainly carry in it the risk of sexual abuse and oppression by a physically stronger man (*īsj*)".

[171] The text is formulated descriptively (this is how it will be in future) rather than prescriptively (this is how it should be). In both creation stories, humanity's ruling over the animals is perceived as being part of the creation order, while (the) man's ruling over (the) woman is seen as a consequence of the 'Fall' (as indeed is all 'unpleasant' ruling).

some reference to hunting in the early phase of that development. The implicit reference to arable farming and cereal cultivation is certainly not the principal significance of the judgement pronounced on humanity. The passage is not concerned with the 'introduction' of agriculture. Rather, its main thrust is the *change* humans will experience in their relationship with the earth and in the activities involved in obtaining food from it. The harmony in the original, natural relationship has been disturbed.

Human toil is to cease only on the body's return to the earth from which he was taken, for God pronounces: "you are dust, and to dust you shall return". Over the centuries it is this passage that has served as such a pregnant expression of the transient nature of human existence.

Exegetically, it is interesting to inquire whether death is here viewed as a punishment. Gispen holds that it is: "The threat of 2:17 is thus indeed implemented, even though the punishment is not carried out on the day he eats the forbidden fruit".[172] Although this is the line taken by many commentators, it cannot be derived directly from the text.[173] There are, in essence, two possible readings of this passage:

1. Human life is proclaimed finite; death appears for the first time in (human) existence.
2. This death is a punishment for human disobedience.

The first explanation can be derived only implicitly from the text. The passage in question evokes more of a picture of a 'punishment' that will hold for 'the span of a human life', with the finiteness of life presupposed, than one of the end of life being introduced as a new phenomenon. First, with respect to (the) man, we read in verse 17: "in toil you shall eat of it [the ground] all the days of your life" (the same phrasing as that at the end of verse 14, where the serpent is told: "dust you shall eat all the days of your life"), followed then, in verse 19, by: "in the sweat of your face you shall eat bread till you return to the ground...". Each time, the emphasis is on events

[172] W.H. Gispen 1966, 191.

[173] A reading of this nature is closely tied up with the assumption that there is parallel between the main themes of the creation stories of the HB and those of the *Umwelt*, one such theme undoubtedly being to provide an explanation of the mortality of the human race. Although this is also my basic assumption, there may still certainly be major differences between the various stories.

occurring during life, which are now about to undergo a change. As already discussed, the reference to death in Gen 2:17 ("you will (surely) die") does not support the idea of death still being entirely unknown in whatever form. Finally, there is the tree of life. As yet, the text contains no reference to humankind not being permitted to eat from this tree, but now, in verse 22, God steps in to prevent such an occurence, for otherwise (hu)man(kind) will "live for ever".[174] If, previously, humanity was not familiar with death in the garden and was (already/still) immortal, what would then have been the consequence of eating of this tree? And if those consequences are now (in 3:22) different from before the 'Fall', what is the significance of this tree in the narrative of the garden? One possible interpretation rests on the assumption of human immortality being due (in the garden) to access to the tree of life. In this reading, the other tree, and the commandment not to eat of it, were introduced by the author as a means of introducing human mortality into the story (and thus the real world). With this explanation, we remain close to the basic assumption of comparative religious studies that the creation stories of the ancient Middle East seek to answer the fundamental question of why human beings are mortal. In answering this question, however, the passage concerned does not point explicitly or exclusively in any single direction.

The second interpretation, 'death as a punishment', also begs a number of questions. Nowhere is 'returning to dust' mentioned as an explicit consequence of events; it merely indicates the limits of human toil. If we assume that this phrase refers to death, then the reason for that 'return' is not given as the eating of the tree, but the fact that humankind has been 'taken' from the earth. With the words "you are dust", this state of affairs is sketched evocatively once more in the following phrase. These lines describe '*the why of how*' of human mortality.

It remains anomalous that death, announced earlier with the words "*in the day* that you eat of it you shall die", is now to be inflicted 'sometime in the future' and then only on the human race. If death had been announced as a punishment to 'man' in the previous verses, can humankind then have that punishment retracted by eating of the tree of life, and thus (as a second chance?) gain immortality?

Another approach to explaining the consequences of eating of the

[174] On the tree of life as an archetypal motif, see, *inter alia*, R. Cook 1988.

tree proceeds from the idea that humankind here first gains a new level of self-awareness as mentioned earlier. By eating of the tree, man and woman gain a reflexive consciousness that allows them to reflect on themselves and on their place in reality. This is why they become aware of their nakedness, even though that nakedness was already a fact, and it is also why they now also become aware of their own mortality, even though death already existed as a principle. This explanation is not entirely satisfactory, but it does enable a consistent link to be established between the first, certain, consequence of eating of the tree and a second, possible consequence.

Although it can be concluded from these passages that, from now on, human beings will be mortal, all things taken there are insufficient grounds for assuming that 'death itself' makes its appearance after the 'Fall', by way of punishment. It is perfectly feasible that it is not death as such that is introduced at this juncture of the narrative, but *hostile* death. In keeping with other elements of the story, the 'natural' phenomenon of death also assumes a different character, now acquiring *hostile* features.

Westermann sees death not as a punishment, but as a "Grenze für die Mühsal der menschlichen Arbeit", as a limit to human toil. In his words, there is 'something positive' in what the man and woman are told, because it may be good to return to the earth after a life of work and toil. It is death that occurs when a human being is "alt und lebenssatt". In his view, the narrative provides no full, 'whole' answer to the question of the cause of death and sin. "Die eigentliche, die Erzählung bestimmende Frage ist: Warum ist der von Gott geschaffene Mensch ein von Tod, Leid, Mühe und Sünde begrenzter Mensch?" and this is not an objective question concerning the cause of these limitations, but an existential question posed by human beings confronted with such in their everyday lives. To Westermann, the expulsion from the sight of God is the crux of the story of Genesis 3.[175]

To my mind, Westermann's view that Gen 2-3 is concerned with a conflict between God(s) and humankind does insufficient justice to all the references to nature and natural processes found in the creation narrative. Besides God(s) and humans, there is the 'third party' of nature. In virtually all cultures, creation stories can be considered as cosmologies in which the position of the human race

[175] C. Westermann 1974, 363.

is established in relation to the reality in which it finds itself. As such, they can be taken to address the mutual connections and relationships between the triad of God(s), humankind and nature.[176]

After the curses and pronouncements of verses 14-19, the story picks up the earlier thread again in verse 20. Now it is immediately apparent that relations between the man and his 'wife' have changed; she is given a 'proper' name by the man. In the light of what was said earlier on this point, this passage can be taken as constituting the first expression of male domination. According to verse 20, the meaning of that name ('Eve') refers to the future of woman as "mother of all living" but, as said earlier, it may also be related to the Aramaic for snake or serpent.

In verse 21, the makeshift aprons of fig leaves are replaced by 'garments of skins'. These garments are made by God Himself. It is to be queried whether it has not been too easily assumed, in the exegeses and translations familiar today, that the passage refers to animal skins.[177] The Hebrew text employs a status constructus, *kotnôt 'ôr*, and this admits of two translations: 'garments of skin(s)' and 'garments for the skin'.[178] In early-Christian and rabbinic literature 'garments of skin' is read as referring to the fact that, here, humans for the first time acquire a 'body with a skin'.[179] In his commentary on Genesis, Rashi also reports the view that the garments originated 'from the skin', as with the wool of hares.[180] The explanation of these garments being made of animal hides is thus less self-evident

[176] For further elaboration of this triad concept in Western culture, see C.J. Glacken 1966. For a broad treatment of this theme in the HB (particularly in the portions attributed to J), see Th. Hiebert 1996. For a similar approach to Genesis, see, *inter alia*, E.J. van Wolde (1989, 1994).

[177] C. Westermann (1974, 366,367) describes God making garments from 'available materials' and continues with an extensive treatment of the whys and wherefores. Likewise, A. van Selms (1984, 76), U. Cassuto 1989 and H. Jagersma 1995 simply assume, without any further argumentation, that the reference is to animal skins.

[178] M. Reisel (1966, 29) opts for the translation 'garments for the skin', adding "i.e. garments to cover the skin". The word *kotnôt* is generally employed for a long shirt-like cloak that is usually, though not always, made of linen or cotton (our word 'cotton' may be derived from the Hebrew). The word *'ôr* denotes the skin of a human or of an animal.

[179] See, *inter alia*, L. Ginzberg (1909/1937, Vol. 5, 93); Philo *Questiones in Genesin* I, 4.

[180] Pinchas Doron 2000, 16.

than is generally assumed. If we follow the RSV and many other translations in 'garments of skins', we must assume that these skins came from a dead animal or from dead animals. In treating this passage, different scholars have placed different emphasis. Some see it as indicating the altered status of animals, because this is the first (implicit) reference to the killing of animals for human ends. To kill animals is now legitimized is then the reading, with God Himself setting the example. Other exegetes hold the passage to symbolize God's caring for humankind. Although barren times will now be encountered outside the garden, God does not let the man and his wife unprotected. He wishes to shield them from the cold and, in passing, so to speak, to let it be known that they are free to dispose over animals and the lives thereof.[181] However, this 'cold' has then appeared out of the blue, as it were, and there is no explicit reference in the passage to any such 'free disposal' over the animals. Such disposal at this point seems fairly unlikely, moreover, in the light of Genesis 9. It is more plausible to read this verse, in tandem with the foregoing account, as stating that the covering of nakedness is now a permanent necessity for humankind. The focus is not on the material of which the clothes are made, but on the need to be clothed. It is God Himself who confirms, and indeed in a certain sense resolves, this issue by actually making the clothes.[182] The switch from fig leaves to skins may perhaps indicate a changed attitude towards the use of animals. Whether (the) animals were killed specially for this purpose or whether use was made of skins from animals that had died a 'natural' death, or whether the author was not, indeed, referring to dead animals at all but to the wool or linen cloaks he knew, about all this the narrative has nothing to say. Here, too, one should beware of reading too much into the text we have.

The conclusion of Genesis 3 has already been mentioned. God observes that human beings have become "like one of us, knowing good and evil". This is apparently at odds with 'living for ever', and humankind is therefore deprived of the possibility of taking and eating

[181] W.H. Gispen 1966, 192.

[182] The text may reflect the view that clothing, like other elements of culture, is to be considered a gift from the gods (C. Westermann 1974, 366,367; W. Robertson Smith 1889/1894, 437). The Gilgamesh epic, too, establishes a linkage between the three elements of knowledge, clothing and food. Enkidu, it may be added, probably (the text is very incomplete here) acquired his clothing from the temple prostitute Shamkat (Tablet II, 2,3).

from the tree of life. The man and the woman are expelled from the garden, to till the ground. This 'punishment' is to take effect immediately. And the way (back) to the tree is cut off, guarded by cherubs with flaming swords.

CHAPTER THREE

CLEAN AND UNCLEAN

3.1 Introduction

In comparison with the creation stories, over the centuries the 'dietary laws' of the Hebrew Bible have played but a small role in the Christian theological and exegetic tradition. Christianity, with the exception of several minor denominations, is one of the few world religions that has no canon of rules setting out what may and may not be eaten.[1] Although Catholicism does have a tradition on this point, in the form of fasting and fish-on-Friday, not a single animal or animal product is explicitly prohibited.[2] This section of the law is clearly viewed as being typical of the *old covenant.* Christians find arguments for a different interpretation of the dietary laws in the New Testament. In the gospels (Matthew 15 and Mark 7, for example) Jesus casts new light on the distinction between clean and unclean.[3] In the book of Acts we read that the apostle Peter, while in a trance, hears a voice that tells him to kill and eat all kinds of animals. Some of these animals are apparently unclean, because Peter protests against the command, arguing that he has never eaten anything "common or unclean". Thrice he is told "What God has cleansed, you must not call common" (Acts 10:9-16). Although the rest of the narrative makes clear that we are here concerned with contacts between Jews and non-Jews, in the Christian tradition this

[1] From the early beginnings of the Christian church there have always been devotees who distinguished themselves from non-Christians by their dietary practice (R. Lane Fox 1989, 87). In the later tradition, too, we find rules concerning foodstuffs, in the penitentials of the early medieval Irish monks, for example (J.M.D. de Waardt 1996). These rules have never been afforded official status in the principal Christian churches. The rules adopted more recently by minor denominations are generally concerned with the use of the stimulants alcohol and tobacco rather than the eating of meat. Individual vegetarianism does have a long tradition among Christians.

[2] Except, perhaps, "what is strangled and from blood", for which the apostle Paul makes an exception (Acts 15:20).

[3] Both these passages can be interpreted as not implying any rejection of the dietary laws, however; see: E.P. Sanders 1990, 28.

has been taken to constitute a breach with the orthodox dietary laws. A new light is also cast on the laws by the apostle Paul, who, in a discussion about whether they should be imposed on fellow believers, declares that "nothing is unclean in itself" (Romans 14:14). These New Testament passages mark a clear reorientation, with regard to the dietary laws themselves as well as the underlying motives. This reorientation in the NT reflects the contemporaneous environs, for in the classical world we also see a certain distance being taken from the Jewish dietary laws, which were moreover not always understood or respected.[4]

Against this background, it is understandable that in the later Christian tradition theological interest in the relevant passages of the HB declined. The difficult question as to which animals are precisely referred to in the lists of Leviticus and Deuteronomy attracted ever less theological interest.[5] The issue was left to other scholars, including anthropologists and students of comparative religion (among others, W. Robertson Smith 1889/1894, J.G. Fraser 1918, F.J. Simoons 1961, E.R. Leach 1964), Old Testament scholars with a biological interest (G.R. Driver 1955, for example) and even a pharmacologist: D.I. Macht 1953.

In the Jewish tradition this is very different, of course. To the present day, preparing and eating everyday meals according to the dietary laws (to 'kosher' standards)[6] is an essential element of orthodoxy and interpretation of the relevant passages is still an ongoing topic of debate.[7] Over the centuries, many exegetes have studied and debated these texts and their meaning, and some of the fruits of these labours have been laid down in the Mishna and the Talmud.[8] In

---

[4] Among Greek philosophers the dietary laws were a cause of amazement as well as some amusement (Plutarch *Moralia* 669E–671C). In the Epistle of Aristeas (128), too, Jewish food protocols are assumed to be viewed by many as a curiosity.

[5] See, *inter alia*, B. Holwerda 1957/1958, 353 and B. Maarsingh 1974, 93,94.

[6] The official Hebrew term for the dietary laws, *kašrut*, is derived from the root *kšr*, which can be translated as 'suitable' or 'fitting'. The three consonants can be found in the English word kosher. It is to be noted that *kšr* occurs only three times in the HB (Esth 8:5; Eccl 10:10, 11:6), and not in connection with food.

[7] A recent book on Jewish housekeeping considers the dietary laws at length, for example (B. Greenberg 1983, 95-119). She writes "*kashrut* is not simply a set of rules about permitted and forbidden food. It's a way of life" (95). For a more scientific review of the Jewish tradition, see: I. Grunfeld, 1972.

[8] Among others, Maimonides *The Guide for the Perplexed* (*Moreh Nebuchim*) Part III, 47,48; Mishna tractate *Tohorot*; Talmud tractates *Ḥullin* and *Tohorot*.

this tradition, interpretation of the texts also gained major practical significance because decisions had to be taken about whether it was permitted to eat 'other' animals not mentioned in the written Torah.[9]

Above all, it has been the work of the cultural anthropologist Mary Douglas (1966, 1972, 1975) that has spurred renewed interest in the dietary laws among non-Jewish scholars. At first her work drew greatest attention from her own anthropological community.[10] The work of the French scholar Soler (1979, 1996) also deserves mention in this context. Independently and, in his own words, ignorant of the studies by Douglas, he has likewise developed an interesting analysis of the passages in question. Hallpike (1979) has demonstrated that in many cultures animals, plants and natural phenomena are classified in a consistent and systematic fashion.[11] In contrast to anthropologists, scholars of the Old Testament and the history of religion have, with the odd exception, expressed little apparent interest in the theme.[12]

The past few years have seen growing interest in the dietary laws among (other) scholars, including Christian exegetes, however. In the more recent commentaries on the books of Leviticus (E.S. Gerstenberger 1993) and Deuteronomy (C.J. Labuschagne 1987/1990) we see greater interest in anthropological opinion. There is also a measure of convergence in the viewpoints expounded. This emerging consensus relates mainly to the agreed importance of dietary rules in religious practice as a whole.[13]

The study by W. Houston (WH in this chapter) is especially informative and comprehensive. He builds on the work of Mary Douglas but disagrees with her on many important issues. Houston sets the dietary laws in the context of the sacrificial temple cult and

---

[9] Good practical and historical reviews are to be found in: I. Grunfeld (1972) and in the *Encyclopaedia Judaica* under the entry *Dietary Laws* (1971, Vol. 6, 26-46).

[10] Among others R. Bulmer 1967, S.J. Tambiah 1969 and M. Harris 1974, 1985.

[11] Thereby building on the classic work *Primitive Classification* by E. Durkheim and M. Mauss 1903. See also J.M. Drummond 1966, L. Dumont 1980 and S. Aran 1990.

[12] For the exception, see G.J. Wenham (1979, 1981, 1982). For the mainstream, see for example K. van der Toorn (1985, 36) and, particularly, R. Albertz' *Religiongeschichte Israels in alttestamentlicher Zeit* 1992, in which the dietary laws are conspicuous in their virtual absence.

[13] See, *inter alia*, E.B. Firmage 1990, 1992, R. Lane Fox 1991, J. Milgrom 1991, C.J. Labuschagne 1992 and W. Houston 1993.

holds that the distinction between clean and unclean animals dates back to the time that the (or some) Israelites were pastoral nomads. His central thesis is that the dietary laws came to play a decisive role in the development of "a monolatrous Yahwism". In his words: "Purity became the guardian of monotheism" (WH 1993, 123). Houston's work contains an excellent historical and thematic review of the various opinions on the dietary laws and their reading. In the following I shall restrict myself to the most relevant of these. Within this context I shall then focus on the differences and the arguments involved, with a view to assessing whether some form of synthesis is feasible.

3.2 Relevance

The question of the relevance of a deeper analysis of the dietary laws in the framework of the present study can be addressed at various levels (see also Chapter 1, Section 1.8: Selection and themes).

First of all, let me state that I fully endorse Mary Douglas' views concerning the significance of rules of diet within the overall body of the religious thought of a people, or tribe, in this case the Israelites.[14] In principle, rules concerning daily (and nightly) activities can have a profound and pervasive influence on the everyday attitudes of a people. This is indeed the case with the laws concerning purity and impurity.[15] In Douglas' view, such rules must be based on, and refer to, central elements in the religion of the people in question. Similarly, then, it must be possible to explain them with reference to those elements. In following this approach, she takes the text as her basic starting point. If her hypothesis is correct, study of the dietary laws is important, if only to gain a better understand-

[14] Douglas analyses the dietary laws from the perspective of holiness and goes on to observe: "Holiness was given a physical expression in every encounter with the animal kingdom and at every meal. Observance of the dietary rules would thus have been a meaningful part of the great liturgical act of recognition and worship which culminated in the sacrifice in the Temple" (Douglas, 1966, 57).

[15] Pure and impure in the broad sense, i.e. not merely in relation to eating (meat). For the 'common man' in ancient Israel, eating meat will not have been an everyday occurrence. It is widely held to have been limited to the sacrificial cult and only later extended to profane life, and then mainly on special occasions. On this point, see J. Milgrom 1991, who explains the differences he observes between Leviticus (slaughter in the temple(s) only) and Deuteronomy (profane slaughter, too) in terms of this extension.

ing of a key element of ancient Israelite cosmology, as expressed in the HB.

A further answer to the question of relevance proceeds from the fact that the dietary laws are among the few texts that expound in concrete terms on the main theme of this book—humanity's relationship with nature—and, in this context, occupy a niche all of their own. The lists and criteria given in Deuteronomy and Leviticus to delineate clean and unclean animals are unique in the literature of the ancient Israelites.

3.3 Origins of the Text

Rules pertaining to diet are to be found in various places in the HB, but the most extensive passages are in Leviticus 11 and Deuteronomy 14. Both these chapters contain a list of animals that may and may not be eaten. Leviticus, a book of priestly codes, is more extensive than Deuteronomy in this respect. As with so many parts of the HB, (comparative) dating of these texts is problematical.[16] Although it is generally accepted that Leviticus, as part of the oeuvre of P, is to be dated later than Deuteronomy, a number of scholars doubt whether this also holds for the portion of Deuteronomy 14 we are concerned with here. Firmage (1990, 208) considers it likely that (a version of) Leviticus 11 served as the original source for Deuteronomy 14. Milgrom (1991, 698-704) is more categorical on this point. He argues that Deuteronomy is an abridged version of Leviticus. He comes to this conclusion against the background of his general views on the relationship between P and D. In his judgement, the similarities and minor differences can be explained by assuming that "D had the entire MT of Lev 11 before him, which he copied, altered, and above all abridged to suit his ideological stance and literary style" (704).[17] From the angle of interpretation, the is-

[16] There is a reasonable consensus that the core of the book of Deuteronomy is late pre-exilic (late 7th century BC). On this point see, *inter alia*, F.R. McCurley (1974, 295-297), H.D. Preuss 1982, K. van der Toorn (1996, 352) and J. Milgrom 1999. C. Houtman 1980 has published a clear and comprehensive review of the origins and composition of the first five books of the OT, the Pentateuch. According to A.D. H. Mayes Deut 14 represents a post exilic priestly insertion in the Deuteronomic law (1994, 181)

[17] The formulation of "D" having an "entire MT" before him is an anachronism, but the substance of Milgrom's position is clear. Even so, his formulation is

sue of dating becomes particularly relevant where there is major discrepancy, or contradiction, between the two texts. This appears to be the case with the 'winged swarming things'. In Deut 14:19 this entire category of animals is termed unclean. Leviticus also includes this general prohibition (11:20), but makes an exception for four kinds of animal (probably locusts). This addition and anomaly can be more readily understood if it is assumed that the existing text or verse had given rise to debate about the status of certain animals that were, in principle, available and attractive as a source of nourishment. An issue of this nature will most certainly have been put to the priest(s) and in Leviticus the debate is once and for all decided by codifying a clear position. Following this line of argument, the Leviticus text must then be of more recent date. Milgrom (thereby following many other scholars) resolves the apparent discrepancy between the two passages in a different manner. He assumes that the following verse of Deuteronomy (14:20, which he translates as "you may eat only pure winged creatures"), by opting for the word *'ôph* rather than *ṣippôr*, refers to flying insects.[18] We shall be returning later to the 'selection criteria' for this particular group of animals.

Besides the issue of the relative dates of these texts, there is also the problem of the age, development and observance of the dietary laws themselves. Did the people of Israel know the laws as we do today and, if so, in what form and from what time onwards? Were they ever practised in this form?[19] Which animals were eaten and which were not? Did there already exist a certain practice for which the laws formed a kind of *post factum* legitimization, or did the rules (help) mould the practice? In specific cases the mention of certain animal species in the HB begs questions regarding historical dating. For example, on archeological grounds it is to be queried whether, at the time of Abraham,[20] camels were kept in the area he is re-

---

far too categorical in my opinion. He supposes a knowledge of the development of the text that is simply lacking, lost as it is in the distant past.

[18] This reading can also be found, *inter alia*, in S.R. Driver: "By "winged swarming things" (*šereṣ hā'ôph*) are meant winged insects" (1895/1902, 163), B. Holwerda (1958, 358) and C.J. Labuschagne (1990, 69).

[19] These questions are of particular importance for the pre-exilic era. With respect to the latter days of the Israelites (the final centuries BC), we know from the book of Daniel and from non-biblical sources that the dietary laws played an important part in Jewish religious practice, sometimes in far stricter form than laid down in the Pentateuch.

[20] We hereby assume that Abraham is a historical figure who must have lived around 1500 BC (? reported dates have a range of ± 800 years). For decades, his

ported to have kept them according to the narrative of Genesis 24:10.[21] Scholars such as Firmage and Houston, and with them many others, assume that these texts represent a version of the laws that has been subjected to substantial editing by the priestly scribes. They hold the view that the people of Israel will certainly have followed certain dietary rules and customs, possibly even in written form, but to their mind the ultimate rendition and argumentation are of a later date.[22] This 'later date' is then held to be the period of exile, or shortly afterwards. In that period, so they assume, there was a strong desire to lay down the Israelite religious heritage, thereby attending to every last detail in the ideological and religious sense. The presumed systematic nature of the texts is therefore more characteristic of the author(s) than of the Israelites as a people. Van der Toorn also subscribes to this view (1985, 36). On the basis of the

---

historicity has been an issue of considerable debate and study. Following original doubts, voiced, *inter alia*, by Wellhausen in his *Prolegomena* (1878), the tide begins to turn in the 1940s and 1950s. Owing in part to new archeological finds, there emerges a reasonable consensus on the significance of these data, which are seen as "supporting the substantial historicity of patriarchal tradition" (W.F. Albright 1950, 3). There also appeared to be general agreement on the fact that these finds did not perhaps confirm the existence of the person Abraham himself, but that "His life and times, as reflected in the stories about him, fit perfectly within the early second millennium" (G.E. Wright 1957, 40). Albright and Wright's work is reprinted, summarised and discussed in E.F. Campbell, Jr. & D.N. Freedman (eds.) 1983. Over the last few decades this position has been further examined and strongly criticized. There is now considered to be no solid proof of the historicity of the patriarchs and the general consensus today tends towards a later dating (exilic or slightly earlier) of the (Abraham) stories (see J. van Seters, 1975; for a recent review of the early history of Israel, see K. van der Toorn 1996, 183-205).

[21] See D. Hakker-Orion 1984. Archeological finds indicate that the domesticated 'camel', here a reference to the dromedary (*Camelus dromedarius*), most probably did not appear in Israel until the end of the second millennium BC, and then only in the Negev region. The Genesis narrative (12:16) also states that Abraham received his camels 'from outside Israel', namely from the Egyptian Pharaoh. The story of Jacob (Gen 30:25-43) also refers to camels being employed in the days of the patriarchs as beasts of burden for journeys through the desert. Hakker-Orion holds the view that the first historically reliable biblical reference dates from the era of the judges (1200-1000 BC). Even during this period, however, the camels still came 'from outside Israel'. We read in Judg 6:4-5, for example, that the Israelites were robbed by Midian, Amalek and the tribes of the east, who were as numerous as grasshoppers: they could not be counted, like their camels (on which they were presumably mounted).

[22] The historian R. Lane Fox (1991, 82-88) takes the same line. He deems it logical for the priests to have started with the sacrificial animals when drawing up rules and categories. The sacrifice of cattle, sheep and goats was customary, so the question was: what do these animals have in common? The issue of precisely why these animals were sacrificed is not addressed by Fox.

available material, it is difficult to arrive at a final pronouncement on these issues. It is entirely feasible that editors left their mark on these texts in a later phase of history, but there is no unambiguous evidence of this. (On this point, the reader is referred to what has been said earlier of the creation stories and the Pentateuch as a whole.) If there has indeed been pronounced editorial abridgment of these passages, then to me it seems reasonable to assume that the purpose of the priestly scribes would have been to lay down, tighten up and religiously ground already existing practices (and possibly also text fragments) rather than design a system all of their own. Even in the unlikely event of the priests designing a system and associated rules of their very own, the question: what system and why that particular system? remains extremely relevant. After all, it is that system that has been codified and handed down, constituting "the Law pertaining to clean and unclean" for the Jewish people for over 2500 years.

Relevant material published to date on 'practice' is in concordance with the notion that the priestly scribes did not venture far from everyday practice or that they helped shape this practice. This conclusion is based largely on archeological finds at ancient settlements in combination with analyses of animal remains from (domestic) refuse or from sacrificial sites. With such simultaneous evidence, and given the traces of certain usage indicated by analysis, it can assumed that these animals were used and/or eaten, or at least may have been so. By whom the animals were then sacrificed and/or eaten cannot always be unequivocally stated, though, given the virtually permanently mixed population of ancient Israel. This is particularly important for interpreting the finds of (remains) of wild and domesticated pigs.[23] Houston provides a thorough review on this point. With respect to the issue of this non-written evidence he arrives at the following conclusion: "The position is, then, that the biblical system of rules arose in a setting that was eminently compatible with it: it required no sharp changes in habitual dietary and cultic practices

[23] On this point see R. de Vaux 1973 and U. Hübner 1989. De Vaux holds that there is archeological evidence for the sacrifice of pigs among and by the Israelites. Hübner posits that the idea of pigs being unclean was a relatively late addition to Jewish law. Houston disagrees with both these views. Relevant in this discussion are the passages in Isaiah (65:3-5 and 66:3,17), which refer to the eating of pig meat in a cultic context. Houston considers that the passages concern "obscure and perhaps often secret cults offered to the dead or the deities of the underworld, or both" (WH, 168).

general in the land and its environs at least since the beginning of the Middle Bronze Age" (WH 177).[24] This conclusion leaves unanswered the question of the origins of these "habitual dietary and cultic practices". It is therefore entirely feasible that older texts and traditional customs gave rise to a certain practice and that in a later era the priests were keen to again lay down the details of this practice, making use of these older text fragments. In the view of Houston, and with him others, the origins of the (oldest) dietary rules are to be sought in the sacrificial cult practiced by the nomadic pastoralist people of Israel.[25] The questions arising in this context—what may and may not be sacrificed?, what is fitting for the altar?—were later extended to the realm of the table—what may be eaten?, what is fitting for the table?—and to social life—what may be touched and what are the consequences? The depth of study in the fields of textual analysis, religious history and undoubtedly archeology required to address these questions in any fundamental way is beyond the scope of the present work. It is not only new research on existing sources that appears necessary to really advance our understanding of these issues, but above all new materials and new finds.

Following the line adopted with regard to the creation stories in Chapter 2, in this chapter I shall follow primarily the Revised Standard Version (RSV) of the Bible when citing the English and the Masoretic text (MT) for the purpose of analysis. Some reference will also be made to Dutch-language translations.

3.4 Theoretical Framework

In interpreting and understanding the biblical texts, we shall approach them from the following, interrelated angles:

A. The literal text.
B. The systematics of the text and the underlying rationale of the rules. The reasons for these particular systematics.
C. The function of the text and the rules.

[24] D. Hakker-Orion 1984 assigns the following dates BC: 3150-2200 Early Bronze Age, 2200-1550 Middle Bronze Age, 1550-1200 Late Bronze Age. Houston further distinguishes the period 2300-2200 BC as the "Early/Middle Bronze Age".

[25] Besides Houston, also W. Burkert 1983 and E.B. Firmage (1990, 1992), for example.

D. The connection with other parts of the HB and, particularly, the Torah.

The theoretical framework within which I shall analyse the texts has already been discussed in the previous chapter. This framework hinges on the following points, here phrased as postulates:
1. The dietary laws of the Bible do not form an arbitrary set of dietary guidelines but a coherent corpus of rules.
2. These rules are such that they can be explained 'necessarily and sufficiently', i.e. without any loose ends remaining. They are in accordance with all the (biological) information in the text or, less strictly, the biological data do not run counter the stated rules.
3. The structural systematics and underlying rationale should be distinguished from the function of (enforcing) the rules and dietary laws.
4. The key to why these rules and their internal systematics are as they are should be sought primarily within, and with reference to, the text.
5. These rules are rooted in a conception of humanity's relationship with nature within the overall context of creation: in our terms, a cosmology.
6. This cosmology dovetails with major concepts in the Torah and therefore reflects the key elements of the religion of the Israelites.
7. Two main functions of the dietary laws can be identified: they serve firstly as a daily reminder of the key elements of the Israelites' religion, and secondly as a mark of distinction.

In this way the upholding of the dietary laws may certainly have led to the Israelites being able to distinguish themselves from the other peoples in their environs and thus become a separate people in a religious as well as practical sense. In this respect, the laws may indeed have acted as a "guardian of monotheism" (WH 123).

The first of these points is of an axiomatic nature: it cannot be derived, constituting the basic premise that justifies the quest for coherence. It is indeed a stimulating point of departure, because it encourages scholars to seek such coherence. Before Mary Douglas presented her systematic interpretation, as far as is known a quest of this nature had scarcely ever been undertaken.[26] Although coher-

[26] This is obviously not to say that the dietary laws have attracted no interest

ent explanations have been given in the past, by the Church Fathers among others, these were concerned specifically with the function of the dietary laws (see point three).[27] Firmage already considers it a merit of Douglas' work that it questions traditional opinions on the subject. In the theological tradition the dominant view is

> that no single reason could be operative, that the real reasons were no longer understood when the law in its present form was written, and that the present criteria are secondary .... If it has done nothing more than to shake our confidence in these assumptions, Douglas's work merits recognition.[28]

---

at all. In this connection the work of J. Döller 1917 deserves mention. He discusses numerous motifs, some of them for each individual animal mentioned, considering these from the perspective of the comparative history of religion. He does not assume a 'monocausal' explanation, however, and consequently does not posit a single, overriding motif. One of the few theologists to have done so is the Dutch scholar B. Holwerda. In his lectures on Deuteronomy, Holwerda treated the dietary laws (and the work of Döller) extensively. He discusses over fourteen different motifs, but rejects them all and goes on to formulate a position of his own (see 3.6.1). His lecture notes were published posthumously by his brother D. Holwerda in 1957/1958. They have (unfortunately) not been widely disseminated.

[27] Such is the case with the *Epistle of Aristeas* (128-171), in which the dietary laws are discussed. Aristeas, a fairly unknown Jew living in Egypt (probably in the 2nd century BC), assumed that the dietary laws contained moral lessons for humanity. In this view, that birds of prey were not be eaten was a reminder that human beings should not kill one another. The *Epistle of Barnabas*, an apologetic text (i.e. contradictory to Jewish interpretation of the HB) probably dating from the 2nd century AD, contains a similar moral and allegorical explanation of the dietary laws. *That it was forbidden to eat certain animals was intended by Moses in a figurative sense and is not based on divine mandate. When Moses mentioned 'swine' his intention was to say: we may not associate ourselves with those who have forgotten the Lord and wallow in luxury (Barnabas, 10, freely translated).*

In this allegorical interpretation, unclean animals are seen as exhibiting objectionable behaviour that is to be discouraged and banished from human society, with clean animals possessing favourable traits that are to be encouraged. In this reading the hare, for instance, may not be eaten because it forms a new anus each year and is therefore symbolic of those who engage in the vice of pederasty. Why do we eat animals with a cloven hoof? the epistle asks, and the answer is: "because the righteous man does not remain on earth all his days, but also anticipates eternal life. From this we see how good a law-giver Moses was" (*Barnabas*, 10). In this explanation the animals themselves disappear into the background.

[28] E.B. Firmage 1990, 177. The same credit can be given to Douglas' cultural theory (1985, 1986; M. Douglas & A. Wildavky 1982). Moreover do her ideas play a significant role in environmental discourse (see K. Milton 1996).

In the Jewish tradition, too, the predominant attitude is that no reason(s) can be given for the dietary laws.[29]

The second point is related to the first and also follows from it. The explanation given for the coherence within and among the rules must be 'necessary and sufficient' and in accordance with the biological facts reported in or related to the text. The explanation must have no loose ends, as it were. Exceptions of whatever sort do not prove the rule, but tend to weaken it. Explanations that appeal to a specific (different) rule for each animal basically explain nothing and are therefore unsatisfactory. If this criterion is applied unremittingly, however, difficulties are encountered. In both lists we find terms that occur just once. This kind of *hapax legomenon* obviously causes problems because it is difficult to establish an accurate meaning with any degree of certainty. Consequently, such terms cannot be used to assess the consistency of the rules. Given the overall introductory texts, in our case these unique terms undoubtedly refer to certain animal species and, more specifically, in Lev 11:13-19 and Deut 14:20-20, to certain birds. There can be no complete certainty as to their exact reference, however.[30] Whether, and to what extent, this stands in the way of drawing certain conclusions is a matter we shall consider later. As a minimum, the text should at any rate not contain any biological information that is clearly at odds with the rules.

The third point makes a distinction between explanations of function and those of substance. A great many of the exegeses on the dietary laws make insufficient, or no, distinction between these two

[29] This is with reference to the fact that the text itself provides no (explicit) explanation, even though numerous elements of such an explanation can obviously be found in the eyes of Jewish exegetes, too. The *Encyclopaedia Judaica* (vol 6, 42) states: "The Pentateuch classifies the dietary laws as *ḥukkim*, "divine statutes", which by definition are not explained in the text (Yoma 67b)". In the same vein Blu Greenberg, in *How to run a Traditional Jewish Household* (1983, 95), writes: "The laws of kashrut fall into the category of statutes, *chukkim*, Biblical laws for which no reason is given. In other words, we don't know why this food is forbidden and that one not".

[30] There are numerous publications on the flora and fauna of the Bible (see chapter 1 note 97), varying from glossy editions with full-colour photographs to straightforward nomenclatural lists. With regard to animals, the list of A. Shulov 1975 is extremely accurate and valuable for its reporting of sites. Unfortunately, though, the domesticated animals are omitted from the list. A more recent publication is Peter Riede's *Glossar der hebräischen und aramäischen Tiernamen/Tierbezeichnungen* 1993. This glossary is virtually comprehensive as well as accurate.

types of explanation.[31] While the first type is concerned with establishing the effect of (upholding) the rules (wherefore?), the second seeks to explain the systemic substance and the underlying rationale (why these particular animals and rules?). In the classical theological literature the functional explanations predominate, probably unconsciously, because no difference was perceived between the two, or at least not made. The dietary laws were seen primarily as a mark of distinction: by upholding these laws the people of Israel distinguished themselves from the peoples around them and were thus able to become and remain a people in their own right. The question of the 'why' of the laws is then considered satisfactorily answered by indicating their function, the 'wherefore'. By focusing so strongly on the issue of functionality, exegetes of an earlier era omitted to consider other important questions, or answered these only very unsatisfactorily. The major question in this context is obviously why it is this particular set of rules that must (or can) fulfil this purpose. If and when arguments can be found among such functional explanations for deeming certain species unclean, then those arguments are very diverse and in some tautological cases hardly convincing (for example, the swine posited as unclean because of its 'dirty' way of life; see also the remarks on the Epistle of Barnabas in note 27).

The fourth point in fact embodies the classic exegetic attitude: the text is taken as the point of departure and used to construct an argument. This does not imply a restriction to, but the primacy of the text in question.

The final three points of the theoretical framework indicate the potential role that might be played by the dietary laws in the overall context of the present volume. The interpretative framework employed by Douglas and other cultural anthropologists, including her notion of the 'order of creation', nowhere appear to contradict the key concepts or doctrines of the HB.

Although like any interpretative model the above theoretical framework is not binding, it does offer an attractive perspective with which to clarify the status of the dietary laws in the HB as a whole. From

[31] A recent example is E.S. Gerstenberger (1993, 119), who writes in his commentary on Leviticus that: "Die Speisegebote gehören also zu den Normen, welche die Gemeinschaft konstituieren. In diesem Bestreben, Gemeinschaft zu begründen und abzusichern, liegt der Hauptantrieb zur Zusammenstellung der vorliegenden Speiseregeln. Die früheren Motive und Zwecke sind darin aufgehoben". This is confusing as well as arguable.

forming an enigma to the biologically interested exegete, they can become a constituting element of the religion of Israel—provided, that is, the texts can indeed be satisfactorily interpreted within that framework.

3.5 Text and Interpretation

The passages of central importance in the present context are to be found in chapter 11 of the book of Leviticus and in Deuteronomy 14:1-21.[32] In our treatment we shall follow the text of Leviticus, but give due consideration to both texts where Deuteronomy differs. The two texts are reproduced below in the RSV translation:

Lev 11
1. And the LORD said to Moses and Aaron, 2. Say to the people of Israel,
These are the living things which you may eat among all the beasts that are on the earth. 3. Whatever parts the hoof and is cloven-footed and chews the cud, among the animals, you may eat.
4. Nevertheless among those that chew the cud or part the hoof, you shall not eat these: The camel, because it chews the cud but does not part the hoof, is unclean to you. 5. And the rock badger, because it chews the cud but does not part the hoof, is unclean to you. 6. And the hare, because it chews the cud but does not part the hoof, is unclean to you. 7. And the swine, because it parts the hoof and is cloven-footed but does not chew the cud, is unclean to you. 8. Of their flesh you shall not eat, and their carcasses you shall not touch; they are unclean to you.
9. These you may eat, of all that are in the waters. Everything in the waters that has fins and scales, whether in the seas or in the rivers, you may eat. 10. But anything in the seas or the rivers that has not fins and scales, of the swarming creatures in the waters and of the living creatures that are in the waters, is an abomination to you. 11. They shall remain an abomination to you; of their flesh you shall not eat, and their carcasses you shall have in abomination. 12. Everything in the waters that has not fins and scales is an abomination to you.
13. And these you shall have in abomination among the birds, they shall not be eaten, they are an abomination: the eagle, the vulture, the osprey, 14. the kite, the falcon according to its kind, 15. every raven according to its kind, 16. the ostrich, the nighthawk, the sea gull, the hawk according to its kind, 17. the owl, the cormorant, the ibis, 18. the water hen, the pelican, the carrion

[32] For the arguments for reading Deuteronomy 14:1-21 as a single body of text, see C.J. Labuschagne 1990, 59 *et seq.*

vulture, 19. the stork, the heron according to its kind, the hoopoe, and the bat.

20. All winged insects that go upon all fours are an abomination to you. 21. Yet among the winged insects that go on all fours you may eat those which have legs above their feet, with which to leap on the earth. 22. Of them you may eat: the locust according to its kind, the bald locust according to its kind, the cricket according to its kind, and the grasshopper according to its kind. 23. But all other winged insects which have four feet are an abomination to you. 24. And by these you shall become unclean; whoever touches their carcass shall be unclean until the evening, 25. and whoever carries any part of their carcass shall wash his clothes and be unclean until the evening. 26. Every animal which parts the hoof but is not clovenfooted or does not chew the cud is unclean to you; every one who touches them shall be unclean.

27. And all that go on their paws, among the animals that go on all fours, are unclean to you; whoever touches their carcass shall be unclean until the evening, 28. and he who carries their carcass shall wash his clothes and be unclean until the evening; they are unclean to you.

29. And these are unclean to you among the swarming things that swarm upon the earth: the weasel, the mouse, the great lizard according to its kind, 30. the gecko, the land crocodile, the lizard, the sand lizard, and the chameleon. 31. These are unclean to you among all that swarm; whoever touches them when they are dead shall be unclean until the evening. 32. And anything upon which any of them falls when they are dead shall be unclean, whether it is an article of wood or a garment or a skin or a sack, any vessel that is used for any purpose; it must be put into water, and it shall be unclean until the evening; then it shall be clean. 33. And if any of them falls into any earthen vessel, all that is in it shall be unclean, and you shall break it. 34. Any food in it which may be eaten, upon which water may come, shall be unclean; and all drink which may be drunk from every such vessel shall be unclean. 35. And everything upon which any part of their carcass falls shall be unclean; whether oven or stove, it shall be broken in pieces; they are unclean, and shall be unclean to you. 36. Nevertheless a spring or a cistern holding water shall be clean; but whatever touches their carcass shall be unclean. 37. And if any part of their carcass falls upon any seed for sowing that is to be sown, it is clean; 38. but if water is put on the seed and any part of their carcass falls on it, it is unclean to you.

39. And if any animal of which you may eat dies, he who touches its carcass shall be unclean until the evening, 40. and he who eats of its carcass shall wash his clothes and be unclean until the evening; he also who carries the carcass shall wash his clothes and be unclean until the evening.

41. Every swarming thing that swarms upon the earth is an abomination; it shall not be eaten. 42. Whatever goes on its belly, and whatever goes on all fours, or whatever has many feet, all the swarming things that swarm upon the earth, you shall not eat; for they are an abomination. 43. You shall not

make yourselves abominable with any swarming thing that swarms; and you shall not defile yourselves with them, lest you become unclean. 44. For I am the LORD your God; consecrate yourselves therefore, and be holy, for I am holy. You shall not defile yourselves with any swarming thing that crawls upon the earth. 45. For I am the LORD who brought you up out of the land of Egypt, to be your God; you shall therefore be holy, for I am holy.
46. This is the law pertaining to beast and bird and every living creature that moves through the waters and every creature that swarms upon the earth, 47. to make a distinction between the unclean and the clean and between the living creature that may be eaten and the living creature that may not be eaten.

Deut 14:3-21
3. You shall not eat any abominable thing. 4. These are the animals you may eat: the ox, the sheep, the goat, 5. the hart, the gazelle, the roebuck, the wild goat, the ibex, the antelope, and the mountain-sheep. 6. Every animal that parts the hoof and has the hoof cloven in two, and chews the cud, among the animals, you may eat. 7. Yet of those that chew the cud or have the hoof cloven you shall not eat these: the camel, the hare, and the rock badger, because they chew the cud but do not part the hoof, are unclean for you. 8. And the swine, because it parts the hoof but does not chew the cud, is unclean for you. Their flesh you shall not eat, and their carcasses you shall not touch.
9. Of all that are in the waters you may eat these: whatever has fins and scales you may eat. 10. And whatever does not have fins and scales you shall not eat; it is unclean for you.
11. You may eat all clean birds. 12. But these are the ones which you shall not eat: the eagle, the vulture, the osprey, 13. the buzzard, the kite, after their kinds; 14. every raven after its kind; 15. the ostrich, the nighthawk, the sea gull, the hawk, after their kinds; 16. the little owl and the great owl, the water hen 17. and the pelican, the carrion vulture and the cormorant, 18. the stork, the heron, after their kinds; the hoopoe and the bat. 19. And all winged insects are unclean for you; they shall not be eaten.
20. All clean winged things you may eat.
21. You shall not eat anything that dies of itself; you may give it to the alien who is within your towns, that he may eat it, or you may sell it to a foreigner; for you are a people holy to the LORD your God. You shall not boil a kid in its mother's milk.

3.5.1 *Leviticus 11:1-8*

Leviticus begins by declaring the origins of the Law: "And the Lord said to Moses and Aaron: Say to the people of Israel...". This is the first mention of these two prominent figures in one and the same text and it underscores the importance of the following passage. According to Houston, the mention of Aaron together with Moses implies a priestly instruction, "with ritual implications". This then applies particularly to the latter part of the chapter (verses 24-27). In the first part there is no direct ritual link. Here, the focus is on the rules and criteria, and on animals that may or may not be consumed by the Israelites. These animals are grouped according to the segment of the environment in which they are found: animals of the land, water and air. Means of locomotion appears to be a second important distinctive and ordering criterion. One is immediately struck by the parallel with the first creation story.

"These are the living things (*ḥayyāh)* which you may eat among all the beasts (*b^e^hēmāh*) that are on the earth", Leviticus continues. In the HB *ḥayyāh* is generally employed to denote living creatures and sometimes, as we saw in Genesis 1, as a generic term for land animals. *B^e^hēmāh*, in contrast, is used mainly to denote four-footed mammals and, more specifically, livestock. Houston (1993) consequently argues for dividing this sentence in two, with the general reference to living creatures serving as a title to the first part of the chapter. This requires a change in punctuation and in reading of the MT, however, and conflicts with the structure of verse 9[a]. It seems unnecessary, moreover. It is immediately clear, from what follows, that the focus is indeed on the quadrupeds, and within this group, on those that are domesticated, an important class of animals for a people of shepherds and farmers. The reference is in fact to these (domesticated) animals along with their non-domesticated, wild relatives and members of the Deer family (which were also hunted), and also such smaller mammals as the Hare and the Rock Badger. These are the categories to which the stated criteria are relevant. When it comes to the 'wild' mammal fauna of Israel and its environs, large predators for example, their uncleanliness is apparently clear. This latter group of animals is treated, albeit implicitly as 'going on all fours', further on in the text (verse 27). The same holds for the crawling and swarming animals, the *šereṣ* of the land, which are also missing from this list, but are later explicitly mentioned as being unclean (verses 29 and 44). Houston sees in this initial absence and later explicit mention arguments for assuming that the second part of the chapter is a later addition.

Of hooves and cud-chewing

With the aid of two criteria: having cloven hooves and chewing the cud, a distinction is to be made between the clean and unclean species of animal among "the beasts that are on earth". For an animal to be classified as clean, both criteria must be satisfied.

Positive fulfilment of the two criteria yields a zoologically recognizable taxon: the sub-order Ruminantia of the order Artiodactyla, as has indeed been pointed out by others including Hunn (1979). Besides having a divided hoof (or being 'even-toed', as the zoological nomenclature has it), these animals have several other features in common, such as horns or antlers and upper-jaw incisors. While not constituting definite proof, this is at least an indication that the criteria were not chosen arbitrarily. Milgrom assumes that the codifier of the laws was concerned with limiting the consumption of meat to the flesh of the livestock kept by the nomadic people of Israel (cattle, sheep and goats). The cloven hoof would in itself have been a sufficient criterion, were it not that pigs would then be classified as fit for consumption. To avoid this, the criterion of 'cud-chewing' was added (Milgrom 733).

In verses 4-8 the criteria are applied to a series of species whose status was possibly the subject of debate, or about which confusion was to be avoided. None of the animals identified may be eaten, on the grounds of their meeting only one of the two criteria. Both the animals themselves and their flesh and their carcasses are declared to be unclean. The camel appears to be problematical.[33] Although it is not a member of the aforementioned taxon, lacking horns for example, it appears to be excluded here on incorrect grounds and assigned to the category of unclean animals. The camel does in fact have cloven hooves. In an attempt to explain this 'error', it has been pointed out that the hooves rest on a thick, elastic sole, partly hiding the divided hoof. The codifier was apparently thrown off guard

[33] As far as is known, the two-humped Camel (*Camelus bactrianus*) is not and never has been native to this region; it is a Central Asian species. The Camel was not entirely unknown in the Middle East, though, witness a depiction of the species on the Obelisk of Shalmaneser in the British Museum. In the HB the term 'camel' is almost certainly used to denote its close Middle Eastern relative: the single-humped Dromedary (*Camelus dromedarius*) or Arabian Camel (see also D. Hakker-Orion 1984). Biologically, then, it would be more correct to employ the term 'dromedary' in the English. I shall not do so, though, because the translation 'camel' is so familiar; it is indeed one of the Hebrew-derived words (*gāmāl)* to have been absorbed in the English language.

by this situation, and consequently classified the camel erroneously (and thus wholly unjustifiably) as unclean. In other explanations it is assumed that the camel was not to be eaten because it was a very valuable animal as well as a symbol of affluence. For this long-standing—and, in economic terms, very sensible—tradition, justification had to be found in the law. The slightly anomalous hoof provided such justification, and it was henceforth the camel's fate to be unclean. If economic motives were the prime concern, it might be added, how is it then to be explained that the camel was not to be eaten under any circumstances at all? One would expect an exception, as it were, for the slaughter and consumption of old animals that had come to the end of their working life.

The first explanation is in line with the overall notion that the criteria predate their application, while the second implies that the criteria were adapted to match existing practice. Milgrom follows Douglas in her view that the criteria were earlier, and considers the discussion closed: "Thus the verdict is clear and decisive: the criteria came first and only afterward four anomalies were found". His principal argument rests on the completeness of the list of anomalies. As far as is known, there are no 'missing species' that should be considered anomalous and that are moreover known (or can be assumed) to have lived in the Mediterranean region in the era in question.[34]

Throughout the passages in question the criterion of the hooves is formulated in a rather enigmatic fashion, owing partly to the use of two different verbs (*prs* and *šsʿ*) and their derivatives. This leads to problems both in translation and in applying possible translations

[34] Besides the four mentioned, the following anomalies are involved. Outside the Mediterranean region, the Llama is the only known exception. The Llama and its relative the Vicuna are natives of South America. They are ruminants, but their hooves are not fully cloven. That the Llama is not included in the list is hardly surprising. Milgrom does note the absence of the Hippopotamus, which (on the basis of archeological evidence) can be assumed to have lived in the (environs of) Israel. The Hippopotamus is a herbivore, but it is not a ruminant. It has four clearly discernable toes and is a member of the order Artiodactyla. According to Milgrom, the cleft in the animals's hoof is so weakly developed that "it was missed by the ancients and even omitted by Aristotle in his *Historia Animalis*". He consequently takes it that it was the animal's habits and weakly cloven hoof that kept the Hippopotamus off the list and sees this as constituting proof for Douglas' position: that the rules predate the criteria (Milgrom 1991, 727,728). However, Milgrom does an injustice to Aristotle here, because the latter most certainly does assign the Hippopotamus to the class of the διαχιδῆ animals with a hoof divided in two (*Historia Animalium* 499 b9).

to concrete animals, or in both. Let us first consider the translations.

The most usual meanings of *prs* are: to break or distribute (of bread, for example) and to divide or split, and of *šs'* : to split, cleave and tear to pieces. If we take 'divide' for *prs* and 'cleave' for *šs'*, the MT text of the verses in question translates literally as follows:

3. Consumable: All (beasts with) divided hoof and cloven cleft (of the) hoof
4. Non-consumable of the beasts that chew the cud or have a divided hoof:
   the camel: undivided hoof
   the hyrax: undivided hoof
   the hare: undivided hoof
7. the pig: divided hoof (and) its hoof with cloven cleft, but not chewing the cud
26. All beasts with divided hoof (but) without cloven cleft.

Over the ages, translators as well as exegetes have endeavoured to resolve the evident difficulties of these lines. In the English language both the classical King James' Version and the newer Revised Standard Version start out by translating *prs* as 'part' and *šs'* as 'cleave', giving 'cloven-footed'. The Dutch NBG translation, in verse 3, opts to make the second clause subordinate to the first, as: (having) 'parted hooves, that is fully cloven hooves'. The same phrasing is adopted in verse 26, but with 'yes even' in place of 'that is'.

These translations are not only over-elaborate; they also beg a number of questions regarding the real-world animal species that are denoted. Apart from the fact of the camel having cloven or divided hooves, already discussed, there is the additional problem that there are no known species satisfying the criteria of verse 26 as they stand in the above translation.

Having closely analysed the terms used in verse 3 for "that parts the hoof" (*maphreset parsāh*) and "cloven-footed" (*šôsa'at šesa' p^erāsôt*), Houston opts for a radically different translation (WH 36). He rejects the generally adopted rendering of *maphreset parsāh* as 'that cleaves/parts the hoof', because in his view this "makes v.3 tautologous, and, worse, it makes v.26 nonsense". He takes a different stance, thereby following Ibn Ezra, Hirsch (1958, 268) and (in part) Milgrom (1991, 646). His approach is grounded in a closer analysis of the verb *prs* and, in particular, the analogy with Ps 69:32 (RSV: "...an ox or bull that hath horns and hoofs"), the only other place in the HB where the verb occurs in the hiphil form. He considers this sufficient justification for adopting the translation "with hooves" or, as the Jewish Publication Society (JPS, 1985) has it, "that has true hoofs". Houston assumes that this qualification of having 'true' hooves was added to exclude the camel. Biologically, such hooves can be taken to mean *hard, horny extremities (on the legs) on which to walk.* In Houston's approach, *šôsa'at šesa' p^erāsôt* does not constitute a further specification of *maphreset parsāh* but a separate, albeit secondary, criterion. The clean animals must have ('true') hooves (1) and these hooves must be (fully) cloven (2).

In this reading, verse 3 loses its tautological nature and verse 26 gains meaning. At the same time, this solution ensnares Houston in problems in interpreting verse 4. There, four animals are excepted from "among those that chew the cud or part the hoof". If we here read only "(having a) hoof", the question arises why the ass and horse are not mentioned (WH 37). On the basis of the LXX text, Houston assumes that the phrase "(fully) cloven", although intended, has perhaps been lost or omitted.

In this verse Houston is fully in line with Milgrom. The latter translates verse 3 as "Any quadruped that has hoofs, with clefts through the hoofs", likewise giving arguments for translating *maphreset parsāh* simply as 'hoof'. He takes the hiphil as a denominative, to be rendered literally as "that grows a hoof" (Milgrom 646). Since the possession of hooves is in itself an inadequate criterion for the clean land animals, verse 3 then adds the further condition *šôsa'at šesa' p^erāsôt.* The phrase 'clefts through the hoofs' ('cloven cleft') can then be read as 'hooves divided into two'.[35] Partly in view of the parallel text in Deuteronomy 14:6, which has *šôsa'at šesa' š^etê p^erāsôt* here, it can perhaps be assumed that the word *š^etê* has been lost in the Leviticus text.[36] Although the RSV translation "the hoof cloven in two" is better than the NBG's "both hooves fully cloven" (dealing as we are with quadrupeds), it is still illogically repetitive, as it follows the condition "that parts the hoof". What the verse surely describes is the *šôsa'at šesa'* in each hoof creating *š^etê p^erāsôt*: two hooves, as it were. Given the anatomy of the Artiodactyla, this is an entirely adequate description. Anyone who have ever observed hoove prints of these animals is able to confirm that.

---

[35] See also C.J. Labuschagne 1990, 66.

[36] The LXX, the Peshitta and several other manuscripts also have this reading in the Leviticus text.

Of the animals in principle sanctioned for consumption, several species are subsequently rejected because they do not satisfy one or more of the appropriate criteria. Three species may not be eaten because, although they chew the cud, they do not have (proper, true) hooves. The apparent problem of the camel is thus resolved, for like the hare and the hyrax it has no hard, horny extremities on which to walk.[37] Milgrom's approach requires a different reading of verse 4, too, for here animals are pronounced unclean for possessing the 'wrong' kind of hooves (like the camel), characterized as *ûpharsāh ' ênennû (lô') maphrîs*. Milgrom rejects the traditional translation "(because) it does not part the hoof" ("for its feet are clearly cloven"), reading instead "it has no hoofs".[38] To reflect the verbal form, one might say: 'it does not hoof'. This reading is supported by verse 26, for if *maphreset parsāh* is read as 'parts the hoof', we would be confronted with a self-contradictory text. The fourth animal species that is taboo, the swine, does have hooves, even hooves 'divided in two' and therefore 'correct', but it does not chew the cud.[39] In this the pig is unique, being the only cloven-hooved animal that is not a ruminant. Verse 26 refers to those animals that have hooves, but not cloven hooves, and that do not chew the cud, a description to which horses, asses and mules conform.

Given the arguments, I would hold the translations proposed by Milgrom and Houston (or the JPS variant "true hoofs") superior to those of the RSV, NBG and other such renderings. They are linguistically defensible and can be consistently matched to biological data.

The second criterion, 'chewing the cud', is a literal translation of the Hebrew and poses no immediate difficulties. Nor is identification of the relevant animals controversial. What is to be queried, however, is whether these animals are all true ruminants. The *šāphān* belong to the Order of Hyraxes (Hyracoidea), herbivores the size

[37] The 'unhoofly' nature of what we term the camel's 'hoof' is apparent when the animal walks, the elastic sole accommodating every unevenness encountered. This can be observed in zoos and is confirmed by those who have ridden camels.

[38] WH (1993, 38 note 1) states: "Milgrom on the other hand (1991, 646) simply asserts that the text is correct in saying that the camel is hooved.", but he has not read Milgrom accurately on this point.

[39] The reference is to the Wild boar (*Sus scrofa*) and domesticated breeds of swine. Hybrids are also found, as domesticated animals revert effortlessly to the wild and interbreed readily with wild pigs.

of a Hare.[40] Grzimek's standard animal encyclopaedia (Volume XII, Mammals 3, 600) describes them as ruminants. They have a very long gut and two so-called caecae, or 'blind guts', and go through 'ruminating' motions as they chew. This quality is also ascribed to the third animal, the Hare. It seems likely that these observations led to these two non-domesticated species being termed ruminants, or 'regurgitators'. As a description of their digestive behaviour, this suffices.[41] In terms of biological taxonomy, though, the Hare is not a ruminant.[42] However this may be, use of the twofold criterion thus leads, intentionally or otherwise, to the correct biological classification of the clean animals remaining intact. The main focus of the debate with respect to 'chewing the cud', then, is why it is employed as a criterion. Milgrom, like Firmage, holds the view that it was added to exclude pigs. Harris (1985, 79), on the other hand, assumes that it was the 'hoof' criterion that was added to exclude the camel. However, the text provides insufficient grounds for concluding that the reference is to two independent and complementary criteria. If the intention was merely to exclude pigs, it would have been far more straightforward to simply prohibit their consumption. Why, then, the choice for such a general, indirect criterion as 'chewing the cud'? And if the camel was the only ruminant that was not to be eaten, why such convolutions about the hooves?

Further on in the text, in reference to birds, it will become clear that feeding habits form an important criterion, and this also appears to be the case with the land animals. The codifier of the laws was keen to ensure that these animals were not carnivores. They should not feed on other animals, consume blood or associate themselves with 'death' in any way, as this would conflict with the ideal. The quality of being a ruminant constitutes a very good guarantee

[40] It is not known exactly which (sub)species is referred to here. It lives in rocky terrain (Prov 30:26). English translations generally render the species as 'rock badger' or 'daman' (*Hyrax capensis*), also known as the Syrian coney (*Procavia syriacus*; United Bible Societies 1972/1980, 69) or the Rock hyrax (*Procavia capensis syriaca*; P. France 1986, 43). On the basis of its geographical distribution, the Ethiopian rock badger (*Procavia habessinica*) might also be eligible.

[41] See also *Brehms Tierleben*, Saugetiere Band III, 1890/1893, 355.

[42] There are three possible (sub)species intended: the Syrian hare (*Lepus syriacus*), the Ethiopian Highland Hare (*Lepus starcki*) and the Arabian hare (*Lepus arabicus*). Like rabbits, besides excreting pellets hares also produce soft droppings which they reconsume. These 'caecotrophs' are retained in the stomach for a while before being further digested. They are extremely rich in vitamins and their redigestion gives hares an extremely efficient digestive system.

that the animal in question is herbivorous and thus a 'vegetarian'.

In her analysis, Douglas is concerned primarily with the linkage between the environmental segment (land, water, air) in which the animals live and their means of locomotion. To my mind, this is too narrow an approach. The emphasis on chewing the cud indicates that the ideal, 'pure' animal should at any rate be vegetarian. In this ideal, there is evidently no place for death. In this context, 'death' should be taken as a premature end to life (cf. Chapter 2, Sections 2.5.2 and 2.6). The hooves prove to provide perfect support and delineation for this criterion. Some animals use their hooves to fight. It may be the case that animals with powerful, undivided hooves such as the horse were 'mistrusted' for this reason, too. Support for this thesis can be found in the fact that the horse not only uses its hooves to kick, but is also, more generally, an animal which is associated in the HB with aggression and war. As a general rule, though, this 'weapon' motif is inadequate, as is demonstrated by the example of the ass (an undivided hoof, but not used 'offensively' as with the horse).

Although there remains a degree of uncertainty in the arguments concerning the clean (land) animals, the positive upshot of the passage seems entirely clear. The intention of the author is readily surmised. He was seeking 'sufficient' criteria for the clean quadrupeds: criteria that were in accordance with the ideal picture held of these animals and at the same time practicable to use. As was fitting for a society of nomadic pastoralists, the basic point of departure was the livestock so familiar to everyone. It is certainly feasible that the criteria were designed to justify existing sacrificial practices, but then the question becomes: what religious motive is used for such justification, and what underlying ideas may have played a contributory role? In both cases it is difficult to ignore the creation narrative as an appropriate analytical framework. The ideal can be derived from the account of creation. According to the codifier, all ruminants with hooves that are visibly cloven meet the norm. This norm can be seen as the 'ideal' of a clean quadruped. Feeding habits and mode of locomotion are seen as important criteria, both independently and in combination. Both qualities can be readily observed and for the land animals—a key group—the twofold test yields an unambiguous result.

3.5.2 *Deuteronomy 14:1-6*

This chapter of Deuteronomy starts by prohibiting two mourning customs. Mourning customs were not generally forbidden to the Israelites. The question of why these two customs should be mentioned here we shall leave undiscussed.[43] From our perspective, what is important is the motivation for these commandments found in verses 1[a] and 2. As the text makes clear, the people of Israel, as children of YHWH, must conduct themselves in an appropriate fashion. They are a sacred, chosen people and this must be reflected in their behaviour, both in times of death and mourning and in everyday life, and particularly when life and death confront one another. In this sense, verse 2 forms a bridge between what precedes and what follows. The reiteration of the 'holiness' motif in verse 21 makes clear that Deuteronomy considers this extremely important. I shall be returning to this matter later on, in discussing Leviticus 11:44,45.

In verse 3 the text continues by indicating which animals are fit for consumption and which are not. In terms of argumentation and criteria, the text does not deviate from Leviticus. The 'problematical cases' are discussed in virtually identical fashion. The minor interesting differences have already been treated in the previous section. However, while Leviticus merely provides general rules as to which land animals may be eaten, Deuteronomy first provides a concrete list of animal species and concludes with the rules. The species named in verses 4-6, three domesticated and seven wild, all belong to the aforementioned zoological taxon. Identification of the domesticated species poses no problems. These are the ox (i.e. cattle), the sheep and the goat: the animals traditionally kept by pastoral tribes in the region. With the undomesticated species, identification is slightly more problematical, but there is no doubt that the reference is to deer, gazelles, antelopes and wild species of cattle, sheep and goats. These species recur, although not always consistently, in numerous translations. The badger, the translation of *dîšon* given in the NBG, does not belong in this list, however. The badger is not a member of the order Artiodactyla, eats insects and small reptiles, and cannot therefore be included in the list of clean animals. The *dîšon* is probably a species of antelope or a mountain goat,

[43] In my view, C.J. Labuschagne (1990, 62-64) provides a satisfactory explanation.

or ibex, as in the RSV. The list appears to be reasonably complete, but the text leaves the door slightly ajar, because verse 6 may also be translated as: 'and (further) every other animal (or quadruped) that has a divided hoof and chews the cud'.[44]

3.5.3 *Leviticus 11:9-23*

Fish

Verses 9-12 of the Leviticus text explain the rules for "all that are in the waters". It has already been noted that the HB has little terminological differentiation among the main inhabitants of the waters: the fish. There are no names for individual species of fish. The only aquatic animals afforded separate mention are the "great sea monsters", encountered in our treatment of Genesis 1:21, and the *Liwyātān,* mentioned elsewhere. The Israelites were most probably well aware that fish were not the only inhabitants of the water. The criteria deciding which aquatic animals may be consumed are quite clear. All creatures with fins and scales may be eaten, while others may not. All non-piscine species such as crustaceans, molluscs and marine mammals are thus excluded. The combination of fins and scales is seen as representing the 'ideal' for aquatic animals. As with the land animals, the mode of locomotion also plays an important role. In water, animals should swim using fins and scales. Anything lacking fins and scales "is an abomination to you", so warns verse 10. Their carcasses are also "an abomination". The use of the word abomination, *šeqeṣ*, (three times) for this category is noteworthy. In reference to the land animals that are taboo because they satisfy but one of the two criteria, the only term used is 'unclean' (*ṭāmē'*). This difference in terminology indicates that there are degrees of (im)purity. The general term used frequently in the HB for unclean, root *ṭm'*, is used adjectivally mostly in a ritual context. The religious meaning evolved from (or alongside) the general meaning. Consequently, *ṭāmē'* is "the most widely used technical term in the priestly writings for everything that is ritually defiling" (WH 18). The connotation of 'dirtiness', sometimes encountered, is therefore less accurate for the passages in question and misses this deeper significance.

---

[44] There are indeed species that satisfy these criteria, such as the giraffe and the okapi, which are both ruminants, have cloven hooves and also horn-like protuberances. Their status in the tradition is not entirely clear, however (*Encyclopaedia Judaica* 1971, vol. 6, 26-46).

According to Houston, 'uncleanliness' arises when the integrity of someone or something is breached. This may be the result of intrusion by, or mixing with, something or someone not belonging to the same 'realm'. Such realms are to be found at many different levels: the household, life and death, the natural segments of water, soil and air, sexuality, the temple cult and so on. These realms are in fact concerned with upholding overall social structure and order, the order of creation. On this important issue Houston sides with Mary Douglas, who in her *Purity and Danger* concludes that the breaching of the order, the transgression of limits and the mixing of elements from different realms, is the pivotal element in the notion of uncleanliness. The notion is also captured in both authors' use of the term 'pollution', and this line of thinking provides an important key to understanding the many different rules concerning purity and impurity as a coherent whole. I shall be returning to this broader issue later.

Birds

Verses 13-19 sum up the species of birds that are not to be consumed. It is this list, in particular, that has caused translators such difficulties over the ages. For many species, the list is the sole reference in the HB. The sequence of species in Leviticus is slightly different from that in Deuteronomy. Leviticus names twenty species, Deuteronomy twenty-one. The two lists have nineteen species in common. The fourth word in Deuteronomy, *rā'āh*, is probably identical to the word *dā'āh* in Leviticus. Transposition of the initial letters *daleth* and *resh* is not uncommon, particularly in names (in Gen 10:4 and 1 Chr 1:7, for example, and in Num 1:14 and 2:14). The sixth word in the Deuteronomy list, *dayyāh*, may be a variant of the *'ayyāh* found in both lists, but this would mean that the same bird occurs twice in the Deuteronomy. In the plural, the term *dayyāh* is encountered once more, in Isa 34:15 and in the RSV is there translated as 'kites' (and in the NBG and KJV as 'vultures'). This species (name) is lacking in the Deuteronomy texts of the Septuagint and the Textus Samaritanus. It is virtually certain that these three (two?, four?) similar terms denote birds of prey, and most translators opt for species of vulture, kite or falcon. Etymologically, the terms have been associated with 'keen eyesight', but the status of these derivations is very unclear. Given the general sequence of the list (with the larger raptors already mentioned), the RSV translation as 'kite' would appear to be the most appropriate choice. Both the Black kite and the Red kite are still commonly observed in Israel today.

According to Houston, textual analysis and comparison have demonstrated that there may have existed a shorter 'original list' lacking systematic structure. Some exegetes see the fact that Leviticus here mentions twenty creatures as being connected with the numerical structure of the numbers of creatures named throughout the chapter. These are always multiples of four. First there are four unclean land animals, then five times four unclean winged creatures, then four clean animals and finally, in verses 29 and 30, two times four unclean 'swarming' creatures. From this perspective, then, in the final version of Leviticus the author was apparently concerned with improving the systematics, by reordering the species, and with getting the numbers right, by adding or omitting species (WH 44 *et seq.*). I here leave this matter for what it is, because it does not detract from my own line of argument concerning the background to the classification, though I shall return to it later since it may constitute indirect evidence.

Although we cannot unequivocally identify each individual species, the text indicates that verses 13-19 at any rate refer to creatures that are carnivorous, or 'non-vegetarian'.[45] They eat small terrestrial animals (all birds of prey, owls), fish (osprey, 'divers', cormo-

[45] G.R. Driver 1955 holds the opinion that the (original) Leviticus list, with the exception of the last two species (the hoopoe and bat), referred only to birds of prey, including owls, and that this list was structured in order of decreasing size. On these premises, he arrives at a translation of his own that deviates markedly on a number of counts. Driver distinguishes no less than eight species of owl. Because of his collaboration with the leading ornithologist David Lack, this 'Driver list' has gained considerable authority, although nowhere in the article is the precise contribution of Lack made clear. Driver's zoological identifications are generally within the overall 'range of uncertainty', although his translation of *Rāḥāmāh* by 'Osprey' seems unlikely to me. Most translators opt here for the Egyptian (or 'carrion') vulture. The argument Driver gives for favouring the Osprey (the black and white colouring, because the stem of the word, *rḥm*, is held to refer to such) is far more applicable to the Egyptian vulture. And if size is important, this, the smallest of the vultures is here better placed on the list than the far larger Osprey. Driver's arguments for translating *Bat hayya'ᵃnāh*—daughter of the desert—by Eagle Owl are also disputable. This creature is often mentioned together with the *tannîm* and is generally situated in an inhospitable environment. It must have been an easily recognizable bird that appealed to the imagination, given the relatively large number of references in the HB. Following the LXX and Vulgate, then, I would hold 'ostrich' to be the most plausible translation. The information over the bird's behaviour, such as the 'mourning' sound reported in Mic 1:8, does not contradict this identification, since the nocturnal cry of the Ostrich can also be thus described.

rant, pelican, heron), amphibians (stork), reptiles (stork, hoopoe) or insects (stork, hoopoe, bat) or are omnivores (ravens and crows, ostrich). Besides feeding habits, other motives may have also played a role in classing a particular species as unclean, either as a secondary or primary argument. It has been pointed out, for example, that owls are found in desolate places such as ruins and rubble, while certain other species such as the ostrich live in inhospitable areas such as deserts.[46] In the case of the *šāḥaph* (seagull?), the *ʾanāphāh* (heron?, cormorant?), the *šālāk* ('diver', cormorant?) and the *qāʾāt* (bittern?, pelican?) we have creatures of the air that swim and dive in water and thus do not always remain in the right 'realm'. In exceeding the bounds of their domain, they breach the natural order. The same argument may hold for the creatures of the dusk, the various species of owl (and nighthawk) identified in translations: the *kôs* (small owl, Little owl?), the *taḥmās* (Nighthawk?, night owl?, Tawny owl?), the *yanšûph* (eared owl: Eagle owl?, Long-eared owl?), the *tinšemet* (white owl?, Barn owl?) and the only non-avian species in the list, the *ʿatallēph* (bat). The last of these creatures, with its characteristic appearance and behaviour—flying without 'real' wings, at dusk, between night and day—can be classed as unclean for several reasons. In general, these arguments are well in line with the overall hypothesis that animals breaching the 'order of creation' are to be deemed unclean. However, none of the various breaches of order encountered with particular species (groups) is in itself sufficient to explain the entire list.

In the literature there have indeed been many attempts to argue that there are good reasons for considering the various creatures 'abominable'.[47] They have been characterized as having a terrible

---

[46] The reference is to the Syrian ostrich (*Struthio camelus syriacus*, literally: 'Syrian camel bird'), which used to be common in the Middle East. It was hunted extensively for its feathers and has been extinct for a number of decades. Reports on the bird's extinction are contradictory and unreliable. According to A. Schouten van der Velde (1992, 138) the last individual was shot in 1948. J.S. Morton (1979, 138) has the subspecies becoming extinct "in the nineteen-forties", while P. France (1986, 118) holds that the last remaining specimens drowned in 1966 "in a flood in Southern Jordan". The ostrich has always been considered an odd bird: primarily because it cannot fly, but for other reasons, too. Aristotle (*De Part. Anim.* IV, XIII, 14) even held the view that the ostrich stands intermediate between the quadrupeds and the birds, one reason being that it has hair on its eyelids, so that the bird appears to have eyelashes.

[47] In the popular zoological literature we also encounter this kind of argument. "That the carrion vulture was counted among the unclean birds will surprise

taste (bat, gull) or containing toxins,[48] or as demonstrating disgusting behaviour (vultures) or dirty habits (the hoopoe, which fouls its own nest). To the extent that these are convincing arguments, they are selective and are not applicable to all the creatures mentioned. This kind of reading seems too influenced by contemporary associations with certain kinds of behaviour. The case can be made that such negative connotations did not always exist in ancient Israel. The carrion vulture would not otherwise have been given the name *rāḥāmāh* (related to *rḥm*: to love, take pity on), or the stork the name *ḥasîdāh* (related to *ḥsd*: friendliness, tenderness, piety).[49]

As a first ordering principle we should therefore, I feel, return to the motif of feeding habits discussed earlier.[50] Although identification is in many instances by no means certain and some groups of birds, such as the ravens and owls, are not classed as raptors in modern zoological classification schemes, they all consume living animal prey or remains, or both. This also holds for the last two creatures referred to in the Leviticus list: the *dûkîphat* (hoopoe) and the *'atallēph* (bat).[51] The 'mixing' referred to earlier then plays a

---

nobody. Its mere name evokes disgust" (A. Schouten van der Velden 1992, 104).

[48] The pharmacologist David Macht 1953 held the conviction that the unclean animals are poisonous. He therefore prepared flesh extracts from the muscles of clean and unclean animals (specifically, the land animals) and applied this to seedlings of the white lupin. The extracts of the unclean animals proved to be toxic to the plants, and the clean non-toxic. He also gives an explanation for the fact that the quails in the desert (Exodus 16, Numbers 11) were not to be eaten: while migrating, these birds eat poisonous berries and plants such as Monkshood and Rough chervil. To my knowledge, these curious experiments have never been followed up. They are congruent with a wholly materialist explanation of the dietary laws, but their value is as yet unclear.

[49] Due caution is required in following this line of argument, however, because Hebrew also has 'exorcising euphemisms', such as "house of the living" for a graveyard.

[50] This was and is also the view of the Jewish exegetes who classed all the birds listed as *dôrēs* (from 'to grab' or 'to beat'), a term referring to the 'bloodthirsty' manner of eating (still reflected in the German *greifvogel* for bird of prey). This position is already stated in the Epistle of Aristeas (146) and is reiterated in the Mishna (*Ḥullin* 3.6, "any fowl which seizes is unclean"). The same line is taken by WH (46) and Milgrom (1991, 661).

[51] J. Milgrom (1991, 662), who takes Driver's line that we are here concerned basically with carnivores, has the following to say on these species: "except for the last two in the list (19-20), the hoopoe and the bat, which, however, are tabooed because of their dirty habits and inedibility." This seems untenable with respect to exceptionality (both eat living prey: the hoopoe reptiles, and bats mainly insects) and disputable with respect to the qualifications 'dirty habits' and 'inedibility'. Such qualifications are very much culturally determined, and it is difficult

supplementary and supportive role. Both the eating of prey—and the associated risk of consuming blood—and the mixing constitute a breach of the creation order. As I shall argue below, the latter is the prime principle underlying the dietary laws.

There are many 'non-vegetarian' species (songbirds, tits, fly-catchers, kingfishers, etc.) that are not included in the list. If the systematics I propose is correct, the list is by no means a complete review of every unclean bird. Rejection of this systemic proposal and acceptance of the idea that the list is 'complete' implies that unmentioned birds are (or may sometimes be) clean. On this line of reasoning, additional evidence as to the correctness of the systematics may therefore be gleaned by seeking, in the HB and relevant non-biblical sources:

1. mention of consumption of 'non-vegetarian' species, or species that otherwise breach the order of creation;
2. mention of clean species and/or criteria for such.

Let us now therefore briefly review the available data on these two points. The HB has no references to the consumption of bird species which should be unclean according to the proposed systematic but are omitted in the lists of Leviticus and Deuteronomy. From the available archaeological evidence it cannot be deduced whether unclean birds were eaten by the Israelites. There is some evidence, however, for particular birds being associated with certain kinds of 'unfavourable' place, like ruins. Literary, non-biblical sources such as the Sfire inscription and the Balaam inscription of Tell Deir 'Alla show considerable similarity (with regard to the unclean species mentioned in Leviticus) when it comes to creatures living in such desolate, 'eery' places. Aside from the list of Deuteronomy and Leviticus, we encounter characteristic species of such habitats in several other passages of the HB (Isa 2:20, where the bat is mentioned, Isa 13:21,22, 34:11-15, Jer 50:39, Mic 1:8, Zeph 2:14, Job 30:29 and Ps 102:7).[52] Houston considers it likely that the extension

to see why they should be applied so selectively to these two (groups of) animals. They behave no more 'dirtily' and are not, objectively speaking, any more 'inedible' for humans than other species (groups). It is true, though, that animals which are conspicuous—and the hop certainly is—are avoided by potential predators because of their apparent unpalatability. This phenomenon sometimes acts to protect species which resemble conspicuous species, but which are not (that) unpalatable.

[52] For the Balaam inscription and a comparison of biblical and non-biblical

of the original list of unclean birds was inspired by this non-biblical literary tradition.[53]

The correctness of the systematic might also be reflected in information on 'clean' birds. Again, data are scarce, however. Criteria for establishing which birds are to be deemed clean are lacking, and must be deduced from our knowledge of the birds cited. In the HB there is scarcely any reference to the consumption of (clean) birds.[54] Given the many passages describing meals and the food thereby eaten, we may assume that birds did not constitute a major element of the Israelite diet. Exodus 16 describes how the people of Israel 'murmured' in the Sin desert and expressed a longing for the "fleshpots" of Egypt. Moses was told that the people "shall eat flesh" at twilight. "Quails" (*śᵉlāw*) then appear over the encampment in massive numbers and are apparently easily caught. It is generally assumed that the reference is indeed to quails, even though the story has symbolic features and its historicity has been called into doubt. The quail is a bird that can be found in enormous numbers in this region during the migratory season. In Numbers 11 there is another reference to eating quails in the desert, there in a negative context, almost as a plague. Finally, Ps 105:40 also describes the consumption of quails in the desert. The second species referred to elsewhere in the HB occurs in the extravagant daily menu of the court of Solomon reported in 1 Kgs 4:22,23 (MT: 1 Kgs 5:2,3).[55] The birds referred to are probably domesticated, given the use of the term *barurîm 'abhûsîm*: "fatted fowl", and the Dutch NBG opts here for the translation 'geese'. These are the only explicit references in the HB to the eating of birds.

The peacocks and apes on the 'Tarshish fleet' (1 Kgs 10:22) were probably decorative. There are several indirect references from which it can be deduced that the *qôrē'* (the Rock partridge, or Chukar) was

---

lists, see: J. Hoftijzer and G. van der Kooij 1976, particularly 206-207.

[53] W. Houston 1993, 197.

[54] The consumption of flesh by birds is described more frequently than the consumption of the flesh of birds. Carrion-eaters were part of everyday life in the Middle East. The oft-described situation of "the birds of the air" eating human flesh (corpses) was thus readily conceivable as an image of repulsion (e.g. Gen 40:19, Deut 28:26, 1 Sam 17:44).

[55] There is good reason to assume that the quantitative data on Solomon are highly exaggerated, to underscore the king's greatness. The offering of *twenty-thousand* oxen and *one-hundred-and-twenty-thousand sheep* reported in 1 Kgs 8:63 is a case in point.

also eaten. In 1 Sam 26:20 we read that this partridge was hunted, presumably with a view to consuming it. It is not reported who was hunting the bird, so it may have been non-Israelites. The other references to *qôrē'* provide no clues as to consumption of this species.

Finally, the dove or pigeon should be mentioned. In the HB there is frequent reference to this bird, but never as food. That the dove ('turtledoves and young pigeons') was considered clean is reflected in its being the only bird named as a sacrificial animal in the HB.[56] Although other species may have been offered, too, there are no explicit references.[57] The archaeological evidence shows that pigeons most probably were indeed eaten. The remains of geese, ducks and chickens have also been found, from which it can be deduced that these birds were also consumed. Finally, the excavations at Tell Sweyhat and at Lachis prove that chickens were to be found in contemporary Israel as early as the Early Bronze Age (prior to 2000 BC).[58]

This information on clean and unclean species of birds appears to support rather than contradict the proposed systematic. The species consumed are primarily seed- and grass-eaters and the birds to be deemed unclean according to the criteria were not eaten.[59] Many of the latter are described in 'negative' terms, as being associated with desolate places. These descriptions indicate that this may also constitute the reason for their uncleanliness. In verse 13 the

[56] See, for example, Lev 1:14, 5:7, 12:6; in Gen 15:9 we also encounter the 'turtledove and young pigeon', in a ritual confirming a promise to Abraham.

[57] Gen 8:20 reports that Noah sacrificed "every clean bird". The phrasing certainly suggests that a number of other birds besides the pigeon were also sacrificed. In the eyes of the narrator the distinction between clean and unclean animals had existed since time immemorial and it was therefore permitted to sacrifice these birds.

[58] See: D. Ussishkin (1978, 88-89) for these early chicken finds; L.K. Horwitz & E. Tchernov 1989 for the later chicken remains (Iron Age) and those of geese and ducks; J. Boessneck & A. von der Driesch (1978, 285) for the doves and pigeons.

[59] Strictly speaking, insects and molluscs sometimes feature in the diet of some seed-eaters such as the partridge. I do not consider this to detract from the ultimate validity of the rule, however. There remains an unmistakable difference between the (visible) consumption of larger animals, such as lizards by a Hoopoe, and the feeding behaviour of seed-eaters, whereby (as we now know) minute animals are also consumed. This is reflected in the use of the (modern) Hebrew word for these minute animals, *bāṭēl* meaning 'nothing' 'null'. On this point, too, we should not read the rules scientifically or technically, but against their historical and religious background.

unclean birds are termed 'an abomination'. There are to be abhorred (in the ritual/cultic sense).[60]

In addition to specific species, four times mention is made of a certain bird "according to its kind". *L^e^mînāh* echoes Genesis 1, where it is also translated as 'according to its/their kind'. Here, it refers to the various (sub)species to be assigned to a given zoological family on the grounds of their outward features and/or common behavioural characteristics. Unfortunately, explicit characteristics or criteria are lacking; these must be derived from our knowledge of the cited species. With respect to the families or groups mentioned, we know that various different (sub)species were and still are found in the Mediterranean region. *'ayyāh* may refer to kites or vultures, *'ōrēbh* to crows and ravens, *nēṣ* to falcons and *'ᵃnāphāh* to herons.

Flight

In the eyes of the writer, the possession of wings (and feathers) is characteristic of creatures of the air. From this perspective, these physical features form the best guarantee that these animals indeed have the correct mode of locomotion through the air, i.e. flight.[61]

---

[60] E.S. Gerstenberger (1993, 118 *et seq.*) holds the view that the term "abscheulich" should be avoided because it led to the extermination of "ekelhaften Unwesen" in later eras. He translates this term as "sollt ihr meiden" (to be avoided). Both his translation and its motivation I hold to be erroneous. Any abuse in later eras cannot constitute a reason for employing a 'kinder' translation than is linguistically and historically warranted. We are here concerned with a cultic term. That there is a connection between the term "abscheulich" in the dietary laws and the (later) qualification "Un-wesen" or "pest" has not been demonstrated. Biologically, too, these terms are not congruent. In Greek culture, with no dietary laws or 'unclean' animals, it was ethically justified (for humans) to exterminate animals that caused damage or were a nuisance. In this context, see also Chapter 4.

[61] Elsewhere in the HB (Prov 1:17 and Eccl 10:20) birds are denoted by the splendid name *ba'al kānāph*: 'lord of the wing' or 'feathered one' ('lord of the feather'). If only for this reason, flightless birds such as the ostrich were deemed unclean. The chicken forms an interesting case. These birds are not mentioned in the HB. E.L. Greenstein (1984, 93) writes: "The chicken would clearly be tainted and impermissible in the biblical scheme for the same reason. In its postbiblical development, rabbinic Judaism found the chicken "kosher", but debated whether the chicken is meat or *pareve* (a condition neither meat nor dairy)". Although Greenstein's reasoning is probably correct, the correctness of his conclusion with regard to the (hypothetical) biblical chicken is to be doubted. The chickens with which he was familiar and on which his conclusion was based do not apparently fly. At best they flap their wings, but never manage to become airborne. In contrast, the 'primeval' chicken, from the forests of Burma, roosted in trees and was certainly capable of flight. Descendants of these Red Burmese junglefowl still have the

Among other grounds, the following passage of Leviticus lends credence to this motif.

Having treated the winged creatures, verse 20 states that "all winged insects that go upon all fours" (in many translations, there is also the connotation of 'teeming') are to be an abomination. These creatures have characteristics associated with two different realms: the air and the land. They have wings and they 'go on four' (as the original text reads). This 'hybrid' character apparently renders them unclean. They breach their bounds and can consequently not be classified purely or properly. That this uncertain classification is an important issue is also reflected in the exceptions listed in Leviticus at this point. Of the winged creatures with "legs above their feet, with which to leap the earth", four may be consumed, each 'according to its kind'. It is not clear whether the latter phrase denotes allied species or subspecies of the cited insects. It is generally assumed that the reference is to various representatives of the order of crickets and grasshoppers. Translators and exegetes have wrestled with these terms, and doubts have often given rise to inconsistency or illogicality, as in the NBG, for example, where categories of 'locusts' and 'grasshoppers' of varying size evidently overlap. The annotated Dutch *Statenvertaling* is more consistent on this point, translating the first creature (*'arbeh*) as grasshopper (as it does throughout the HB) and leaving the other three untranslated. Although the latter does have translations for the other 'tricky' animals encountered earlier in the text, here the margins of uncertainty were apparently considered too great and simple transcription was preferred. The commentary on this verse speaks of "kinds of animal that the Latins call *insecta*". Whatever the difficulties involved in accurate translation of these terms, the basic thrust can be deduced from the context. The passage is concerned with animals that have both wings and legs, one of which features is characteristic of creatures of the air, the other of beasts of the land.[62] It is this hybrid nature that is the root of the problem. Clean animals must have an unambiguous status in the

---

capacity for flight and also perch in trees if they have the opportunity. With a wink in the direction of the titel of a novel by the Dutch novelist Frans Pointl 1989, the conclusion must therefore be: free-range chickens are not kosher, only 'chickens that fly over the soup'.

[62] The text in Lev 11:23: "But all other winged insects which have four feet are an abomination to you" also indicates a conflict between 'air' (ideal type: wings) and 'land' (ideal type: legs). Insects, it should be added, generally have six, never four legs.

order of creation, which is why Deuteronomy classes the entire category as unclean. For Deuteronomy these are 'birds' with the wrong method of locomotion, 'teeming winged creatures'. In conclusion, and in contrast, Deuteronomy then continues, in verse 20: "All clean winged things you may eat" (a better translation of *'ôph* than the 'fowls' of the KJV and NBG). This rule may have been added to include the locusts in the list of clean winged creatures.

Leviticus has the same general rule, but makes an exception for the four species (groups, including subspecies/varieties?). Despite all the uncertainty regarding the exact zoological reference here, the general consensus is that various kinds of locust and grasshopper are intended. With the somewhat complicated phrasing of Lev 11:21, these creatures are implicitly categorized as land animals. The intended meaning of the text appears to be: all things considered, the conclusion is that these difficult cases have the 'correct' mode of locomotion. Literally translated, these creatures 'go on four', and this is the 'correct' way to move on the environmental segment land. Creatures of the earth run and jump, but they should never fly. In this sense, the text of Leviticus can perhaps be read as a priestly decision in a (long-standing) debate. The question arises why such a decision should have focused so specifically on these animals. There are plenty of other animals of debatable status, certainly from the vantage point of modern biological understanding. With the available data, there is no unequivocal answer to this question. Employing our present biological knowledge to interpret the text is not the same as reading the list through Linnean eyes.[63] Certainly in the form at our disposal, it is a religious codification designed to be practicable for everyday use and therefore alluding to that practice in its phrasing. Locusts (and allied species) are a tasty morsel that, in the Middle East, can provide an over-abundant source of food. These insects themselves consume only vegetable matter, and so could not be excluded on the grounds of their diet alone. It is feasible that this verse embodies a concession to extant practice among the Israelites.[64]

[63] Our modern-day classification of animals and plants, which is based on anatomical features, is largely the work of the Swedish naturalist Carolus Linnaeus (1707-1778). In his *Systema Naturae* (1735) he divided nature into three kingdoms: mineral, vegetable and animal, subdividing the latter two into classes, orders, families, species and varieties.

[64] In the later Jewish tradition, too, locusts form a 'difficult' category. I. Grunfeld (1972, 51), following Hirsch, considers it better not to eat these animals at all: "but

Does this 'exception' betray nomadic, desert origins to the rules or to the customs on which the rules are based? Whatever the case, in our present context this 'controversy' between Deuteronomy and Leviticus is relevant mainly for the argumentation developed in the text. In Leviticus, verse 23 reiterates the general rule of verse 20 and closes the first section of the chapter.

3.5.4 *Deuteronomy 14:21*

"You shall not eat anything that dies of itself"
In verse 21 Deuteronomy concludes the edicts on clean and unclean animals with two more general rules. The first is concerned with animal carcasses and remains. The people of Israel are not permitted to eat such flesh, although they may offer it to "the alien who is within your towns" (i.e. the *gēr*, the familiar stranger[65]) or sell it to "a foreigner" (a *nokrî*, an unfamiliar stranger, comparable to our 'tourist'). The text is concerned with the carcasses of clean animals, since unclean animals may not be consumed under any circumstances and, according to verse 8, their carcasses may not be touched.

In the English language the term 'carcass' can be used to denote either the flesh of animals that have been slaughtered (and in the Israelite context, slaughtered according to the prescribed rules), or animals that have died of natural causes or been killed by other animals. Although the Hebrew term *nᵉbēlāh* is translated as 'carcass' in various passages of Leviticus and Deuteronomy, the more accurate rendering is that of Deut 14:21: "You shall not eat anything that dies of itself", i.e. of natural causes. For animals that have been killed by other animals, Hebrew has the word *ṭᵉrēphāh*, from the root *ṭrph*, to tear or rip to pieces. The phenomenon of a wild animal attacking their flocks and ripping an animal to pieces was certainly familiar to the Israelites, and the question of what to do with this (clean) meat naturally had a very practical significance. In Exod 22:31 we read that this 'torn' flesh may not be consumed, for: "You shall be men consecrated to me". It is to be cast to the dogs.

The reason for the taboo on eating this category of carcass is clear:

---

for us, who lack experience and tradition concerning this, every locust is forbidden".

[65] Although a *gēr* is usually a non-Israelite living more or less permanently among the Israelites, except for Lev the term was also used to refer to an Israelite with no land of his own and 'living in' somewhere outside Israel.

with such flesh, the blood has not been removed, or appropriately removed. Deuteronomy is short and sharp in its formulation of the rule, proceeding immediately to the motive: the sacredness of the people of Israel (of which the alien and the foreigner are not a part). Leviticus treats the issue more extensively, and the text also differs in its emphasis. In these priestly writings, the focus is on ritual purity and uncleanliness. As we shall see, in considering the text of Leviticus 11, contact with the carcass of a clean animal renders a person unclean until the evening of the day.

The taboo on consuming blood is described at length in Lev 17:10-16. From this passage it can be deduced that this taboo also held for the 'familiar strangers' living with the Israelites: the *gêrîm*. Ritual uncleanliness was 'contagious'. Verse 13 of chapter 17 describes the appropriate procedure when hunting (clean) game animals.[66] The blood must be 'poured out' of the carcass and then 'covered with dust'. Blood itself may on no account be consumed, for as verses 10 and 14 state: "whoever eats it shall be cut off" (from their people).[67] Leviticus does, however, allow for the possibility of eating meat in which there may still be blood present.[68] The passage in question does not explicitly prohibit the consumption of such meat. It merely indicates that it is to be avoided, why it is to be avoided and how it can be avoided. The text then states that such action causes the same ritual uncleanliness in both Israelites and 'strangers'. In verse 15 we read that if a native Israelite or a stranger eats an animal that has died a natural death (*n^ebēlāh*) or been 'torn by beasts' (*ṭ^erēphāh*) —and here the reference is obviously to a clean animal—the per-

[66] Against this background, hunting methods whereby the animal is not immediately killed were to be preferred. In this sense nets, traps, pitfalls and so on are superior to spears or bow and arrow, even though the latter were sometimes used, as is clear from the story of Esau (Gen 27:3). It is to be doubted whether the Israelites hunted with any regularity. There is only sparse and terse reference to hunting in the HB, while hunters (Nimrod, Gen 10:9 and Esau) were not described in positive terms. On this point, see also B. Janowski *et al.* 1993, 150-154).

[67] This explicit denunciation, in combination with the text of Acts 15:20 (see note 2), was the reason why the consumption of blood and blood-containing products (such as the sausage variously known as black pudding or blood sausage) remained taboo for some Christians in later times.

[68] In a strict sense some blood always remains present in meat. The dietary laws are concerned with blood that is visibly present in recognizable form: blood that flows through the body and in biological terms forms an independent organ. In modern orthodoxy the use of salt is common practice to remove the blood of the meat.

son consuming it will remain unclean until the evening. Although the consumption of animals dying of natural causes is prohibited in Deuteronomy, and the consumption of 'torn' animals likewise in Exodus, Leviticus posits no explicit taboo, but concentrates on the ritual cleansing that must follow such behaviour. Leviticus 11:39,40 contains a similar passage on the uncleanliness resulting from contact with the carcasses of clean animals.

If, in the text of Deuteronomy, there is contact with the dead bodies of clean animals then, according to the priestly code of Leviticus, both the Israelite who gives away or sells the meat and the alien (i.e. the 'familiar stranger') who eats it will remain unclean until the evening. Whether the same holds for the 'unfamiliar stranger', or 'foreigner', is debatable. In this context, the status of the foreigner is less clear; as a transient in the community, he or she would not have been bound to the same ritual obligations. On the ritual status of those consuming or handling such meat the verse has nothing to say, however. We must therefore assume that this passage from Deuteronomy lays down a clear rule for Israelites on the consumption of meat from such carcasses, with a different view holding with respect to the ritual implications of handling such meat. Fully in line with the ethics of the Torah, the 'familiar stranger' is then better off than the true foreigner. The latter must pay for the meat, while the former is given it free.

The passage then continues by stating the reason for dealing with the flesh from such carcasses in this way: "for you are a people holy to the Lord your God". This motif of 'holiness', which we also encounter in verse 2 and very explicitly in Lev 11:44,45, will be discussed separately below.

"You shall not boil a kid in its mother's milk"

The second general rule in this concluding verse, 21, states that a young goat may not be boiled in it's mother's milk. At first sight, this sentence appears a little out of place at this particular juncture. Although its literal meaning may be clear, moreover, in this context it is difficult to interpret. What is the connection with the other rules concerning clean and unclean animals? Is this taboo a specific one, and if so, to what exactly does it refer? Or should we follow certain Targumim and read these lines as constituting an overall taboo on preparing and consuming meat and milk in combination, and if so, why?

The rule, which we encounter in exactly the same phrasing in Exodus 23:19 and 34:26, has long been a bone of contention among both biblical scholars and anthropologists. Generally speaking, we can distinguish 'broad' and 'narrow' interpretations. In the former, the domain to which the rule is held to apply is sometimes broadened substantially. For example, S.R. Driver (1896, 166) suggest that the taboo may have been intended to discourage the consumption and use of milk, and thus in fact discourage the keeping of herds and flocks. This then implies an encouragement of arable farming, of making the fields and orchards fertile. The anthropologist J.G. Frazer devotes a chapter to the issue in his *Folklore in the Old Testament* (1919). He sets the taboo in a broad cultural and historical context and arrives at an entirely different conclusion to Driver. In Frazer's view the rule is designed precisely to ensure that responsible use continue to be made of milk. Taboos concerning the boiling of milk are to be encountered among many different peoples. These can be traced back to a fear of harm thus being done to the animal from which the milk has been taken. In this connection Frazer mentions "a belief that a cow, whose milk has been boiled, will yield no more milk and that the animal may even die of the injury done to it". In several Targumim we also find a broad interpretation. Here we read: You shall not eat meat with/in milk (Targum Onkelos) and: it is forbidden to boil or eat meat with milk (Targum Pseudo-Jonathan). This reading goes further than the original text and is in fact based on an exemplary interpretation thereof. Having the status of *midrash halakhāh*, these Targumim lay down the basic rules for kosher consumption of meat and milk in the later Jewish tradition. It is already encountered at an early date.[69]

Standing in contrast to these broad interpretations are a variety of 'narrow' readings, which likewise have a long history. St. Augustine, for example, and later Luther and numerous other theologists, held the view that the reference is to a temporary restriction: the kid-goat may not be eaten as long as it is being nursed by its mother. The passage, whose meaning seemed clear on first reading, then runs: "You shall not boil a kid that is still drinking its mother's milk", another way of stating: "You shall not eat a very young goat".

In this reading, we are concerned with a taboo grounded in gen-

[69] J. Milgrom (1991, 742) traces this position back to the Septuagint and Philo, and therefore holds that the interpretation originated in Alexandria.

eral ethical and humane principles: this is something that is just not done, fresh young lives are to be respected. Support for this interpretation is sought in texts that appear to have a similar background. The two passages mentioned in this context are Leviticus 22:28, where we read "And whether the mother is a cow or a ewe, you shall not kill both her and her young in one day", and Deuteronomy 22:6, where the taking of both a mother bird and her young is similarly forbidden. In this 'humane' interpretation, cooking a 'kid that is still in (on) it's mother's milk' is equivalent to 'killing' or 'slaughtering', for killing normally precedes cooking. Elsewhere in the HB the killing or slaughter of very young animals (after their seventh day of life, but still being nursed) is permitted, however: in Ex 22:30 and Lev 22:27, for example. In the case of a bird's nest, taking only the young (which is allowed, according to verse 7 of the text) appears to be seen as just as 'inhumane' as, if not crueller than, the taking of both the mother bird and her young. The most down-to-earth and practical explanation for this passage in Deuteronomy is that it is better to leave the mother animal unharmed if her young are taken (for consumption), because this leaves intact the potential for continuation of the source of food (new eggs/young). There is also an alternative explanation, however, which to my mind establishes a better link between these two texts on 'mother and young' and which I shall return to below.

The latter, more functional explanation is in line with the interpretation of the text of Deut 20:19, in which the besiegers of a city are prohibited from felling the fruit-trees. Here, too, the issue of a food supply and the scope for continuation thereof is in all probability more important than the ethical motive proposed by some scholars. Modern exegetes who take this passage as expressing an early form of environmental awareness also make too much of the text as it stands.[70] Verse 19 cannot be read in isolation from verse 20. The text simply advocates adopting a rational, functional stance towards nature and natural resources, even in time of war. It is better to leave fruit-trees standing, as a potential food source (with some fruit possibly remaining for those besieged), so it is stated, and fell other trees for building siegeworks. A modern interpretation might read that 'scorched earth' tactics—the senseless burning and destruc-

[70] Moreover, even if a particular rule may have a positive impact on the environment, such a potentially favourable effect should not be too easily taken as constituting the motive for that rule.

tion of large expanses of inhabited land—are prohibited. Means and ends are to remain in balance.[71]

The introduction of arguments of 'humaneness' is also risky in itself, because what is and is not considered humane is very much culturally and historically determined. Particularly when it comes to evaluating human attitudes towards nature, plants and animals, it must be remembered that the cultures of antiquity held entirely different standards from our own. For the Israelites and others, the slaughter and sacrifice of animals was part of everyday life,[72] while most modern northern Europeans have never seen the inside of a slaughterhouse with their own eyes, let alone observed a ritual slaughter. By adopting a relatively modern concept such as 'humaneness', one runs the risk of presupposing contemporary feelings and associations on the part of the author(s) of the text.

The broad and the narrow interpretations are both also unsatisfactory because they consider the text too much in isolation from the preceding verses. If a plausible explanation for this particular rule can be found that is in line with the text and connects it with the preceding passage, then this is to be preferred.

This is exactly what J. Soler, a French scholar of the structuralist (Lèvi Strauss) school, does in an article from 1979. In a sense he reads the text literally. In his view the mother's milk represents, *pars pro toto*, the mother herself, and what is prohibited is to cook the mother animal and her young in one pot, for that would be *culinary incest*. The taboo in this verse is then in line with the basic idea that uncleanliness is a question of 'disorder'. A mixing of realms is to be avoided: mother and son belong neither in bed (Lev 18:6) nor in the pot together.[73]

The explanation of Labuschagne (1992) also merits consideration.

---

[71] On this point, see C. Houtman 1982, who arrives the same conclusion and supports it with different examples.

[72] It may not be deduced from this that the Israelites were great meat-eaters, though. On the contrary, meat consumption was a luxury in that era and meat will certainly not have been on every Israelite's daily menu. Nonetheless, they will have been familiar with slaughtering animals, both in their everyday surroundings and at the sacrificial sites. This is not to detract from the fact that Milgrom may be right in suggesting that originally the consumption (and therefore slaughter) of animals was *exclusively* a ritual affair and that the practice of 'home slaughter' and the consumption of wild animals such as deer was only sanctioned at a later date.

[73] J. Soler 1979. See also his contribution to J.-L. Flandrin & M. Montanari (eds.) 1996.

He discusses classical and recent interpretations and observes that none of them satisfactorily answers the basic question: "why the boiling (seething) of a kid *in its mother's milk* should be regarded as malpractice". Along with a number of other exegetes, he assumes that the reference is to a newborn kid and that the text must therefore be interpreted against the background of the vernal and autumnal sacrificial feasts.[74] On such occasions the newborn young was taken (together with its mother according to L.) to the temple on the eighth day to be sacrificed to the Lord as a burned offering (as described in Exod 22:30 and Lev 22;27). The meat would then be prepared in the tastiest fashion: in milk. And what milk was most readily available? That of its own mother! (1992, 14) Having thus reconstructed an appropriate context, Labuschagne comes with his 'solution'. The milk produced just after giving birth, known as colostrum or beestings, has a slightly reddish colour because of the blood proteins it contains. In this reading of the text, then, the reddish colour indicates the presence of blood or is at any rate associated with blood. This would then appear to constitute a link with the preceding passage and a sound motive for the taboo. Flesh and blood are to be kept separate (Gen 9:4, Lev 17:13,14 and Deut 12:23). After its seventh day of life, then, a very young kid may be slaughtered (in the correct fashion) and eaten, but not cooked in the 'bloody' colostrum. Support for this interpretation is to be found in both of the Exodus texts cited. There the rule is preceded by several others, all of which establish that the sacred may not become polluted, with blood cited as a source of desecration and ritual contamination.

---

[74] M. Haran 1979 sets the taboo in the context of the harvest festivities (November), because it is around this time that goats give birth to their first young. He rejects a magical or cultic motif in the taboo and gives his support to the 'humane' interpretation: "it is rather a deliberate reminder of humane behaviour even in the midst of general jollity" (35). E.A. Kauf 1988 considers the issue in the broader context of (festive) meals and draws the following conclusion: "Zusammenfassend ist zum Böckchen, das in Milch gekocht wird, festzustellen, dass es sich hierbei um ein spezifisch als Festessen ausgewiesenes Gericht der nicht-urbanen Bevölkerung Palästinas in der letzten 5000 Jahren handelt..." (166). This was then transformed into a general dietary law by the compiler of Deuteronomy, so that it might historically serve as an "Abgrenzung gegen die "Völker".." (168). Neither Haran nor Knauf explain why it was prohibited to prepare the kid in *its mother's* milk, however. Rightly, Labuschagne sees this as a shortcoming of these interpretations—a shortcoming that would have been avoided, it might be added, if these two authors had followed the reading: "You shall not eat a kid that is still being nursed", for this holds no reference to 'boiling in (mother's) milk'.

Labuschagne's explanation is intriguing and the presence of blood in colostrum has indeed been demonstrated.[75] But what is the precise situation on the eighth day, with which we are here concerned? Contrary to what Labuschagne holds, by this stage of her nursing the mother animal is no longer producing colostrum, and here milk is unlikely to have an anomalous composition, let alone a reddish colour.[76] The question arises whether a kid-goat may be cooked after the seventh day in its mother's milk if the colostrum phase is over. Although the verses in Deuteronomy[77] appear to prohibit this, Labuschagne's explanation does not. If there are no longer any traces of blood in the milk after the seventh day, then there is no question of any 'mixing' of flesh and blood. In his article, however, Labuschagne suggests that the mother's milk remained taboo for a longer period of time (the whole nursing period?, under all circumstances?) "on the account of the reddish colour of the beestings, in ancient Israel the mother's milk was considered to contain blood" (14).[78]

[75] The two proteins albumin and globulin cited by Labuschagne as being responsible for the reddish colour are in fact colourless, however. He also suggests that there might be traces of blood present, and this indeed proves to be the case. The reddish colour of (initial) mother's milk is connected directly with blood, for colostrum contains red blood cells. The protein involved is thus the 'normal' blood protein haemoglobin (P. Walstra & R. Jenness 1984, 152).

[76] "The first secretion after parturition has a very different composition and is called colostrum or beestings," write Walstra and Jennings. The percentage of protein declines in the first few days after birth and in cattle the reddish colour has vanished completely after four of five days (pers. comm. H. Kooistra, Dutch cattle farmer from Lellens, 29/08/96). "Colostrum changes into more or less normal milk in about three days, however, this varies widely among cows, and some show a far less marked transition ...others an even sharper one" (Walstra & Jennings 1984, 152). C.J. Labuschagne (1992, 14) assumes a gradual transition and therefore holds that "nobody can say exactly when the milk becomes 'ordinary'".

[77] In the parallel texts Exod 22:30 and Lev 22:27 there is reference to killing and sacrificing after the seventh day, but not to consuming or cooking in the mother's milk.

[78] In his *Lectures on the Religion of the Semites* W. Robertson Smith writes: "Mothers milk is simply goat's milk, which was that generally used (Prov. xxvii, 27)". In explaining the taboo he also links milk with blood: "Many primitive peoples regard milk as a kind of equivalent for blood, and thus to eat a kid seethed in its mother's milk might be taken as equivalent to eating "with the blood," and be forbidden to the Hebrews along with the bloody sacraments of the heathen" (Smith 1894, 220/221). However, there is no support for Labuschagne's view that the people of ancient Israel continued to regard mother's milk as suspect because of the assumed tainting with blood, nor for Smith's implicit assumption that in Israel, too, milk was automatically associated with blood. Both views would lead to the conclusion that all milk (being derived from mothers) is suspect and should therefore never be consumed because of the ('knock-on') effect of the (colostrum) blood.

More generally, Labuschagne holds the opinion that "the blood motif" is the key theme of the dietary laws, and he concludes his article thus:

> Indeed the blood aspect is the red thread that runs through the history of our prohibition. We have observed the following development: initially, in the Book of Covenant, it was a prohibition directed against the practice of cooking a kid in its mother's milk, which in early stages was considered to contain blood; then, in Deuteronomy, it became a dietary law, still ultimately based on the prohibition on the consumption of blood. After this, in its further *Wirkungsgeschichte*, the prohibition was subsumed in Kosher cooking: no consumption of blood and no simultaneous consumption of meat and milk (1992, 17).

His general motif (blood as the 'red thread') we shall be returning to below, so at this stage let us restrict ourselves to the question: to what extent does his analysis explain the stated prohibition? As we have seen, the solution he proposes is not without its problems. It rests on a number of assumptions and not all the questions raised by the text are satisfactorily answered.[79]

The basic assumption of both Soler and Labuschagne—the text prohibits a certain action because it involves 'pollution', the mixing of things that should remain separate—I would hold to be correct, however, and it also ties up well with the general rule encountered earlier.[80] The question, though, is what exactly is being mixed here? Milgrom, who rejects the 'culinary incest' line taken by Soler (although in my view on erroneous grounds[81]), makes a good suggestion here, inspired by Keel (1980). However, I feel that a combination (and in part an exchange) of elements from the theories of Labuschagne and Soler can also put us on the right track here. What is at stake is the mixing of life and death in the same place and at the same time. In this reading (contrary to what Labuschagne and

[79] For example, the question of why a goat-kid may not be slaughtered at all before the eighth day, regardless of whether it is subsequently sacrificed, simply eaten or prepared in (its) mother's milk.

[80] In the Jewish tradition there is the 'Law of mixtures' (*ta'ᵃrûbhôt*); see: I. Grunfeld (1972, 140-149). There is also a specific Hebrew term for the mixing of two different entities: *kil'ayim*.

[81] Here J. Milgrom (1991, 740) finds himself unable to follow this, to his mind fascinating, theory, on the grounds that the text employs the neuter word *gᵉdhî*. Although he subscribes to the general idea of 'disorder through mixing' put forward by Soler, he considers this particular example incorrect. The reference is then not to a male kid and its mother, but to a young and its mother. He thus implies that incest can only occur between different sexes, a debatable viewpoint.

Soler hold), mother's milk is a symbol of life, and this is accentuated by the milk being from the goat-kid's own mother. It is the sole source of nutrition available to the young while it is being nursed. Keel, who has studied numerous ancient Middle Eastern depictions of animals suckling their young, speaks of "eine Gestalt, die in besonderem Masse das Gewahrwerden göttlicher Lebensmacht und Lebenslust erlaubte und gleichsam den Blick auf die freudige Zärtlichkeit und Liebe freigab, die den Fluss des Lebens in Gang hält" (Keel, 1980, 142). The slaughtered kid may not be simultaneously associated in time and space with this (symbol of) life, for that would lead to a mixing of realms, to confusion and to new uncleanliness. This explanation is fully in line with the texts forbidding the killing of a mother animal and her young on the same day (Lev 22:28) and the taking of a mother bird from its nest together with its young (Deut 22:6). It is permitted to sacrifice a young animal, but not until after it is seven days' old and no longer being constantly suckled.[82] Although all milk is produced by mother animals, it is conceivable that once its own young have been weaned the description 'mother's milk' might have been superseded simply by 'milk'. The taboo in our text is concerned with the essence of the entities that are not to be mixed. The source of life: mother's milk, is not to be mixed with the flesh of the young it was designed to nourish: it would mean putting life and death in one single pot.

The later Talmudic rules concerning milk and meat and the strict domestic segregation of implements used for these two categories of food in the later Jewish tradition go beyond the letter of the text.[83]

[82] This taboo may also be practically motivated. A newborn animal does not have much meat, nor is that meat particularly firm. Neither is it generally held to be tasty, although we do not know whether the ancient Israelites were of the same opinion. By the time the young is a week old, the meat is already firmer. Lactation begins shortly after birth, induced partly by the young animal's frequent attempts at suckling and drinking. Although in the first few days this yields little reward in the form of milk (colostrum), the essential antibodies contained in this colostrum are all-important to the young's survival. If the young is taken away from its mother soon after birth, lactation can be artificially induced and sustained, but this requires "a terrific amount of time and patience" (pers. comm. H. Kooistra, Dutch farmer from Lellens, 29/8/96). If it is the milk that is important, then, it is unwise to remove a newborn kid from its mother. This may constitute a practical motive for not slaughtering a young in the first week of its life, but it still provides no explanation for the simultaneous slaughter of mother and kid-goat.

[83] For a review of the rules in question, see I. Grunfeld 1972, 115-139. A difficult passage to interpret within the framework of these strict rules is Genesis 18,

Given the underlying motif, however, the general taboo on handling or eating meat and milk at one and the same time is readily defensible and understandable.

3.5.5 *Leviticus 11:23-47*

In verses 23 to 47 of Leviticus 11 further specific details of the rules concerning clean and unclean animals are elaborated, summarized and also motivated. The first and principal elaboration relates to the ritual uncleanliness that arises through physical contact with carcasses. With reference to numerous examples, this section of the text sets forth the circumstances under which such uncleanliness is 'contagious' and what actions are to be taken vis-à-vis the persons or objects affected. As a general rule, uncleanliness lasts for the remainder of the day, that is, until the evening. Those who have become unclean must wash their clothes. Objects must be placed in water and also remain unclean until the evening. An exception is formed by earthenware vessels and their contents, for stoves and for ovens (or pots): these must all be "broken in pieces". There are degrees of contagiousness.

According to verse 26, mere physical contact with certain living animals is even sufficient to become unclean. The text refers here to "every animal which parts the hoof but is not cloven-footed or does not chew the cud", and the fact that the latter two criteria are not satisfied makes these animals especially unclean, as it were, and they are thus a source of uncleanliness even when still alive. Horses, asses and mules satisfy this description.[84] Identification of these particular animals immediately gives rise to problems, though, for why are they so unclean that they can transfer uncleanliness? And even if the rule is indeed applicable to these animals, could it be applied in practice? Horses were used to wage war, and people rode on asses and mules and used them frequently as beasts of burden.[85]

---

where Abraham serves his three male visitors both veal and milk (18:8). One of these men is described as being God himself.

[84] In his commentary on Leviticus, B. Maarsingh (1974, 98) includes horses and asses in the category of plantigrades, animals "that go on their paws" (verses 27, 28): a zoological classification that can certainly be termed unique. On the question of what animals are referred to in verse 26 he remains (wisely) silent.

[85] As a cross between a horse and an ass, the mule was already evidently unclean. It was forbidden for the Israelites to breed mules (Lev 19:19). They were used, however, not only as pack animals but also as mounts. They were ridden on by the sons of David (2 Sam 13:29) and by King Solomon (1 Kgs 1:33). The

Those employing these animals would thus be almost permanently unclean, which is hard to imagine. Which animals are then intended? A more likely candidate for an 'especially unclean' animal would be the dog. This has a greater degree of ritual uncleanliness than other animals, as is reflected in many passages in the HB. The dog comes under the category of 'pawed four-footers' covered by verse 27, however.

Taking verse 26 as a reference to the horse, ass and mule is obviously a consequence of the translation I have adopted, following Houston and Milgrom, i.e. taking the first phrase to mean simply 'hoof'. The RSV translation "which parts the hoof but is not cloven-footed" in combination with 'non-cud-chewing' certainly makes life no easier, though. On the contrary, the problems become worse, because there are no animal species at all satisfying such criteria. Perhaps the dog is eligible, if we ignore the soles of its feet and take its toes to be 'divided hooves'. In itself this may be an illogical interpretation, but its consequences make it even more unlikely. If we consider toed paws to be divided 'hooves', this opens the door not only for dogs, but for countless other species, too, including rodents and all the predatory mammals. No representatives of these two groups are mentioned subsequently, however. Most such predators are covered by the term 'pawed four-footers' of verse 27, or, in zoological terms, 'quadruped plantigrades'. We encounter the mouse as exemplary rodent in verse 29. It is theoretically possible that the text focuses solely on the domesticated animals, because the risk of physical contact with living animals is greatest within this group. In that case the cat should also perhaps be included, but because this animal is mentioned nowhere in the HB, only the dog remains.[86] All in all, this is an artificial and implausible interpretation of the verses in question.

One explanation that is defensible if we proceed from the classical translation of the text ('parts the hoof but is not cloven-footed') is that the author, precisely by not naming any animal specifically by name (and perhaps not even thinking of any specific animal or

---

Israelites probably purchased their mules from passing foreigners or from adjacent tribes such as the Canaanites.

[86] It is even to be queried whether the dog was found as a fully domesticated animal in ancient Israel, or whether there were only feral specimens running wild (cf. Ps 59:6,14). For a review of the status of dogs in the Near East of antiquity, see Firmage 1992, 1143,1144.

animals), was concerned with once more emphasizing the importance of the two 'cleanliness criteria'. The absence of either one criterion already led to ambiguity and to uncleanliness; how much more, then, the absence of both? Within the category of animals referred to here, the (larger) land mammals, this represents the ultimate in uncleanliness. This is then the reason why (hypothetical) contact with these animals is sufficient to render a person unclean. In ritual texts this possibility—that no practical area of application exists, nor such intended—cannot be excluded. The fact that the law is generally of an 'illustrative' character and contains many rules for everyday implementation renders this interpretation less likely, however. If we accept the classical translation, this brings us back into problems with the earlier descriptions of the hooves, moreover, and those appeared to have been satisfactorily resolved.

An acceptable as well as more straightforward solution may be found by assuming that the reference in verse 26 is to contact with dead animals. This assumes the omission of the words 'the carcass of', or a reading of the text in tandem with the other verses, which are all concerned with dead animals. This is the reading found in the Septuagint. There is then nothing new in the verse, for all carcasses are unclean, regardless of species, but the verse constitutes a logical conclusion to the laws of cleanliness regarding the hoofed animals.[87]

For all the animals mentioned in verses 27-30 it holds that any person coming into contact with their carcasses is rendered unclean until the evening. They must then wash their clothes. Verse 27 deals with "all that go on their paws, among the animals that go on all fours".[88] The zoological systematic as well as the identification of the eight animals mentioned here, in verses 29 and 30, are not entirely clear.[89] The text speaks of "swarming things that swarm upon

[87] If we term the two characteristics 'cud-chewer' and 'parted hoof' A and B, we arrive at the following scheme for the hooved animals: clean (A and B), unclean (A but not B), unclean (B but not A) and, finally, unclean (neither A nor B).

[88] M. Douglas (1966, 56) makes a connection between the animals in verses 29 and 30 and the description of verse 27. On the basis of her translation of the latter verse, she holds that the uncleanliness of the animals in verses 29-30 is due to the fact that they use their 'hands' to walk. The word in verse 27 that she translates as 'hand', *kaph*, means (hand) palm, however, and, because it is also used for the foot, sole. The reference in verses 27 is thus to so-called plantigrades, and this does not necessarily include the animals mentioned in verses 29-30.

[89] Douglas is surprised that the frog is not included in the list and she assumes

the earth". It is only on the identity of the second animal, *'akhbār*, the mouse, that unanimity of opinion exists. All the animals mentioned here are land animals that can be termed unclean on the grounds of their mode of locomotion. The term for 'swarming things', *šereṣ*, probably refers not only to their mode of locomotion but also to their numbers.[90] The term is also employed for the (clean) fish of the sea. 'Swarming' is a good translation in which this double meaning is reflected; 'teeming' would be a second possibility. Swarming and teeming creatures are seldom seen alone; they live in groups or 'swarms'. Here in Leviticus it is the mode of locomotion that is at issue, however (compare Lev 22:5). The cited animals can also transfer uncleanliness to objects with which their carcasses come into contact. The description of the mode of locomotion makes the renderings 'toad' and 'snail' in the NBG and elsewhere, rather unlikely. It may be sometimes 'be teeming with snails', but neither snails nor toads themselves 'swarm'. Reptiles such as varans, geckos or lizards are eligible candidates, as is the Chameleon found in many translations. The last of these creatures matches the description less well, however, because it does not 'swarm' over the ground, but is generally found in trees and bushes. Another obvious choice are small mammals, such as burrowers like the Mole rat or the Ratel for *ḥoled* and also mice, rats, weasels and hedgehogs. Many translations take this general line and have animals from these groups. These animals are fairly common in Israel and are sometimes found dead in the vicinity of human habitation, moreover. This last fact, particularly, gives this list a practical significance. It is of note that the *tinšemet* comes last on the list. This term is identical to the bird name in verse

that this is because the frog hops (and does not therefore walk on its 'hands'). For this reason she holds the frog to be a clean animal and she makes (indirect) reference to a Mishna passage in which it is stated that the frog does not transfer uncleanliness. In the Jewish tradition the frog is most definitely unclean, however. In the *Encyclopaedia Judaica* vol. 6 (35,36) the animal is termed unclean, along with all other amphibians. However, there is a note referring to the Mishna tractate Tohorot, which is said to state that a dead frog does not transfer uncleanliness on contact. The text of the Mishna tractate in question, Tohorot 5:1 and 5:4 (Jacob Neusner translation, 1988, 1045,1046), is rather ambiguous on this point, however. That the frog is not included in the list of Leviticus may also be associated with the fact that this animal was very rarely encountered in Israel. The references to 'frogs' in the HB all concern the plague in Egypt (Exod 7:27 MT, Ps 78:45 and Ps 105:30).

[90] The Hebrew also uses the term *remeš* (for land animals), which is generally translated as 'creeping things'.

18. Most translators take the reference as being to the Chameleon, but the Mole rat is also an option. If we accept this last choice, all eight animals referred to in verse 29 and 20 are then identified as mammals.

Verse 36 indicates the practical limits to the uncleanliness that the carcasses of unclean animals can give rise to: a spring or cistern cannot become unclean. If this were not so, there would be no opportunity for ritual cleansing.[91] The rules of verses 37 and 38 concerning 'seed for sowing' are difficult to interpret. 'Dry' seed is not rendered unclean by any part of a carcass falling upon it, but this is not the case with seed on which water has been put, so says the text. Why, then? The purpose of wetting seeds may have been to achieve (preliminary) germination. In that case there is a good chance of the (part-)carcass falling on seed that has already germinated and consequently on the 'product' of that seed: the seedling that will ultimately produce food. Dead matter thus comes into contact with 'life', causing uncleanliness. This contrasts with 'dry' seed, which will die if left to its own devices and, because it has not germinated, cannot transfer the uncleanliness to products intended for consumption.

That the mode of locomotion constitutes an important criterion for distinguishing between clean and unclean animals becomes clear in the reiterations of verses 41 to 47. Twice we read of "swarming things that swarm upon the earth" and twice, in verses 43 and 44, of "any swarming thing", while the concluding verses 46 and 47 refer to "every creature that swarms upon the earth". The word 'swarm' occurs no less than seven times, to emphasize its importance.

Holy

The pivotal passage of this part of the chapter is in verses 44 and 45. Here we read why the Israelites must respect the "the law pertaining to beast and bird..." (verse 46). The concept of 'holiness', already encountered in Deuteronomy 14, here constitutes the central motive for keeping the dietary laws. The passage in Leviticus is emphatic on this point: verse 44 commands: "consecrate yourselves therefore, and be holy, for I am holy", and verse 45 ends with "you shall therefore be holy, for I am holy". This argumentation is re-

[91] It is difficult to imagine that every cistern or well into which a dead animal fell would have been clean. The intention of this verse therefore appears to be to keep open the option of cleansing with water (from the 'root source').

peated regularly in Leviticus, in connection both with the laws of cleanliness and, in Lev 20:25,26, with the commandment to distinguish between clean and unclean animals. Holiness constitutes a pivotal notion in understanding the motives for the dietary laws and this holiness is brought into direct relation with the holiness of God himself.[92]

An additional motive is provided in the words: "For I am the Lord who brought you up out of then land of Egypt, to be your God". This second motive also recurs regularly in Leviticus (22:33, 23:43). The exodus from Egypt, the deliverance from bondage and the passage through the Reed ('Red') Sea, marked the birth of the people of Israel. The phrasing of this act of God, which is described in Isaiah 43 as giving rise to the creation of Israel, is encountered frequently in the HB and cited as the basis for the special relationship between God and His chosen people. In Exodus 20 and in Deuteronomy 5 the same words are used to introduce the Ten Commandments (or, more literally, 'words'), and are indeed intimately connected with the first: "You shall have no other gods before me".

In Deuteronomy the phrase recurs in the concluding motivation of the fourth 'word', on respecting and celebrating the sabbath. Exodus motivates this fourth 'word' with a reference to the Creation in six days with rest on the seventh, a day that must therefore be blessed and hallowed. Keeping and celebrating the sabbath has its deeper motivation and foundation in both the creation of the world and the creation of Israel as a people.[93]

As we have seen, in their structure, content and phrasing the dietary laws also refer to the creation story. The fact that here, in

[92] In his treatment of the dietary laws E. Firmage (1990, 1992) also takes this as his basic point of departure, but he makes no connection with the Creation.

[93] The outlined connections in the (*Letztgestalt* of the) Pentateuch leave out of consideration the exact origins of these views, but they do support the idea of relatively strong, priestly (final) editing having taken place. These views, particularly those concerning the Exodus and the role of (a non-historical) Moses as lawgiver, may have arisen in so-called Deuteronomist circles. For a rather categorical defence of this line of thought, see K. van Toorn (1996, 352-362) and for a treatment allowing greater scope for a historical basis for the stories of the Exodus and Moses, J.C. de Moor 1990 and W.W. Hallo (1996, 314-320). C. Houtman (1980, 252) is very outspoken on the presumed insignificance of dating for interpreting the text: "the intention of this work [the Pentateuch] can be derived from the text itself and is not tied up with a dating of the material in any given period. From our perspective, dating of the material is of less importance for interpretation".

the motivation given in verse 45, there is such emphatic reference to the creation of the people of Israel in the Exodus, introduces a powerful parallel with the sabbath. This parallel is reinforced through use of the terms 'holy' and 'consecrate'. I shall return to both in the following section.

Leviticus concludes with two summarizing verses that lay down, in sober wording, the practical purpose of the preceding rules: to distinguish between the clean and unclean animals, and between the animals that may be eaten and those that may not. It is the *Tôrāh* regarding the consumption of animals.

Chapters 12, 13, 14 and 15 continue with the purity laws. The subject is now not diet, however, but birth (13), skin rashes and leprosy (13 and 14) and ejaculation and menstruation (15). Literally as well as metaphorically, they form the core of the Book of Leviticus, which has 27 chapters. As a priestly code of law, Leviticus as a whole in turn forms the core of the Pentateuch. In so far as this is relevant for interpreting the dietary laws, this broader context will be discussed in the following section.

3.6 The Rationale of the Dietary Laws

Against the background of the theoretical premises described above and proceeding from the discussion and analysis of the relevant texts, the question now arises as to the rationale of the dietary laws and, subsequently, the role of these laws in the everyday religious life of the Israelites. In the quest for a central motif in the dietary laws, I shall start by considering those authors who have indeed proposed such a motif.[94]

3.6.1 *Several motifs*

Hygiene

The hygiene motif occupies a position all of its own. At the heart of this explanation stands the assumption that the dietary laws were necessary to prevent people consuming certain animals that are unhealthy or possibly even unsafe to eat.[95] In the case of the birds

[94] I am concerned here with the central motif put forward by these authors, although it is frequently not the sole motif proposed in their work.

[95] This view is stated very explicitly by Maimonides: "I maintain that the food which is forbidden by the Law is unwholesome" (Maimonides *The Guide for the*

listed as taboo, the laws were "for Israel's welfare".[96] As a motif this is in line with the view, already discussed, that certain animals were deemed unclean because of their 'dirty', unhygienic behaviour or habits. Historically, the hygiene motif has been put forward with particular reference to the taboo on eating pork. Pig-meat may contain trichinae (nematode worms), and those eating meat infected with these parasites may become afflicted with a dangerous infection known as trichinosis. However, contracting an illness through consumption of infected meat is not restricted or specific to pigs or other 'unclean' animals. Infection can also occur when 'clean' meat starts rotting. Provided the meat is properly prepared, moreover, the consumption of pork is not in itself a health hazard.

The hygiene motif forms a special case in two respects. In the first place, as far as I know it has never been put forward as an explanation of the dietary laws in their entirety. As a part-motif it is found in the work of many authors (including Döller, Gispen and König). As a general motif it has been mentioned in several encyclopedias and has, above all, always been the popular view.[97] It seems likely that the 'general public' saw, and sees, pigs as representing, *pars pro toto*, all unclean animals. This is understandable, because pork is eaten in many cultures and its non-consumption has become more or less symbolic of the 'anomalous' dietary habits of the Jewish people. The hygiene motif is a special case, too, because although its unlikeliness (certainly as an overall motif) has been demonstrated repeatedly it has remained an extremely popular explanation.

Blood

The first scholar we shall consider who proposes a dominant motif is Labuschagne, who in his commentary on Deuteronomy, and particularly in his treatment of the dietary laws and purity laws,

Perplexed, 370). Later scientific developments have cast doubt on this hygiene motif (see, *inter alia*, Schapiro, 1928), although the occasional scholar thought he had found due confirmation (D.I. Macht 1953). In our day there are few advocates of this position. In this context, see also J. Cooper 1993, 27.

[96] M.S. Miller & J.L. Miller 1944, 321.

[97] As a 'generally accepted idea' it has also gained a place in a publication by the well-known Dutch fictional author K. van het Reve (1978, 109-112). In this collection of essays originally published under the pseudonym Henk Broekhuis, he considers all kinds of statements and opinions that have nestled into our modern Western culture as *idées reçues* but which, on further consideration (by Broekhuis, naturally), often prove untenable.

arrives at the following evocative conclusion: "Blood runs like a red thread through the texts".[98] In this light, the dietary laws are to be explained as a set of rules aimed at preventing the consumption of blood.[99]

It is clear that blood plays an extremely important role in the dietary laws as well as elsewhere in the HB. The consumption of blood is explicitly prohibited; there is a clear answer as to why it is not to be consumed and where blood is involved, uncleanliness lies in waiting. The pivotal significance of blood is also reflected in the laws on menstruation and on childbirth and also, of course, in the fact that the taboo on consuming blood is already mentioned in the primeval myths of the post-deluvian narrative and that it is the only dietary law that appears to survive intact in the NT. The question, though, is whether this motif is sufficient in itself, that is, whether it provides a full explanation. This is not the case. The main argument against blood constituting the central motif can be derived from those situations in which there is uncleanliness but no blood. Why should the hyrax be unclean, or the horse? Also left unexplained is the uncleanliness that arises through contact with dead animals and people, even though they have died a natural (bloodless) death.

Death

Partly for this reason, Holwerda assumes that death itself is the key motif of the dietary laws. In this reading, anything in which death is directly or indirectly involved is unclean.[100] For Holwerda this also makes clear why there is no longer a place for the dietary laws in the NT: Christ has conquered death once and for all.[101] Although

---

[98] C.J. Labuschagne (1990, 60), reiterated in 1992, 17.

[99] C.J. Labuschagne 1990 does not apply his blood/death motif to all the animals, though. He in fact leaves a surprising amount of scope for other views and motifs, for example for the (to my mind very implausible) view of W.F. Albright 1968 that the unclean animals are to be considered as spreaders of germs. In addition, writes Labuschagne: "It is very debatable whether the deuteronomist was still aware of the underlying motifs, even if these were known" (67). I would tend to disagree with this position. There is no solid proof that the deuteronomist was no longer 'aware' of certain (whatever) motifs. On this point, too, at any rate, the Book of Deuteronomy makes a very 'aware' impression.

[100] "In our opinion the general principle is this: that one should refrain from taking any food that is connected in any way to death and everything that belongs to the realm of the dead. This is why it is referred to as *tô' ēbâ*" (B. Holwerda 1958, 378 translation JJB).

[101] B. Holwerda (1958, 383) considers this motif already present in the Torah:

the death motif explains why contact with the bodies of humans or animals that have died of natural causes leads to uncleanliness, it begs new questions, the main one being why uncleanliness is then associated with affairs that are totally unrelated to death in any form, as in the case of childbirth (Leviticus 12). Neither does it adequately explain why herbivorous animals such as the camel, hare and ass were unclean and why it was permitted to slaughter, sacrifice and eat the clean animals. If 'death' were such a crucial issue, a different list of animals would have resulted, and possibly even rules for a vegetarian diet. Although blood and death are (and remain) important motifs, then, they cannot in themselves provide a watertight explanation.

Creation order

To what extent does the (creation) order motif proposed by Douglas and Soler succeed in this aim, then? Both scholars have worked to lay bare, on anthropological grounds, an organizational principle behind the dietary laws. From this perspective, the information in the text refers back to the creation narrative of Genesis 1 and the status of the animals then created according to the tripartite division of land, water and air. Soler states that: "The Hebrew order of the world is based on taxonomy" and "Uncleanliness is simply disorder wherever it may occur. Man must not mix that which God has separated". Douglas adopts the same position. In her view the word *tebel*, rendered in the RSV in Lev 18:23 as "perversion" in connection with sexual practice, should be read as "mixing" or "confusion". She explains the ordering embodied in the dietary laws with reference to the mode of locomotion associated with the three environmental media. In regarding the animal world, great significance was undeniably attached to mode of locomotion. This is reflected in the descriptions of the various (groups of) animals and also proved pertinent, above, with reference to the status of locusts. The various texts appear to point to the authors having an idealized ordering of the animal kingdom. There are three readily distinguishable media: land, air and water. This spatial subdivision of Genesis 1 is fundamental, for it is the result of the separating activities of the Creator. The animals on land have four legs with hooves

---

"And thus God was able, by introducing this separation (*between clean and unclean*, JJB) into his law, thus imprint into the minds of His people that they had been delivered from death in order that they might expect life in Christ".

on which to walk: they are 'four-footers'. Similarly, the aquatic animals have scales and fins with which to swim, and the creatures of the air feathers and wings with which to fly. This is the basic pattern of the order of creation and this is in itself sufficient to render animals that are at odds with this order unclean, or even an abomination.

In itself, though, the mode of locomotion is still inadequate as an overall motif. As has already been pointed out in a previous section, Douglas, in seeking animals with the 'wrong' mode of locomotion, is herself in error in her translation of Lev 11:27 and interpretation of 11:29,30. Milgrom uses the same argument against Douglas, adding that her theory does not adequately explain why the hoof of unclean land animals should be divided. The same holds for species of bird that move 'correctly' in their medium, the air, by means of feathers and wings, but are still unclean.[102] Here, too, her 'classical' theory breaks down.[103]

For the time being at any rate, then, the conclusion must be that breaching of the order (of creation) is a powerful and important motif, but that Douglas' elaboration of this motif is not entirely satisfactory.

Reverence for life

Milgrom observes that the dietary laws not only regulate the nature of meat consumption, but also set limits on it. His ultimate reconstruction of the motif of the codex of dietary laws reads thus: "Its purpose is to teach the Israelite reverence for life". This motif is elaborated in the laws in three ways: first by limiting the number of species that may be eaten, then by prescribing 'humane' slaughtering procedures and, finally, by prohibiting the consumption of blood.

---

[102] E. Firmage (1992, 1124) considers Douglas' theory and her criteria (means of locomotion) incorrect, because all birds should then be unclean. After all, all birds can walk and by far the majority also fly. Personally, I do not consider this a major objection, however. Flight remains the dominant mode of locomotion for birds, and air is the medium in which they fly. A true breach of the order only arises with birds for which walking is the principal mode of locomotion (such as the ostrich) or which spend a large amount of their time in a medium other than the air (such as members of the heron family). Douglas' criteria are not in principle wrong, but are nonetheless insufficient.

[103] In Douglas' article *Deciphering a meal* 1972, she expands her theory and answers the criticisms of her 1966 work voiced by the anthropologists Tambiah and Bulmer. The 'mixing of realms' remains a pivotal motif for Douglas, but she extends the scope of its application to cover, for instance, the uncleanliness of the pig.

Milgrom then goes on to discuss two obvious objections to this overall scheme. In the first place, there is the question of why the numbers of clean animals are not limited if a restriction on meat consumption was the aim. The answer, in his view, lies in the economic reality of the day. The average Israelite could not permit any losses of his flock, if indeed he had one. But—the second objection—what about the temple rituals then? Many an animal was slaughtered in such rituals, and blood flowed copiously. Here, the answer must be that God can 'recall' anything He has created and that humankind is not acting contrary to the law by eating the flesh of sacrificial animals once the blood has been returned to the Creator. Milgrom indeed considers the dietary laws to be "rudely pragmatic": they do not lose themselves in abstraction, but simply state in concrete terms what is and what is not permitted.[104] The rules and regulations were intended for the Israelites, just as the Ten Words of Exodus 10 and Deuteronomy 5. There is only one exception: the taboo on consuming blood is to be respected by all people.

Milgrom's conclusions are based on a careful reading of the texts and an extensive treatment of all the relevant arguments. As stated earlier, I can well follow the main thrust of his approach and his choices. That also holds for his general "reverence for life" motif, which can certainly be termed a powerful one. At first sight it appears to represent an 'inverse' death motif, but because of its positive elaboration it is more pervasive in its influence. It has always remained a powerful motif and exerted a major influence on later Jewry.[105] It is a functional motif, however, which still leaves us with the question: why precisely this set of animals? Milgrom appears to answer this question from the perspective of the nomadic, pastoral and religious history of the people of Israel (734/735). The exten-

[104] Because of the concrete approach adopted in the priestly laws, abuse also becomes more difficult since the rules could not then be interpreted and applied to serve one's own design. On this point J. Milgrom (1991, 736) sees a clear difference between the priestly laws and the views held by A. Schweitzer. At the beginning of the century the latter advocated the principle of "reverence for life", but omitted to elaborate this principle in any detail. His rule that 'life may only be destroyed if it serves a higher life' could therefore be abused to defend the decimation of certain birds of plumage (such as the Syrian ostrich) for the purpose of putting their plumes on women's fancy hats.

[105] The term 'reverence for life' not only refers to the boundary between life and death, of course, but also implies that animals are to be well-treated in life. Noah Cohen 1959 has treated this issue at length in his *Tsa'ar Ba'ale Ḥayim*. The principle is popular in christian tradition as well, see e.g. Jay B. McDaniel 1989.

sion of the context in which meat consumption was permitted, from ritual sacrifice to 'profane' slaughter, also plays an important role in that history and explains, in his view, the different emphasis of the Deuteronomist and the Priestly authors. To my mind, though, in his "reconstruction of the rationale for the dietary system" Milgrom does too little to incorporate either the parallels between the dietary laws and the story of creation or the motif of 'mixing', the breaching of the creation order.

Sacrificial animals / Food for the gods

Firmage, who has also undertaken an extensive study of the dietary laws, takes Lev 11:44 as the point of departure for his explication, which can be summarized as follows. The priests were faced with the problem of what kinds of food should be deemed clean and unclean. Their solution was simple: they compared the Israelite diet with that of YHWH. The latter was by definition congruent with what was sacrificially offered at the altar, which constituted the table of YHWH.[106] The sacrificial animals thus became the 'paradigm' (in the sense of exemplary case) of cleanliness in the realm of meat consumption.[107] Firmage's explanation is unsatisfactory on two accounts, I feel. In the first place his theory explains the paradigm but not the entire list. Why is it permitted to eat deer, for example? A second, and more important, objection is that his explanation does not resolve the problem, but merely shifts it. The question now becomes: why is it that precisely these animals may be sacrificed, but not others?[108] And to my mind the answer implied by Firma-

[106] Compare the expression "bread of their God" (*leḥem *ělôhîm*) in Lev 21:8.

[107] E. Firmage 1992, 1124.

[108] The question of why the Israelites did not sacrifice animals taken as game is also interesting in the light of the explanation of ritual sacrifice given by Burkert 1983, who follows the line of Meuli 1946. Both trace the practice of sacrifice back to hunting, because only a phenomenon as old as hunting can explain the fact that everywhere in the world of antiquity animal sacrifice constituted the ritual practice *par excellence*. Once certain animals had been domesticated, these became the most eligible animals for slaughter and sacrifice, with game losing its value in this respect. To Meuli, sacrifice was little more than ritual slaughter. As an additional factor, Burkert has emphasized the fact that the shared aggression attending ritual slaughter helps establish a community. On this point he follows the theories of the ethologist Konrad Lorenz. In the approach taken by Burkert, guilt and fear also play a prominent role. To my mind his (and Meuli's) answers are unsatisfactory, however (on this point, see also J. Gould 1985, 16-19). Although it is perfectly feasible that ritual sacrifice developed from hunting practices, in both Greece and Israel it became deeply embedded in religion. In his treatment of animal

ge—that the Israelites simply sacrificed the animals they kept and herded—does not succeed in explaining the complexity of the dietary laws in their totality.[109]

Purity

Houston, finally, sees the central motif of the dietary laws as lying in maintenance and achievement of purity. With regard to the origins of the laws, he follows Firmage. The classification of animals into clean and unclean developed around the places of worship, he holds, to guarantee the "purity of the worshippers". Classification therefore proceeded from the animals that were sacrificed, and the basis of selection of these animals was their general pattern of usage in a pastoral economy. In that economy, pigs were of minor significance and the hunting of mammals was limited to deer and gazelle. As these passages demonstrate, only when "monolatrous Yahwism" began to develop was the distinction made absolute in the commandment to refrain completely from consuming unclean food, as a mark of distinction. In Houston's words, "Purity became the guardian of monotheism" (WH 123).

Once again the proposed rationale is functional. In contrast to Milgrom and Firmage, however, Houston establishes a link with Genesis, situating Leviticus 11 in the context of the entire body of priestly writings. He posits a three-stage development whereby the vegetarian ideal of Gen 1:29 is followed by unconditional meat consumption, in Gen 9:3, to end in restricted meat-eating accord-

---

sacrifice in the antique (and particularly the Greek) world, J.N. Bremmer (1994, 43) concludes: "Finally, whereas the Greeks themselves did represent gods in the act of sacrifice, the protagonists in the modern debate feel apparently ill at ease with the religious functions of sacrifice and approach the subject in a strikingly secular manner. For Meuli, it was nothing but ritual slaughter; for Burkert the shared aggression of the sacrificial killing leads primarily to the forging of a community, and for Vernant sacrifice is, fundamentally, killing for eating. Clearly though, this act which stands at the centre of Greek ritual, is far richer than these reductive formulas suggest." This is indeed the position taken vis-à-vis the dietary laws in the present chapter.

[109] E. Firmage (1992, 1125) himself already recognizes a number of problems. The fish lack a "temple paradigm", as is true, it may be added, of honey (which, in Lev 2:11, is in fact explicitly excluded from the altar!) and vegetable products such as fruits and nuts. Why are these categories clean? Firmage finds himself caught up in tautologies here when he 'explains' the cleanliness of fish with reference to other food products which may likewise be eaten but not sacrificed, such as the aforementioned fruits and nuts. In Firmage's view, the mention of fins and scales in connection with fish is necessary to distinguish them from snakes.

ing to the laws of Leviticus. The vegetarian ideal permits humanity to rule on earth as God's representative, as the prince of peace. As the emergence of 'sin' shows, this is not a state of equilibrium. The situation following the Flood implies only a minimum of laws. There is so little restraint on the destructive elements in humankind that God cannot live in the world and we cannot rule in peace. The solution to this dilemma consists in God's electing Israel to be the chosen people and in that people's respecting the restrictions imposed upon them. In this way sacrifice becomes possible and institutions can be developed, thus enabling God to dwell among His people. In this arrangement the restriction on meat consumption is obviously pivotal, and this can be seen as an "echo of the schism in the cosmos brought about by the presence of violence in the animal world". It should be added that the 'three stages' referred to are not rooted in distinct historical events. "Both the primitive vegetarianism and the omnivorousness are of course myths. They enable the dilemma to which the priestly institutions are the mediating solution to be presented in narrative form" (WH 257).

Like Milgrom, then, Houston proposes a functional motif as quintessential to the dietary laws, but he gives a clearer answer to the question of what their ultimate rationale might be. The elements of ancient Israelite cosmology furnished by Houston in his discourse form useful ingredients with which to work towards a synthesis of ideas.

3.6.2 *Synthesis*

Holiness

"Be holy, for I am holy". This explanatory commandment is given twice in Lev 11:44, 45. By adhering to the dietary laws and not rendering themselves unclean, it was possible for the Israelites to become a holy people, and so they indeed must, for their God was holy to them. The dietary laws thus become charged with religious import; they are brought into connection with the very core of the Israelites' religion and with a pre-eminently divine attribute: holiness. In this arrangement, as laid down in Lev 10:10, there is a relationship between holy and profane, on the one hand, and clean and unclean, on the other. This fact must therefore form the basic point of departure in any attempt to satisfactorily explain the rationale of the dietary laws.

As Ph.P. Jenson has demonstrated in his extensive study, for the priestly authors the notion of holiness was graded.[110] Jenson refers to this as a "Holiness Spectrum", with degrees of holiness from the most sacred to the most profane. By applying such gradation to the various dimensions and aspects of life, following G.J. Wenham, we obtain a 'matrix' describing the priestly view of the world.[111] For some of these dimensions, the gradation might be as follows:

Persons: high priest – priest – levite – clean – (temporarily, slightly) unclean – (permanently, severely) unclean – dead.

Places: holy of holies – holy – the (temple) forecourt – the encampment – outside the encampment – the world of the dead, the *Šᵉʾôl.*

Times: Years: the jubilee year – the sabbath year – other years. Days: the Day of Atonement – the festivals – the sabbath – 'intermediate days' – other days.[112]

Animals: sacrificial animals (which may not be eaten) – sacrificial animals (which may be eaten by the priests) – sacrificial animals (which may be eaten by anyone) – clean animals – unclean animals – animals that are to be abhorred – carcasses of clean animals – carcasses of unclean animals.

From the priestly perspective, it was within this ordering matrix that the Israelites played out and made sense of their religious and social life. It was the task of the priest, and with him the people, to (keep) separate the sacred from the profane, the clean from the unclean (Lev 10:10). To achieve this separation, there were rules and laws concerning all the elements of the sacred order. By respecting and abiding by these laws, the Israelites could become a holy people, and remain so. Only thus would God be able to dwell among His people and could they become His chosen people. For the Israelites this order was also the ever-present context of their relations with the source of all holiness: God himself. Those things and people coming into closest proximity to this source must best match the image in which God is held. From the requirements the Torah sets

---

110 Ph.P. Jenson 1992.

111 G.J. Wenham 1979 and elaborated further in articles from 1981 (on unclean foodstuffs), 1982 (on Jesus' attitude to the laws) and 1983 (on uncleanliness and sexual contact in Lev 15).

112 On the so-called 'intermediate days' associated with the major festivals, see the Mishna *Moed Qatan* (J. Neusner 1988, 324-328).

on people and affairs we can gain a good idea of what that image comprises. Leviticus elaborates at some length on the matter. A case in point form the demands set on the high priest in comparison with the 'normal' priests (Leviticus 21). The sons of Aaron, as the latter are called, do service in the tabernacle/temple and they may not defile themselves by touching a corpse, except that of their parents, children, brothers and virginal sisters. The high priest, however, is not permitted contact with any corpse, not even that of his own mother or father; he may not leave the tabernacle. The priests are not allowed to marry a dishonoured woman, nor a woman who has been divorced by her husband. This obviously also applies to the high priest, but he may additionally not marry a widow: he is to take for his wife "a virgin of his own people".

An example from quite a different realm are the rules on sacrificial animals given in Lev 22:17-25. Here, too, we find a similar subtle gradation. Not all clean animals may be sacrificed, but only the unblemished, male specimens of the oxen, sheep and goats.[113] The text then elaborates on what is meant by 'unblemished': animals that are diseased, mutilated, wounded or blind are not be ritually sacrificed. The rules are very precise. Even a healthy specimen whose legs happen to be too long or too short is 'second-rate'. That is to say, it is not good enough to serve as a peace offering to honour a sacred vow, although it may be sacrificed as a voluntary act.

Leviticus outlines the religious and social model of a holy people. In doing so the dietary laws are subsumed under the purity laws, which are in turn subsumed under the laws of consecration. Implicitly, Leviticus outlines a cosmology: on the one hand because the model refers back to the Creation, and on the other because it makes reference to the Godhead himself who, according to Genesis and Exodus, not only created the world but also, through the deliverance from Egypt and the passage through the Reed Sea, created the people of Israel. This dual creation symbolism constitutes the omnipresent decor for the Book of Leviticus. This book refers ultimately to the situation as it was at the time of Creation, or as it was intended by the Creator (see Chapter 2). However, it is written at a

[113] Apart from these three species, the pigeon may under certain circumstances also be sacrificed. In Leviticus 12 it is stated that a woman who has just given birth must sacrifice a year-old lamb as a burnt offering and a young pigeon or turtle-dove as a sin offering. Doves or pigeons may be brought instead of the sheep if the woman is of insufficient means.

time, and in a world, that has fallen from that ideal.

In the foregoing, we have already identified and discussed some of the main motifs of that ideal. In the first creation story the initial act of the Creator is one of division, creating order out of, and in opposition to chaos. Order represents life, chaos death. Subsequently, everything is given a place of its own and has the characteristics of that place; nothing is holy in itself. Everything functions properly and is good. There is a hierarchy, spanning plant, animal and humankind, but in that arrangement violence is absent and blood is not spilled. The creation has its culmination in the seventh, hallowed day.[114]

The creation order of Genesis 1 has been gravely disturbed by the events in the garden. In the older, second creation story (author: J), which follows the first (author: P) in the text as it stands, the relationship between humanity and nature is disrupted, and chaos and disorder ensue. In referring back to Genesis, Houston makes a solid point. The primeval myths provide an image of the emerging disorder. To bring that image into focus, let us now consider some of the motifs of these early Israelite myths, before returning to our quest for the rationale of the dietary laws.[115]

---

[114] See, *inter alia*, J. Moltmann 1985 for an elaboration of this theme.

[115] The term 'primeval myth' is employed here for the various stories of Genesis 1-11, of which the accounts of the creation and the flood are the best known. To understand these stories it is necessary to consider them in the historical context of the ancient Near East (for a general treatment, see W.G. Lambert 1965, A. Heidel (1971, 1974), R.J. Clifford 1994, Sh. Salmon 1995 and N.C. Habel & S. Wurst (eds.) 2000; more specifically, on the stories of the flood, see G.A. Caduff 1986, N. Cohn 1996, F. García Martínez & G. Luttikhuizen (eds.) 1997 and *Eerdmans Dictionary on the Bible* (entry Flood) 2000.The Genesis narratives show a clear resemblance to the Sumerian and Babylonian creation and deluge myths and much of the former account appears to be derived from these. There are notable differences, however. The significance of the differences and similarities has been the subject of heated debate, which has not subsided since the end of last century (when the respective Akkadian texts were discovered and deciphered) (see: E. Noort 1997, 9 for a literature review). In the narratives of Genesis a Yahwist and a Priestly text tradition are distinguished (see V. Maag 1955 and C. Westermann 1974, 532 *et seq.* for a review of the arguments). In the creation stories these two traditions are clearly juxtaposed, but they are also recognizable in the subsequent text and in the account of the flood (J.A. Emerton 1987/1988 versus G.J. Wenham 1978). In the case of the flood stories, E. Noort 1997 assumes two (J and P) independent sources that were later combined to form a single text. On this question of final editing he writes (with reference, *inter alia*, to P.D. Miller 1978): "Selbstverständlich sind diese beiden Versionen redaktionell zu einer Einheit in der Letztgestalt des Textes geworden, und diese Einheit verdient nicht nur Beachtung wegen ihrer

The primeval myths

In the world outside the garden, violent death soon emerges and appears to dominate human relations. Cain, who kills his brother, fears (rightly) for his life. His somber premonition, "whoever finds me will slay me" (Gen 4:14), is taken seriously. He is assured by God that if anyone kills him his death will be avenged sevenfold. 'If', for he is given a special mark by God "lest any who come upon him shall kill him". Although his tilling of the earth will no longer provide him with a proper harvest, Cain does not do too badly. Through his offspring a grand culture develops. But there is always an undertone of enmity, because his great-great-grandson Lamech also comes to fully exploit the idea of revenge. There is a mixing of entities that should remain separate.[116] Gen 6:1-4 relates how the "sons of God"[117] took daughters of men to be their wives, thus transgressing the boundary between heaven and earth. The Deluge follows, because the world had become wicked in God's sight. The earth is filled with *ḥamas*, violence (6:11). Van Selms inflates this term and sees it as encompassing any form of injustice, not only violence.[118] It afflicts not only the sons of God and the 'mighty men of old' (*gibbôrîm*) born out of their communion with the daughters of men, but eventually spreads to affect almost everything on earth (6:11).

Even the animal world, according to verse 12, had deviated from the straight and narrow. The subject of the second half of this verse, *kol bāśār* (translated literally as "all flesh" in the RSV and as "all that lives" in the NBG) can be read as denoting either human beings and

---

Funktion im Ganzen der Urgeschichte, sondern auch wegen der Tatsache, daß sie den Ausgangspunkt für die weitere Rezeptionsgeschichte bildet." For precisely this reason, in the concluding section of this chapter we shall focus on this *Letztgestalt* (*cf.* also the remarks made in note 93 and the cited position of C. Houtman 1980, 252).

[116] On this point, see C. Houtman 1984.

[117] The term *bh^enē hā^{'e}lōhîm* can be intrepreted in various ways. 'Sons of God' (RSV, NBG) is indeed a possible translation, but an alternative is 'divine sons' or 'divine beings' (as in some translations of Job 1:6). The term *^{'e}lōhîm* can also denote the scarcely conceivable, the unbounded (*cf.* the description of Nineveh as 'exceedingly great' in Jonah 3:3). In this passage in Genesis the earthly daughters stand in clear contrast to entities that are superhuman and virtually not of this world. On this point, see also P.D. Miller (1978, 25,26) as well as K. van der Toorn (1988, 14), who considers these 'mighty men of old' a biblical variant of the Mesopotamian *apkallus*, sages with knowledge of the supernatural, the underworld and so on.

[118] Van Selms (1984, 108) bases himself on the use of the same term in Gen 16:5.

animals collectively or each individually.[119] If we opt for the reading 'human and animal' in verses 12 and 13, in the eyes of numerous exegetes this then implies that there is (also) guilt on the part of the animals. There are (theological) objections to this implication, because the term 'guilt' is connected with sin and animals cannot be afflicted with sin.[120] To my mind, this is a reading adapted to conform with a disputable position. Non-human creation also shared in the consequences of the events in the garden. Without having to take recourse to the notion of the 'sins of animals', it is certainly defensible that the author intended the words "all flesh had corrupted their way upon the earth" to encompass the plant and animal worlds, as well as the human race. In this context, one can think of such phenomena as predation, persistent 'weeds' and poisonous plants and animals. These phenomena indicate that the harmony of creation has been lost.

"But Noah found favour in the eyes of the Lord", we read in Gen 6:8, and the reason follows in verse 9. Among his contemporaries Noah was a *ṣaddîq*, a righteous person consequential in his deeds, and moreover *tāmîm*: a perfect or 'whole' person, as we might say today, or 'blameless' as the RSV and NBG have it. He and his family are saved in the ark, and in how the ark is filled (seven pairs of all clean and one pair of all unclean animals) we see reflected something of the order intended in the creation. In the phrase 'to keep their kind alive' (6:19 and 7:3) there is not only a matter-of-fact justification of the male/female arrangement, but certainly also the motif of not allowing any God-created species to become extinct as a result of the coming deluge. The detailed description of 7:14, with the fourfold repetition of the familiar words 'according to its kind', point in this direction. The phrasing underscores the connection with the first creation story and thus the probable authorship (P) of these verses. The remarkable difference between 6:19 (one pair of all animals) and 7:2 (but seven pairs of all clean animals) is probably also explained by the priestly (final) editing of the text of Chapter 7. That seven pairs of the clean land animals and birds are to be

---

[119] For an extensive review of the use and explanation of this term in the writings of P and in other parts of the HB, see A.R. Hulst 1958.

[120] A. van Selms (1984, 108), for example, takes 'all flesh' in verse 12 to be a reference to humans, while in verse 19 taking it to include the animals. He arrives at this reading because "The Scriptures are not aware of any sins on the part of animals".

taken on board, and only one pair each of the unclean, is perhaps an indication of the priestly scribes' concern with the availability of sufficient sacrificial animals. In my view the use of the number seven reflects the idea that these clean animals are more 'perfect' than the unclean.

This ordered microcosmos enabled a new beginning to be made after the Flood. In many respects the deluge acted as a 'reverse creation' that enabled that new beginning to be made. This motif (a new beginning) occurs in almost all the classical accounts of the deluge, but there are major differences in the status of the animals therein.[121]

The new beginning starts with Noah building an altar on which he offers burnt sacrifices of clean land animals and birds (8:20).[122] These sacrifices move God to say to himself (literally: 'in his heart') that He will never again "destroy every living creature", for (or because, *kî* is explanatory) "man's heart is evil from his youth". The Flood is the first and last such event; the order and rhythm of the

[121] Comparing Genesis with the primeval myths of Mesopotamia (S. Dalley 1989) it is noteworthy that Atraḫasis also takes animals on board his boat in the Mesopotamian account. He selects birds, cattle and wild animals of the open country (*Atraḫasis*, tablet III ii) and invites his family for a feast on board. It is therefore conceivable that the animals were intended for consumption at the feast. There is no mention of animals leaving the boat. The story of the deluge is also related in the Gilgamesh epic (tablet XI ii). Ut-napishtim (probably identical to Atraḫasis) reveals a divine secret to Gilgamesh: Shuruppak is threatened by a great deluge. Gilgamesh is to build a boat in the form of a huge, almost square box (while the ark, referred to as a 'vessel' in Genesis, has the dimensions of a large ship) and he takes on board the following: "seed of living things", "cattle from open country" and "wild beasts from open country". When the ship runs aground on Mount Nimush, after seven days a dove is set free and then a swallow. Both return to the boat, because they find no resting-place. Finally, a raven is released which does not return. Gilgamesh then offers a sacrifice on the top of the mountain. There is no further mention of animals in the story. The motif of 'animals in a lifeboat' is found in both Mesopotamian accounts, but in Genesis (and especially with P) the animals are given greater, and far more systematic attention. In the Greek stories of the deluge, no animals at all are taken into the lifeboat. J.N. Bremmer (1997, 56) refers to this as "a nice illustration of the lack of Greek interest in animals". The systematic and positive focus on animals in the biblical stories is not a chance detail, I would hold, and of greater significance than has yet been noted.

[122] The sacrifice of (clean) animals is thus (because of the strong parallels with the non-biblical myths) a universal element of these myths. The sacrificial practices of the people of Israel are similarly reflected in the account of Abel and Cain, who sacrifice, respectively, the fat (the most valuable part) of the firstborn of his flock and the fruit of his fields.

days, the seasons and the harvests will now be lasting. Noah is blessed and, like the first human beings in Genesis 1:26, is instructed to: "Be fruitful and multiply, and fill the earth".

As the rest of Chapter 9 relates, the 'old order' is not entirely re-established. In the narrative it is God himself who, having giving His blessing, outlines the terms of the new order. In this postdeluvian era, all animals may be eaten. They thus have every reason to live in fear and dread of humanity (9:2,3). However, there are two clear 'constraints', restrictions that characterize the new relationship between humankind and the rest of the natural world and set bounds to chaos.

The first restriction is an explicit prohibition on the consumption of flesh together with blood. Blood is the spirit of life. Although humankind is given disposal over all animal life, as it were, to feed off 'every moving thing', but human dominion may not degenerate (entirely) into consuming the very spirit of life. This restriction on shedding the blood of animals (and fellow human beings) is motivated in verse 6: "For God made man in his own image". Earlier, I cited this passage to support the view that the granting of 'dominion'is intimately bound up with humankind's being created 'in the image of God'. At this juncture, the phrase is reiterated to set bounds on that dominion. In contrast to Genesis 1-3, human beings living outside the garden have (and those after the flood will have) come to rule over one another, but the shedding of human blood will always remain an absolute taboo. Dominion over the animals was already the rule, but animals may now be killed and consumed. Even so, the 'blood of life' may not be consumed along with the flesh. It is forbidden as such, but the prohibition can also be seen as a permanent reminder of the fact that the killing of animals was not originally intended.

The second constraint, described in verses 8-17, is perhaps even more binding. In the 'covenant' established with Noah, God elaborates on the vow he made to himself in 8:21: that there would never again be a deluge to "destroy every living creature". This covenant is established not only with Noah and his descendants, but with every living being. The explicit and repeated mention of this fact is very striking, to say the least. In virtually every verse of this short passage there is reference to nature. At the beginning of verse 10 we find *kol nepheš ḥayyāh*, every living creature, an expression that recurs in verses 12, 15 and 16. In the second half of verse 10 several groups of animals are named: the birds, the cattle, every beast of the earth (wild animals) and, summarizing, "as many as came out of the ark", all that lives on the earth. In verse 11 follows the phrase *kol bāśār*, "all flesh", for all the animals, and we see the same phrase used again on its own (15,17) or in combination with *kol nepheš ḥayyāh* (15,16). That the fish are not mentioned in verse 10 is more or less self-evident. They were not included in the ark's population, were not directly affected by the flood and are moreover subsumed under the title of 'every living creature on the earth'.

We read nothing about plants. The emphasis is all too obviously on life in the form of 'flesh', that is on animals. This is understandable, because compared with the world of the creation myth it is humanity's relationship with the animals that has undergone the greatest change. This is not to say that plant life is excluded from the covenant. The inclusive use of the word 'earth', first negatively in verse 11 ("to destroy the earth") and then positively in verse 13 ("a covenant between me and the earth") indicates the broad reference of the text.

The meaning of this passage can only be that Noah and his descendants (for that is how he is addressed) are told that the covenant is established not only with them, but with all life on earth, the whole of creation. It is a covenant between the Creator and His creation, and for this reason is properly termed the Covenant of the Rainbow. The use of the term covenant attaches an enormous importance to this second 'constraint' on life after the flood. A covenant between God and the earth implies a direct bond, a reciprocal relationship between the Creator and the created. It is a relationship that is shaped outside the human realm, but in which humankind is involved as a 'third' party: we see a 'triangular' motif emerging in the relationship between God, humanity and nature.

Verses 15 and 16 describe how God himself will remember His covenant with the earth on seeing the rainbow. The bow may well (also) serve as a sign for humanity, but the text relates explicitly to the covenant between God and the earth. It is to this side of the triangular relationship that human attention is directed. This becomes even clearer in verse 17, which at first appears to be a reiteration of verse 9. There are two differences, however: in the parties involved, for here the covenant is "between me and all flesh on earth", and in the tense employed, for here the text has the perfect (*haqimôtî*, the covenant which I *have established*). It is now a fact, and one for which

Noah and his descendants will have to make due allowance.

Throughout the history of Western culture, the covenant of the rainbow has been afforded too little significance.

The next part of Genesis 9 describes the emergence of the different peoples of the earth, and in particular the (low) standing of the Canaanites. In the account of the historical development of Israel and specifically during occupation of the promised land, the Canaanites formed a major stumbling-block.[123] Perhaps this was not so much because of their great military superiority, although this was certainly a fact (Deut 7:1), but more because their religion formed a permanent source of temptation and they stood as a model of that which was to be banished. As a people they were more a part of Canaan than any other of the seven peoples cited in Deuteronomy 7. And Canaan is the land that God will show to Abraham and promise him. The required banishment and subjugation of the original inhabitants of that land are justified (or prompted), as it were, by the history of the curse pronounced on Canaan by Noah, for something that his father Ham had in fact done to him.[124]

In Chapter 10 the 'generations' of the sons of Noah are enumerated, with several striking features. In verse 8, for instance, we encounter Nimrod as a descendant of Ham.[125] It is said of him that he began to be a *gibbôr* on earth, someone with power who used force. The RSV translates this as "the first on earth to be a mighty man"—

---

[123] Historically, there are arguments for supposing that the (proto-)Israelites were largely descended from the Canaanites. See, *inter alia*, G.W. Ahlström (1986, 25-36), T.L. Thompson 1992 and K. van den Toorn (1996, 184-187). All the more reason for later scribes to emphasize the difference.

[124] On this point O. Keel & Chr. Uelinger (1992, 474) see a parallel with the later persecution of the Jewish people. They write:

> Die jüdäische Religion hat im 7. und 6. Jh. v. Chr. einen in der Anlage dem christlichen Antijudaismus vergleichbaren, wenngleich in seinen historischen Folgen unvergleichlich harmloseren Sündenfall begangen, als es seine Identität nur finden zu können glaubte, indem es alles Kanaanäische verfemte und zu seiner Vernichtung aufrief, etwa zur Zerstörung heiliger Steine (vgl. Dtn 7,5; 12,5), welche die Väter einst aufgerichtet hatten (Gen 28,18.22). Nicht nur Kultobjekte wurden nun als Kanaanäisch verfemt. Es gehört zur Tragik der jüdisch-christlichen Schuldgeschichte, daß das Volk, das Opfer der Schoa geworden ist, in seinen eigenen religiösen Schriften an zentraler Stelle den Auftrag zur Ausrottung anderer Völker überliefert hat, um dadurch die Reinheit und Heiligkeit der eigenen Religionsgemeinschaft zu sichern (vgl Dtn 7,2-6.16.24-26). Hier sind *alle* zu Demut und Umkehr aufgerufen.

[125] On this legendary Babylonian figure, see: DDD, entry *Nimrod*, and further: K. van der Toorn & P.W. van der Horst 1990.

after the deluge, that is, because he follows in the steps of the *gibbôrîm* of the antediluvian era (6:4). Verse 9 appears to state against whom he uses this force: he is a *gibbôr ṣayid*. Reisel (1966) translates this as "mighty conqueror". This serves as a neat stepping stone to the next verse, in which we read of his conquests and their political outcome: virtually all the major kingdoms to the north-east of Israel[126] He arrives at this rendering by translating the term *ṣayid*, which is generally associated with the hunt, simply as 'seizing' or 'conquering'. The benefit of logic does not detract from the fact, however, that this is quite some way from the original Hebrew. There is a broad consensus that the root of the word, *ṣyd*, is connected with food.[127] As the whole passage from verse 8 to verse 13 relates, however, his activities are not restricted to the hunting of animals. He was not "a man of the field" like Esau (Gen 25:27). Perhaps in Nimrod we have a 'mighty man' who had power over both animals and men. It should be added that that power need not be interpreted immediately in a negative sense. According to the text, he was establishing his rule, in the form of cities and kingdoms. The fact that he is said to be a mighty hunter "before the Lord" can be seen as a superlative. Reisel refers to the classical explanation, *viz.* that this should be understood to mean that he carried out his deeds "knowing that God was opposite him": they had a confrontational, possibly provocative character.[128]

---

[126] The text states that Cush is the father of Nimrod, which appears strange for a Babylonian, because Cush is generally taken as an (eponymic) reference to an African tribe from southern Egypt. It is therefore probable that this passage in Genesis was intended to associate the whole of the contemporary civilized world with Nimrod, both in the north-east (Babylon and Assyria) and in the south (Egypt, Cush) (in the same vein: B. Oded 1986, 18).

The sequence of the names of the cities founded by Nimrod corresponds roughly with the political history of Mesopotamia over a period of over 1500 years. For these and other reasons, it is to be doubted whether Nimrod can be identified with an historical, earthly figure (see: K. van der Toorn and P.W. van der Horst, 1990, who identify Nimrod with the Assyrian god Ninurta).

[127] In Gen 25[27] we read of Esau that he was a *yôdēʿa ṣayid*, a one skilled in *ṣayid* and of Isaac verse 28 relates that he loved Esau: "because he ate of his game". In the Hebrew: *kî ṣayid bᵉphîw*, which can be translated literally as "for the *ṣayid* was in his mouth". From this passage it must therefore be concluded that (hunting) 'catch' is the most obvious translation of *ṣayid*. Other passages in the HB also point to a similar meaning (e.g. Lev 17:13, Hos 9:13). When it is said of Nimrod that he is a *gibbôr ṣayid*, to me it would seem that he was at least a 'mighty plunderer of food'.

[128] M. Reisel (1966, 57); support can be found in the fact that the name Nimrod is generally taken to be derived from the root *mārad*, to rebel. The reading

This enumeration of the generations of the sons of Noah relates what became of them and how, according to the author of this section of Genesis, these *gôyim*, these peoples, were divided over the earth in the era after the flood (10:32).

In this manner, by way of the primeval myths, much of the world around the Israelites was given a place in history. There remained one important phenomenon to be explained, however: the diversity of languages. This was accomplished in Genesis 11:1-9, the account of the building of the Tower of Babel. The striking feature of this story is the 'mixing' it involves, i.e. people's attempt to (once more) establish a link between heaven and earth. It is a link that is to give the people a name, a reputation, and bind them together. In that era, too, unity meant power, but power can lead to a desire for absolute power. Verse 6 indicates that God deems this to be a real danger. The tower is "only the beginning of what they will do", with no further bounds to human aspiration, no dream beyond humanity's grasp. This evokes a response from heaven, in the plural form: "Come, let us go down, and there confuse their language". This might be the 'royal we', but the plural form may also reinforce the idea that heavenly beings are not to intermix with earthly beings and that these beings therefore act in consort to bring such human aspirations to an end. Language, the basis of communication, is 'confused'. This results in the people being scattered and construction of the tower is halted, perhaps directly, because of the 'confusion of tongues', perhaps indirectly, because of insufficient labour. In verse 9 a link is established between the name of the city (*Bābhel*), which was mentioned earlier as being conquered/founded by Nimrod, and this confusion (*bālal*). In doing so, the text distances itself subtly from the (Assyrian) reading of the name as *Bābh-el*, gate of God. This latter meaning may echo in the second half of verse 9, which reads "and from there the Lord scattered them abroad over the face of all the earth". Whatever the case, the response to a mixing of realms is confusion, which leads ultimately to 'unmixing'.

From verse 10 we return to the enumeration of the generations of Shem. Down to Peleg the list is a reiteration of Chapter 10, but

---

with 'opposite' is already encountered in Philo (*Quaestiones in Genesin,* II, 82) and Augustine (*De Civitate Dei*, 16.4). Besides the Hebrew derivation of the name Nimrod, she also refers to the LXX translation of *liphnê* as ἐναντίον which can mean both 'in the face of' and 'opposite'.

the text does not continue with Peleg's brother Joktan, but with his son Re'u. This lineage ends in Abraham and as such forms the link between the primeval myths and the history of the patriarchs, which in turn represents the primeval beginnings of the people of Israel.[129]

God's calling of Abraham from his homeland in the north-east and His promise "to make you a great nation" mark a decisive turning point.

Conclusion

The primeval myths related in the early chapters of Genesis represent a universal cosmology which, embroidering on numerous Mesopotamian motifs,[130] embodies the Israelites' answer to many questions with which people wrestled in the world of antiquity, at the time of writing and presumably in earlier ages.

The stories also highlight the deficiencies of the primeval era and the ways in which the creation ideal had been left behind. The Creator considers it necessary to unleash a deluge. The Flood is 'causal' and, as a narrative, is intimately connected to the Creation and, in many respects, forms a counterpart to it. However, the deluge did not bring about the desired (or, better: the 'presumably desired', for the text leaves this undecided) recovery: in certain respects the chaos remained.[131] From the reader's perspective, this makes it plausible that it required more than mere 'constraints' to keep life adequately within the intended bounds, as it were; it required the election of a chosen people. In this way the primeval myths become linked to the 'genesis' of the people of Israel, and it is within this priestly fabric that the tradition of the Patriarchs and the Exodus are woven.

The Torah makes it clear that the Israelites are a holy people serving a Holy God, with special laws intended to shape that holi-

[129] The 'missing link' between Abraham and the primeval myths, Nimrod, is also established in the *Liber Antiquitatum Biblicarum* of Pseudo-Philo, which dates from the beginning of our era. This Philo describes a battle between Nimrod and Abraham in which the latter is cast into a furnace. Although over 80,000 people do not survive the fire, Abraham is unharmed. For a historical review of the interpretation of the Nimrod texts and the—conjectured—relationship between Nimrod and Abraham, see K. van der Toorn & P.W. van der Horst 1990, 16-29.

[130] See, *inter alia*, W.G. Lambert 1965.

[131] E. Noort (1997, 41) phrases this idea as follows: "Zwar stimmt die schon immer gemachte Beobachtung, daß die Fluterzählungen letztendlich Rettungserzählungen sind, aber die Unheimlichkeit der Rückverwandlung vom Kosmos zum Chaos bleibt."

ness and also serving to form a constant reminder of the source of that holiness. Only by following this path was there still a chance to make visibly manifest some of God's designs in this world.

Against this background, this 'priestly decor', we can now perhaps understand both the function of the (dietary) laws and their underlying rationale. The laws, and adherence to them, are necessary to mould the Israelites into a holy (in the sense of separate) people, in the midst of nations who worshipped false gods: a people that had been chosen by and was dedicated to one God (the "Guardian of Monotheism" function). The laws are also designed to make this people holy (in the sense of radiant) and ensure that they remain so amidst their neighbours.[132] This is a 'referential' function. Both functions serve to constitute a people that is having difficult forging a religious and social identity of its own in an environs that poses an continal threat to such an identity.[133]

The dietary laws can serve these functions because in their substance they are themselves also referential: they refer back to the Creation ideal and to (characteristics of) the Creator in a world that is now some distance removed from that ideal. Let us then attempt to summarize the Creation ideal in terms of a few concise principles.

Life/death/blood Premature, often bloody death is not part of the Creation ideal.[134] Because of the events in the Garden, death has become part and parcel of life and the dominant force outside the garden. Animals are drawn into the chaos that once more reared its head following the 'Fall'. In a sense, they are also its victims. The Covenant of the Rainbow introduces new constraints, but these are not apparently sufficient to maintain the 'quality of life' of the world as a whole. Separate laws for a separate people are laid down to create and guarantee the required quality of life. The eating of animals is accepted under strict conditions. The consumption of

---

132 U. Cassuto (1989, 65) describes holiness (*qᵉdûšāh*) as "elevation and exaltation above the usual level".

133 It may be added that the HB contains numerous references to the fact that such threats were not solely external, but were also regularly posed from within Israel itself. All the more reason, therefore, for the priestly scribes to make this point, in the name of God ('do this, for I am holy/you are holy').

134 'Death', in this context, I understand to mean violent death, the premature ending of life, death whereby blood is generally spilled. On this point, see Section 2.5.2 and 2.6 of the previous chapter.

blood, the quintessence of life, is forbidden and only a limited number of animal species are declared fit for consumption. Only certain herbivorous animals may be eaten, with carnivores and omnivores being declared unclean.

Order/chaos/mixing Creation is separation: thus is order created from chaos. Everything has its place in this Creation Order. 'Mixing' constitutes a breach of this order, creates confusion and is associated with chaos. Animals that breach the order are unclean and sometimes an abomination. If clean and unclean come into too close proximity, the result is (temporary) impurity. Impurity is transferrable: it proves to be contagious. Unclean, too, are those animals that do not clearly belong to one particular domain, such as creatures of the twilight, animals living on the border between land and water, aquatic animals lacking the characteristics of fish, scales and fins, animals that fly but lack the characteristics of birds, feathers and wings, and finally animals with feather and wings that generally walk and scarcely if ever engage in flight. For this reason, the use of two different types of material, wool and linen, in a single item of clothing, simultaneous use of different kinds of draught animals for ploughing and, last but not least, the interbreeding of two different animals, let alone of human and beast, leads to impurity: it is contrary to the Creation Order.[135]

Immaculacy/integrity/deficiency The Creator is a single, undivided entity, immaculate and whole. His Creation was also 'good'. The concept of purity is designed to reflect this. In the context of animal sacrifice and temple ritual, particularly, this led to a differentiated classification embodying 'graded holiness'. There was only limited use for diseased, afflicted or injured animals in sacrificial practices. Similar requirements held for the human members of the temple hierarchy. There is by no means always 'abomination' in-

---

[135] This has always been the conviction in the Jewish tradition, as reflected, for example, in Nachmanides (1194-1270), who commented as follows on the text of Lev 19:19: "He who grafts or interbreeds two different species changes Creation. In doing so he indicates that he holds Creation to be imperfect". This position of Nachmanides is cited approvingly by Rabbi R. Evers in a discussion on the cloning of animals (Dutch newspaper *Trouw*, 7 March, 1997, 17). Rabbi Evers holds the view that scientists involved in cloning projects are not out to change Creation and he opposes Chief Rabbi Meir Lau in the latter's opinion that cloning constitutes a transgression of Talmudic law.

volved; in essence, impurity is a cultic and ritual concept. This is clear from the rules concerning (temporary) illness and the bleeding of women during menstruation and childbirth.

The dietary laws harbour a wealth of symbolic motifs. They stand for the order and peace of civilization as against the disorder and chaos of the wild. By their nature and intensity (graded) they refer back to the nature and source of Holiness. They root the Law in the Creation narratives. In this way they form a constitutive element in the religion of the people of Israel and fulfil a clear function: to constitute a separate people with a single, separate God.

The answer to the question of the rationale of the dietary laws thus lies in the (priestly) ideal of the *Creation and the Creator.*

CHAPTER FOUR

GREEK COSMOLOGIES

4.1 Introduction[1]

This chapter describes some of the principal ideas of the Greek philosophers.[2] Given the vast body of thought that has come down to us, texts have been selected for their relevance to the subject of our enquiry: humanity's relationship with nature. The general historical context and the Greeks' actual dealings with the natural world will be considered only obliquely, as an interpretative framework.[3] In terms of the philosophers considered, I shall be discussing the principal schools of thought as well as a number of 'dissidents'. Here, the reference is to dissidence concerning humanity's relationship with nature rather than cosmology. Several more general motifs can be discerned in Greek thought, and on this note the chapter concludes. The views and beliefs described here epitomize the spiritual heritage of ancient Greece and Rome in this realm. Both directly and indirectly, the Greek philosophers have had an enormous impact on western culture: directly, because of the profound influence of their

---

[1] Some sections of this chapter have been previously published in J.J. Boersema 1994.

In geographical terms the ancient Greek world extended far beyond contemporary Greece to include Western Turkey (Asia Minor), Libya and parts of southern Italy, including Sicily. For several centuries when Greece was at the height of its power, following the conquests of Alexander the Great (4th century BC), its sphere of influence encompassed the whole of the eastern Mediterranean, including much of the Middle East through to the Indus valley.

[2] Use has been made of Diogenes Laertius' *Lives* (in Ferwerda & Eykman's Dutch translation, *Leven en leer van beroemde filosofen,* Ambo, Baarn, 1989) and the following general reviews: A.H. Armstrong 1947/1981, 1972, 1973; S. Sambursky 1956, 1959, 1962; H. de Vos 1963; W.K.C. Guthrie 1965; C.J. de Vogel 1960/1964, 1967; I. Düring 1966; U. Dierauer 1977; D.E. Hahm 1977; M. Pohlenz 1978; A.A. Long 1974; W. Burkert 1985; A.A. Long & D.N. Sedley 1987; D.J. Furley 1987,1989; A.P. Bos 1991; R. Sorabji 1993 and, finally, the controversial work of R. French 1994.

[3] Use has been made of the literature cited in Chapter 1, note 95, especially of R. Sallares (1991), an excellent work with a rather misleading title, as it is concerned mainly with Greek society, agriculture and demography rather than with 'ecology' in today's sense of the term, and besides that of J.D. Hughes 1994.

writings on the development of our culture, and indirectly because they also moulded the intellectual climate in which Christianity evolved from Judaism. As a result, ideas originating in Greece were also passed down by way of early Christianity.[4] We shall be returning to some of these ideas in the following chapter.

4.2 Nature and Cosmos

Before considering these ideas in greater detail, a few linguistic remarks are in order regarding two key concepts: cosmos (κόσμος) and nature (φύσις). The word cosmos has two basic meanings: (1) order, an ordered, harmonious whole, and (2) ornament, adornment. Both these meanings are reflected in the continued philosophical usage of the term for 'the universe', 'the world'. The term thus has a far wider connotation than as used in our term 'cosmology' for the science of the origins and structure of the astronomical universe, embracing the totality of humanity's beliefs regarding the world order.[5] In scientific terms, it was the basic premise on which the Greeks built their philosophical constructs.

A similarly dual meaning is embodied in the Greek term for nature (φύσις), the root meaning of which is 'being' according to some philologists.[6] The root φύ is also found in the Latin *fui* ('I have been'). The medio-passivum φυέσθαι means 'to grow' and can perhaps be regarded as an ingressive, meaning 'to begin to be'. The active φύειν is causative: to make grow or emanate from. From these derived meanings the root meaning of φύσις appears to be: that which emanates, the (true) nature or disposition. Use of the term φύσις in

[4] On this point see C. Clemen 1908/1912; K. Steur 1935; J.N. Sevenster 1961; J. Passmore 1974/1980; M.L. Colish 1979, 1985; P.H. Santmire 1980, 1985; Y. Amir 1983; R.Lane Fox 1988/1989; J. Pelikan 1993; D.T. Runia 1993 and M.Sordi 1994.

[5] In the work *On the Cosmos* (399 a 12-15), attributed to Aristotle, we read: "The one harmony that is the result of all heavenly beings dancing in unison has its beginnings in but one origin and its end in but one purpose, and thus the whole is truthfully termed 'order (cosmos)' and not 'disorder'" (A.P. Bos 1989, 59). Although Bos' designation of this as an Aristotelian work is controversial, the ideas voiced are in line with the ideas of Aristotle.

[6] Extensive argumentation on this point is to be found in the dissertation of D. Holwerda 1955 on the term *phusis*. For my information regarding a number of Greek terms and semantic derivations I am very grateful to him and to his colleague Professor S. Radt (both at the State University of Groningen).

the sense of 'nature overall', for (living) reality, is a later development.[7] Here, too, the dual connotation remains pivotal in later usage. In discussing Greek conceptions of nature, the Dutch scholar De Vogel distinguishes nine aspects, hesitating to reduce them to a single common denominator.[8] She deems the following three aspects characteristic:

1. The Greek philosophers' attitude to nature was contemplative.
2. They regarded nature as a *divine order*.[9]
3. They conceived of nature as a *living being*.[10]

4.3 The Early Beginnings

It is with Thales of Miletus, the oldest of the so-called Ionian natural philosophers (early 6th century BC), that the millennium of Greek philosophy traditionally begins. The era comes to an 'official' close when the Roman emperor Justinian closes the last Athenian schools of philosophy in 529 AD. Thales and the other early philosophers of nature (later referred to as φυσικοί[11]) also mark the beginning of (western) scientific culture. Their ideas constitute a historic break with the past in a variety of ways,[12] the contours of which can be summarized as follows:

[7] The works of the natural philosophers are often prefaced by the epigraph *peri phuseoos*. In the wider literature this is often taken to mean that these writings were concerned with living nature, 'nature overall' (De Vos 1970). D. Holwerda 1955 and C.J. de Vogel 1967 are more cautious, however. They believe these epigraphs were appended to the works of the pre-Socratic philosophers at a later date, because it is only after about 450 BC that the term *phusis* is used in this wider sense.

[8] C.J. de Vogel 1967, 36,37. The three aspects considered characteristic by De Vogel are particularly pertinent to the currents of thought that later predominated in ancient Greece. See also the conclusion of Section 4.3.

[9] On this point only the atomists and the sceptics (followers of Pyrrho of Elis) held dissident views. In the words of the sceptic Carneades, for example (as cited by Cicero in *De Natura Deorum* III 9, 23-24 and III 10, 25-26): "Neither the world nor the universe is divine" and "both were not made by God either".

[10] On this so-called ζῶον concept De Vogel refers, *inter alia*, to the work of S. Sambursky 1956.

[11] For example, by Aristotle: *Phys.* I 2, 184 b 16.

[12] Although it is commonly held that the Greeks laid the foundations of western scientific thought (see, for example, E.J. Dijksterhuis 1950, C.J. Glacken 1967 and D.C. Lindberg 1992), several scholars disagree on this point. R. French 1994, for example, writes: "the motives of ancient writers on nature were rarely 'scientific' and, indeed [...] there was no science at all in the ancient world". French

- They gradually rejected the mythical conception of the world, developing a fundamental criticism of traditional religious thought. They studied 'this world' in terms derived from this world.
- Imbued with a sense of wonder, they sought to discover the nature of the world's coherence, endeavouring to trace all natural phenomena back to a single ruling principle, the ultimate cause (the ἀρχή, a term denoting not only 'origin' but also 'control' or 'grip'). It is by understanding and explaining the phenomena of nature that human beings can 'get to grips' with the world.
- They regarded logos (reason, speech), in the form of logical reasoning, language and mathematics, as the defining human characteristic. They also identified (the order of) the cosmos itself with the logos.[13]
- While basing their ideas on empirical observation, their method was principally that of philosophical reflection. Knowledge was to be acquired through individual effort, employing one's rational faculties of thought. Revelation and ancestral tradition lost their authority as sources of knowledge, in favour of a reasoned explanation of the totality of natural phenomena.

The shift from mythopoeic explanations of the world's creation and humanity's evolution to the new philosophical method of investigation was a gradual one. Traces of mythological thought are still emphatically present in the works of such philosophers as Parmenides and Empedocles. There is a marked difference, however, between how these philosophers, from Thales onwards, began to perceive the world and how the world's beginnings had been described by writers like Hesiod and Pherecydes of Syrus.[14] The crucial difference from the mythopoeic conception of the world is that reality is now held

---

comes to a similar conclusion using six strict criteria to define 'science'. According to his (modernistic) criteria, the Greeks were not pursuing science but "natural philosophy" (French 1994, xi). Lindberg 1992, on the other hand, regards the Greeks as incipiently satisfying many of these criteria (though he himself does not use them explicitly). For him, Greek natural philosophy certainly forms *The Beginnings of Western Science*. From the perspective of the history of science, the latter position seems the more tenable.

[13] Heraclitus (ca 500 BC) explicitly states that the logos is a cosmic law which is congruent with the order of the cosmos (Fragments 62 a, 63 b in: C.J. de Vogel I).

[14] See C.J. de Vogel 1960/1964, I, 1,2. In her view the decisive issue is the clear difference between the methods adopted by the 'philosophers' and the *theologoi*, as they were to be called by Aristotle.

to be fundamentally comprehensible through application of rational thought. It follows that humankind is supremely equipped to understand the cosmos. Proceeding from these basic philosophical and scientific premises, the Greeks went on to construct expanding systems of philosophical thought.

The leading schools of philosophy to emerge were undoubtedly Plato's Academy, Aristotle's Lyceum and the Stoics, the 'philosophers of the porch'. These three schools did not evolve without considerable dispute, however. Furley describes the 'cosmological crisis' that was played out during the evolution of Greek philosophy.[15] He argues convincingly that even in the early phases of this evolution two currents of thought can be distinguished, differing in their basic interpretation of the cosmos. Although in the pre-Socratic era the nuances of neither had yet fully crystallized, in Furley's view after the death of Socrates the two can be clearly distinguished.[16] On the one hand we have the theories of the atomists Leucippus and Democritus, later adopted by Epicurus and his followers.[17] These stand over and against the theories of Plato, adopted and developed by Aristotle and the Stoics. Both sides of the conflict identified by Furley thus embraced several schools of philosophical thought. Although there were also obviously many shades of opinion within these schools, the differences that separated them were fundamental. The Atomists rejected the notion that the cosmos was ordered and intelligible, holding that everything was due to accident rather than design. They did not believe the cosmos was unique, but merely one of many existing in an infinite universe (τὸ πᾶν). They rejected teleological explanations of reality in favour of a mechanistic reading. Over and against the continuous they posited the indivisible, which they termed 'atoms'. Finally, over and against creation and the immutability of species they posited the notion of development and evolution.[18]

It was the Aristotelians, in Furley's terminology, who emerged as

---

[15] D.J. Furley 1989, 223-235.

[16] D.J. Furley (1989, 225,226) holds that these two philosophical tendencies remain clearly distinguishable throughout antiquity, with "very little crossing of the dividing line". There were thus few who held eclectic or 'mixed' positions.

[17] Among his followers was Lucretius (c.100–c.50 BC), whose *De Rerum Natura* is one of the few complete works of the Atomists left to us today.

[18] See D.J. Furley (1989, 225 ff.) for a schematic review and a discussion of the differences.

victors in this cosmological dispute.[19] Within our context, the three major philosophical schools of this tendency can be taken as being the most influential. For the purposes of the present inquiry it is important to examine whether there are any major differences between the Aristotelians and the Atomists with respect to their ideas concerning humanity's relationship with nature and with animals. This proves not to be the case, principally because the Atomists had very little to say on these issues.[20] When the term 'dissident' is used in this chapter, the reference is to philosophers holding dissenting views regarding those matters with which we are here concerned.

The influence of the Atomists' ideas on the later western tradition is a topic to which I shall return in Chapter 5.

4.4 Main Schools of Thought

4.4.1 *Plato*

Plato (428-347) viewed the cosmos as an immaculate whole created in perfect harmony by a master-craftsman, the Demiurge. The world in which we live, though, is merely a reflection of the true world, that of ideal Forms.[21] The dual concept of a macrocosm reflected in a microcosm forms a key element of Plato's cosmology. Three levels are distinguished, of which the macrocosm is the most inclusive, encompassing the cosmos as a whole. The second level is that of human society, embodied for Plato in the city state. The third level is that of the human individual, the domain of Plato's anthropology.

Plato perceived in the cosmos an immanent order. Particularly in the *Timaeus*, his key treatise on cosmology, he stresses the motif of order over and against chaos. The ideal world constitutes the perfect and eternal model, of which the world of the senses is merely the best possible reflection.[22] Plato was not enamoured with the

[19] D.J. Furley (1989, 225) considers Aristotle's cosmology the most representative of this tendency and the difference between his cosmology and that of the Atomists most pronounced.

[20] With the exception of Lucretius. His poetical treatise *De Rerum Natura* makes regular reference to plants, animals and nature in general (see Section 4.5).

[21] The famous allegory of the prisoners in the underground cave who see only a reflection (shadow) of reality rather than reality itself is in *Rep.* VII, 514 a–517 c.

[22] Plato *Timaeus* 28c–29a en 29d–30c.

concept of 'nature overall', in the sense of untamed wilderness.[23] It reminded him too much of the intangible, unmanageable forces of chaos, which had no place in his conception of the perfect cosmos. This attitude probably goes a long way to explain his view, expounded through the mouthpiece of Socrates, that "nature can't teach me anything". The same holds for Socrates' remark at the beginning of the Phaedrus: "Now the country places and the trees won't teach me anything, and the people in the city do."[24] This predominant focus on human society is not peculiar to Plato. Greek philosophy as a whole can justifiably be termed humanist, in the classical sense of the word. This should not be taken as constituting a narrow preoccupation with human interests, or disregard or scorn for the other two elements of the 'triangle of meaning' referred to earlier. Greek 'anthropocentrism' or humanism was frequently imbued with no small measure of religious fervour and a keen sense of responsibility for one's fellow citizens and the world at large.[25] The fact that human beings are the only animals that walk upright, so we read in the *Timaeus*, shows that we are divine beings, for our head, the seat of intelligence, links our divine aspect, our soul, with heaven.[26]

Plato's preoccupation with transcendent reality, the world of ideal Forms, did not withhold him from having astute opinions on 'this world'. Among other instances, this is reflected in his accurate description, in the *Critias*, of the role played by hillside forests and soils

---

[23] In Plato's view the plants were made by the gods for the convenience of humankind. Certain originally wild plants were 'trained' and domesticated by humankind (*Timaeus* 77a). It is because the higher powers created the plants for us inferior beings that we have also been equipped with a digestive system (*Timaeus* 77c,d).

[24] Plato *Phaedrus* 230d.

[25] In the contemporary environmental debate the term anthropocentrism generally has a negative connotation. It is associated with 'nature-unfriendliness' and is often traced back to Greek humanism. In this context the sophist Protagoras ("Man is the measure of all things") is often cited as the chief culprit—to my mind wrongly, for Protagoras is one the last thinkers to come to mind as representing Greek humanism. In his own day his views met with resistance and according to some (unsubstantiated) traditions he came to a sorry end: his books were burned because they were held to be godless. The crucial quotation is often interpreted rather one-sidedly, moreover, as expressing unmitigated pride (for a better explanation, see for example Bakker's introduction to an edition of the works of the Sophists, 1986). For a good review of the historical background of humanism, see C.J. de Vogel 1968 and for the relationship between humanism and the environmental problem, H. Achterhuis 1990 and W. Achterberg 1992.

[26] Plato *Timaeus* 90a.

in the hydrology of lower-lying areas.[27] It is even more clearly visible in his vision of human society, as described in *The Republic* and *The Laws*. In these two works Plato discusses the Greek city state, the *polis*, and how it should be governed. Democracy does not come off well, for Plato believes in aristocracy and meritocracy.[28] These works contain more that is of relevance for the subject of our inquiry, however. In both, Plato displays his appreciation of several crucial parameters in humanity's relationship with the natural environment, namely: the size of the human population, the scarcity of natural resources and the availability of fertile land. Plato is a champion of sobriety, holding that prosperity and abundance keep humankind from attaining true happiness, which is purely spiritual.

To resolve these issues, Plato is not loathe to adopting crass measures. The size of the local population, so he holds, should be tightly controlled by a body of wise men, both qualitatively, in the form of rules regarding sexual relations, and quantitatively, through infanticide. He holds in disdain the commercial, materialistic, urban democracy of Athens. In his view, the ideal society should function as an agricultural, artisanal, small-scale, autarchic economy.[29] For human society, at least the ideal society apparently conceived of by Plato, nature has an innate order and stood in the service of humankind.

---

[27] Plato describes how the Acropolis and the hills around Athens had become denuded, their fertile soils washed away by prehistoric deluges and earthquakes (*Critias* 111, 112). He does not explicitly blame the situation on unsound agricultural practices or deforestation by man, however. Plato was not denouncing anthropogenic erosion, although this suggestion is sometimes made in contemporary historical studies (e.g. C. Ponting 1991, 96 and I.M.B. Wiman 1990).

[28] Although in some respects Plato was more 'conservative' than other Greek philosophers, in other areas he proves remarkably 'progressive' in his views. He advocates the same education being given to men and women, for example. If the training of women is neglected, he holds, the state will be diminished by half (*Leg.* VII, 804d—805b). Given some of Plato's other pronouncements concerning women, this is rather surprising. This example already demonstrates that modern labels like 'conservative' and 'progressive' are not appropriate in such a context.

[29] G. Koolschijn 1990 characterizes Plato's views on the ideal society as a recipe for a "green dictatorship", thus linking them neatly to the contemporary context, although I feel he is interpreting Plato too much in a particular (i.e. modern) light. The term green dictatorship is generally used in reference to views whereby human interests are rendered subordinate to environmental and conservation interests (*cf.* Section 5.4). Nowhere in Plato's writings is there any indication that he was concerned with 'green' interests in the modern sense of the term.

4.4.2 *Aristotle and Theophrastus*

In the cosmology of Aristotle (384-322) the term 'nature' is encountered in both its senses. Aristotle's writings can be divided into three basic categories. His metaphysical works are concerned with general philosophy, while his philosophy of nature is to be found both in his works on physics and in the third category: his extensive writings on biology.[30] No sharp dividing line can be drawn between these various works, though, as Aristotle does not always restrict his discussion of a particular topic to a single category of writings. His metaphysical, physical and biological works indeed reflect an impressive unity of thought.[31]

In Aristotle's view, a phenomenon can only be truly known if we know its cause. In his *Metaphysics* he explains the difference between experience (ἐμπειρία) and art or craft (τέχνη) on the one hand and wisdom (σοφία) and knowledge (ἐπιστήμη) on the other. Experience and craft can only teach us the 'fact' (ὅτι), while wisdom and knowledge can lead to an understanding of the 'wherefore'(διότι).[32] Aristotle is intrigued by the phenomena of growth and motion, and these are important concepts in his philosophy of nature.[33] By analysing motion and change, he contends, the underlying principle of change can be revealed and thus the ultimate cause of reality. Motion is borne out of motion, but the chain of cause and effect stops at the source. Aristotle therefore calls the supreme deity from which everything originates and which first set everything in motion the "first unmoved mover". The deity moves only as the final cause, himself doing nothing, merely setting others and the other in motion.[34] In Aristotle's thinking, the godhead is not conceived of as a (biblical) creator. Aristotle takes matter to be eternal and his supreme deity does not know his world, moreover, but only himself.[35]

---

[30] Not all his works are to be considered authentic, however, in the sense of being personally written by Aristotle (*cf.* note 44).

[31] C.J. de Vogel II, p67.

[32] Aristotle *Metaphys A*, 1, 981 a 5-30.

[33] What, for Aristotle, is φύσις? It is a "principle of motion" *Physica* II 1, 192 b 20 (C.J. de Vogel II, fragment 481 p76).

[34] Aristotle *Physica* VIII 10, 267 a21—b9 and *Metaphys* Λ 7, 1072 b 13-14. A nice simile to illustrate this principle is the effect that an attractive woman may have on nearby men. Even though she may not herself make a move, she can still 'get the men going'.

[35] Aristotle *Metaphys* Λ 7, 1072 b14-30. On this point Thomas of Aquino interpreted Aristotle as follows: "In a direct sense the divine spirit knows only itself,

Growth, as perceived in nature, is a very special form of motion. It is through growth that organisms develop; they grow to fulfil their purpose, 'actualizing' the potential they already hold. For this phenomenon Aristotle employs the term ἐντελέχεια (imbued with innate purpose). It is thus the innate purpose (τέλος) of an organism that constitutes the active force directing its growth. Nature ('the world') has a purpose that it actualizes in accordance with its innate nature ('potentiality'). Aristotle was as convinced as Plato of the inherent order of the cosmos. However, his main concern was with physical nature itself rather than the supreme deity from which it ultimately originated. Nature herself functions in the same way as Plato's demiurge to create order. Nature chooses the best of all possibilities.[36] According to Aristotle growth, motion and change do not give rise to unpredictability but precisely to a functional, hierarchically ordered universe in which everything has its fixed place and purpose. "Nature never creates anything 'in vain'" is a phrase encountered frequently in the writings of Aristotle as well as in those of Theophrastus (*c.* 370-286), who might be termed a landscape ecologist *avant la lettre.*[37]

Aristotle is particularly intrigued by organisms and phenomena that appear to be at odds with the natural order and with the systematics he has developed. He devotes an entire treatise to the ostrich, for example, a bird seen to be a bird but clearly with a number of non-avian characteristics.[38] Aristotle and Theophrastus are very clear about the purpose of nature and the hierarchy embodied therein. The purpose of a plant is to yield fruit and set seed. In a sense, wild fruit-trees fail in their purpose, because their fruit is inferior.[39] This fact in itself betrays the hierarchical order: plants were made for the sake of humankind, either directly, or indirectly as food for animals.[40] The ultimate purpose of the whole of creation and the

---

but in Himself God knows all things" (paraphrased from a quotation in C.J. de Vogel II, p99).

[36] Aristotle *On the Heavens* II, 5, 288a.

[37] The characteristic phrase for "nothing in vain" is οὐθὲν (or οὐδὲν) μάτην, encountered, for example, in Aristotle's *On the Heavens* I 4, 271 a 33 and Theophrastus' *De Causis Plantarum* II, 1, 1, 10.

[38] Aristotle *On Parts of Animals* IV, 697 b 14-26.

[39] Theophrastus *De Causis Plantarum* IV, 4, 2, 3.

[40] A classic *topos* for Aristotle's anthropocentrism is the passage in *Physics* (198 b 16-199 a 8) in which he describes how rain falls on the earth to allow human crops to grow. In the debate on the anthropocentric character of this and other

earth on which we live is to serve humanity, a notion that was widely held in antiquity.[41]

Although Aristotle has a formidable knowledge of the contemporaneous natural world, and particularly of wild animals, that is lacking in Plato's works, nowhere does he allude to the 'intrinsic value' of nature. On the contrary, in one treatise ascribed to Aristotle we read: "Whenever animals gain perfection in their nature, they become tame. Utmost perfection is always tame rather than wild. That which is deficient or not large of stature is generally wild."[42] This is a remarkable idea, which though not found in this wording in Aristotle's classic biological works, should nonetheless not be regarded as 'un-Aristotelian'. It is perfectly feasible that Aristotle, as a knowledgeable and admiring observer of nature and particularly the wild fauna of his region, placed the domesticated animals further up the ladder in his highly hierarchical approach.

Aristotle's systematics were not restricted to the animal world. Human beings also have their allotted place in the *scala naturae*, with (Greek, male) philosophers at the top and slaves and barbarians at the bottom (*cf.* Figure 1, Chapter 5).[43] Aristotle held that slaves were disposed by nature to carry heavy burdens: to this end nature had provided them with strong, broad shoulders. It is therefore not only permissible to treat them as slaves; it is in fact imperative, in order for them to realize their τέλος, their natural purpose.[44] Here, for the first time, we encounter the notion of nature as the norm, later elaborated to become one of the central principles of the Stoics. The fact that Aristotle took nature as the standard does not mean that he held every observed phenomenon to be 'natural', however. In the *Problemata*, for example, he states that human beings generally

---

teleological passages in the works of Aristotle, David Sedley has subjected this and other passages to a close analysis and concludes that Aristotle's teleology is evidently anthropocentric in character (D.N. Sedley 1991).

[41] See, for example, *Politics* I, 1256 b 15-25 and Pliny's *Natural History* VII, 1. Dissenting views were held by the Stoics and the Epicureans.

[42] L.S. Filius 1989, vol. 1, 165.

[43] The view that human beings occupy the top rung of the ladder recurs frequently in Aristotle's works, for example in the *Problemata Physica Arabica* XI, 60 (L.S. Filius 1989, vol. 1, 37). Filius holds that Aristotle's ideas on this point can be traced back to Hippocrates, but that the Arabic translator may also have been influenced by Galenus, who lived many centuries after Aristotle. *Cf.* P. Moraux 1984, 742.

[44] Aristotle *Politics* I, 1254 b 25-40 and 1255 a 1-5.

give birth singly.[45] The fact that humans give birth to twins is anomalous, all the more so if these are of different sex. "In the case of such twins being one male and the other female, then, it is far more obvious to assume that the phenomenon is unnatural".[46] This passage reflects the fact that Aristotle, too, holds certain views about the 'natural order', which he subsequently employs to classify the phenomena he observes.

4.4.3 *The Stoics*

The Stoic philosophers derive their name from the Athenian colonnade where they originally met and taught.[47] Three periods are generally distinguished:

- The early Stoics (*c.* 300-150 BC), including Zeno, Cleanthes and Chrysippus.
- The middle Stoics (150-50 BC), including Panaetius, Posidonius and Cicero.
- The late, or Roman Stoics (25 BC – *c.* 180 AD), including Seneca, Epictetus and Marcus Aurelius.

Given its long history and the many philosophers that played their part in it, Stoic thought has been shaped by a variety of sources. Its most original thinkers were the school's founder Zeno of Citium and Chrysippus, but for the later European tradition it was above all the Roman writers Cicero and Seneca that served as conduits and part-moulders of Stoic thought. Stoicism has had an important influence on the western tradition, although on this point too there is a wide variety of opinion. Hahm perhaps goes furthest in stating: "for half a millennium Stoicism was very likely the most widely accepted worldview in the western world.... it appealed to all classes.....it left its mark on Christianity". Guthrie and Green are more nuanced in their conclusions.[48]

---

[45] Reconstruction of authorship indicates that several works including the *Problemata* may have been written by the Aristotelian school. Although such works may evidently report the words of Aristotle himself, they also contain ideas based on his views as well as the views of others that may have diverged far further from their origins.

[46] L.S. Filius 1989, vol. 1, 158.

[47] The *Stoa Poikilè*, or Painted Colonnade.

[48] D.E. Hahm 1977, xiii; W.K.C. Guthrie 1965, vol. II and the more general study by P. Green 1990, particularly Chapter 36.

The Stoic philosophers saw the world as an ordered and logical whole. In their cosmology the logos plays a key role and with the Stoics the concept acquires something of a religious character. The idea of the logos as a cosmic law can be found earlier but the Stoics put it into a coherent scheme.[49] Because human beings are imbued with the logos, they are able to fathom the divine, the logos that governs the universe.[50] Thus equipped, the human race in principle already participates in the divine, but at the same time his logos is also the ideal channel for increasing his understanding of God. For this reason, human beings are the most perfect of God's creatures.[51] Like earlier philosophers, the Stoics inquired about the relationship between 'the good life' and 'nature' and their answer was clear: the good (moral) life is lived in accordance with (human) nature. Since reason (logos) is the defining human characteristic, to live according to one's reason is to live in accordance with the cosmos.[52] This is elaborated on by Cicero in the *De Officiis*. In his opinion, when the Stoics speak of the highest good ("summum bonum"), of "living in accordance with nature" ("convenienter naturae vivere"), they mean that life should always be lived in accordance with virtue. Of all the other things that are in harmony with nature ("secundum naturam essent") only those should be chosen that do not conflict with virtue ("si ea virtuti non repugnarent").[53] In the Stoics' philosophy, the cosmos, nature and the good merge, with the caveat that 'living according to nature' for them largely meant living according to *human* nature.

Human beings, so the Stoics taught, are driven first and foremost by an urge for self-preservation and therefore attach greatest value

[49] Heraclites is certainly one of the sources, see: De Vogel I, fragment 62a p29

[50] See M. Pohlenz 1978, K. Algra 1988.

[51] See, for example, Cicero *De Natura Deorum* II, 140-153. Among the Stoics, it was only Cleanthes who stopped short of such absolutism, stating that it is not mankind but only God that is perfect. At the same time, though, he was equally convinced that man is superior to all other earthly creatures both physically and spiritually (SVF I, 529).

[52] In the Stoics' terminology: living not only κατὰ φύσιν but also κατὰ λόγον/κόσμον, interpreted to mean living in accordance with nature (ὁμολογουμένως τῇ φύσει ζῆν, Diog. Laert. VII 86-87). Here again Heraclites may have inspired the Stoics for the expression ποιεῖν κατὰ φύσιν can be found in his work as a moral principle see: De Vogel I, fragment 62b p29.

[53] Cicero *De Officiis* III, 13. Elsewhere he also refers explicitly to this Stoic terminology: "Quod cum positum sit in eo quod ὁμολογίαν Stoici, nos appelemus convenientiam, .." (*De Finibus* III 6, 20-21).

to what is most appropriate, or most 'akin' (οἰκεῖος) to them. With the Stoics the question of the kinship of human beings as well as that among humans, animals and plants, an issue that had occupied Aristotle and others before, gains a moral dimension. Aristotle and Theophrastus had observed that many qualities were shared by all people, whatever their way of life, as well as by both human beings and animals. Endowed with reason (logos), human beings obviously distinguished themselves from the ἄλογα, the irrational beasts, and nature had an evidently ordered hierarchy, yet neither Aristotle nor Theophrastus combined these positions to form a coherent ethical system. The Stoics did just this, distinguishing objective, biological phenomena of relatedness (οἰκειότης) from a perception or feeling of kinship (οἰκείωσις). The latter extends by way of what is most akin to oneself (one's own body, offspring and family) to encompass all human beings. It becomes a principle for determining ethical bounds. With animals there is no οἰκείωσις and they are therefore not part of the moral universe.[54] The logical conclusion of this line of reasoning is that animals cannot be 'harmed' in any moral sense of the word.

The Stoic philosophers held that nature, and particularly animals, exist for the benefit of humankind, and were in fact created specifically for that purpose. They based this position on two arguments: the teleological argument that the usefulness of animals is to be seen as reflecting divine intention, and their views on humanity being endowed with a logos.[55]

This thinking in terms of distinct moral domains is rooted in the Stoic doctrine of appropriateness. Theirs was in fact an extremely well-defined moral code, with a sharp dividing line between human and non-human beings, but moral equality among humans. The human community was maximal: besides women it also included slaves and barbarians, a remarkable position in that era. In taking this stance their practical influence was only modest, however. The Stoics' prime interest was in the human condition, leading Colish to state that they did not in fact have a philosophy of nature but rather a philosophy of humanity, even though they wrote extensively

[54] Cicero *De Finibus* III, 20, 67. Elsewhere Cicero writes that animals do not belong to the "civitas deorum atque hominum" (*De Leg.* I, 7, 22). *Cf.* C.O. Brink 1956 for an illuminating discussion of the moral implications of the notion of relatedness.

[55] U. Dierauer 1977, 238-245, particularly 241.

on nature.[56] This accords entirely with the Stoic doctrine that the cosmos was created for the sake of humankind.[57] If the universe was created for the human race, then everything exists to serve humanity, whether directly or indirectly. This view was shared by almost every school of Greek philosophy.[58] The Stoics, however, did dwell on it explicitly and frequently. The belief led to astute and imaginative debates on the purpose of a fly for humankind. That which stands lower on the ladder of creation derives its *raison d'être* from that which is above, it was held. Everything on earth must therefore serve some purpose, and discovering the nature of that purpose leads ultimately to a usefulness for humankind. Nature does nothing in vain, without a purpose.[59]

These ideas may justifiably be termed anthropocentric and utilitarian. However, in line with my earlier remarks (e.g. note 25) on the term 'anthropocentric', the idea of usefulness/purpose should also be interpreted in a broad sense. It also has an ethical dimension, for example, and contemporary thinking should not allow it to become restricted to the purely economic sense of 'utility'.

4.5 Dissidents?

For all their differences, in terms of their ideas about humanity's place in the cosmos, humanity's relationship with nature and, in particular, the dividing line between human and beast, there is still evidently a basic consensus among the philosophical schools considered in Section 4.4. We shall now elaborate on this consensus with reference to the various motifs that concern us here. In the first place, it can be queried whether this consensus extended beyond the principal schools reviewed. Were there no undercurrents of thought, no dissidents worthy of the name? With our brief review of just three major schools, have we not done injustice to the rich tapestry of philosophical beliefs developed by the Greeks? From our present perspective this is not the case, I venture, certainly when these beliefs are measured against their *Wirkungsgeschichte*. In the evolution of

[56] M.L. Colish, personal communication, May 1989.

[57] Cicero *De Natura Deorum* II, 154-162. *Cf.* SVF II 1152 f (Chrysippus) and Diog. Laert. 7, 138 (Poseidonius).

[58] "Der Gedanke, daß Gott die Tiere um der Menschen willen erschaffen habe, liegt also durchaus auf der Linie griechischen Denkens" (U. Dierauer 1977, 240).

[59] Aristotle *On the Heavens* I 4, 271 a 33.

western thought the influence of Aristotle, particularly, can hardly be overestimated. And in terms of their conceptual ideas, the less influential schools are found to entertain no fundamentally anomalous views on humanity's relationship with nature and the hierarchical position of human beings relative to the animals. In this respect the neo-Platonists continued to follow the line laid down by their master, and neither can the Atomists or Sceptics in any way be deemed dissidents.[60]

The same holds for the Epicurean Lucretius (*c.*100-*c.*55 BC) Although in his poetic treatise *De Rerum Natura* he takes issue with the Aristotlean and Stoic view that nature was divinely created for the benefit of humankind, this does not bring him to an explicitly 'nature-friendly' attitude. Like the Atomists, Lucretius did not believe in a godhead or Creator. He held, moreover, that everything in nature points to her being 'human-unfriendly': that much is clear from that fact that human beings are born so helpless and naked. The existence on earth of unyielding and inhospitable places, with their obstinate weeds and menacing wild beasts, contradicts the notion of anthropocentric teleology.[61] Although his beliefs certainly provide certain grounds for a more benign attitude towards nature, Lucretius does not elaborate his anomalous views on this point. His main concern is atomistic cosmology, and it is from that angle that he takes issue with the anthropocentrism of the mainstream philosophers, nature invariably being brought in merely by way of argument in the ensuing anthropological debate. Lucretius, in other words, was putting the 'privileged position' of humankind into some kind of perspective by emphasizing nature's harshness.[62]

There are, nonetheless, certainly a number of Greek philosophers who differed in their beliefs regarding nature itself. The most renowned is Pythagoras, whose name is embedded in our collective

---

[60] Although the Atomists and Sceptics differed in the cosmologies they developed, none of these philosophers elaborated on man's relationship with nature in any detail. Although in their writings the Sceptics make reference to 'wild animals', for example, they do so merely as an argument in discussing providence. To their mind, the existence of wild animals showed there was no such thing as providence, but they were not interested in animals as such. Although a 'sceptic' tone can still often be discerned in western culture, the influence of the Sceptics of antiquity is in fact rather marginal.

[61] Lucretius *De Rerum Natura* 5, 195-221. For a similar Epicurean criticism *cf.* SVF II, 639.

[62] *Cf.* U. Dierauer 1977, 273-279.

psyche thanks to the theorem ascribed to him. He is certainly to be numbered among the 'dissidents' and is in many respects indeed the most original and anomalous of them all. Unfortunately, his life is largely clouded in mystery. Although firm dates of his birth and death are lacking, he is generally held to have lived in the sixth century BC. He is cited by his contemporaries, but such quotations appear only in later sources. Our first direct reports of the scientific principles of the Pythagoreans are the fragmentary texts of Philolaus, probably dating from the fourth century BC.[63] For Pythagoras' own beliefs we are wholly dependent on these fragments, on Diogenes Laertius' collected writings on the lives of the Greek philosophers and on the treatises of Iamblichus and Porphyry on the life and doctrines of Pythagoras. The latter writings are to be dated around the third century AD, some eight centuries after Pythagoras' own lifetime. It may be queried whether these documents constitute reliable records of the philosopher's own beliefs.

In his seminal standard work on Pythagoras, W.Burkert argues convincingly that even in the classical era it was debated whether Pythagoras had himself authored any writings. From the third century AD onwards, documents began to appear that were ascribed to Pythagoras, but none of these earned and retained a status of authenticity.[64] In pursuing the true beliefs of Pythagoras, Burkert assigns decisive importance to Philolaus. He dates the fragmentary texts ascribed to Philolaus to the pre-Socratic era and, with others, considers these to be our earliest source of Pythagorean beliefs.[65] Since its publication, the evidence presented by Burkert in his description and reconstruction of Pythagorean thought has not been seriously disputed.

Pythagoras himself lived with a circle of followers in a religious community in southern Italy. These Pythagoreans lived a rather sober and cheerless life.[66] They were taciturn and abstained from a wide variety of foods, including meat and beans. The community's taboo

---

[63] C.J. de Vogel I, pp14-15.

[64] W. Burkert 1972, 218 ff.

[65] W. Burkert 1972, 277.

[66] In the fourth century BC "Pythagoristae" begin to appear in Attic comedies, figures conspicuous by their anomalous lifestyle and appearance. They clothed themselves in a single garment, rarely laughed, did not wash and therefore probably smelled unpleasant, drunk only water and were vegetarian. In their values and standards they clearly stood apart from the rest of Greek society, being more akin to early Christian monks or hermits (J.N. Bremmer 1992, 205).

on meat consumption is considered related to Pythagoras' belief in metempsychosis, the transmigration of souls: in animals slaughtered for consumption there may reside the soul of another human being, a deceased relative or a friend. In the doctrine of metempsychosis Dierauer sees "eine gleichsam naturphilosophisch-biologische Erklärung für die ewige Kontinuität des Lebens" and he concludes: "In der rein naturphilosophischen Seelenwanderungstheorie wird kein prinzipieller Vorrang der menschlichen gegenüber der tierischen Existenz sichtbar".[67] Iamblichus gives a similar argument: our affinity and fraternal relationship with the animal community.[68] If Iamblichus' picture of Pythagoras and his followers is reliable, we must assume that the debate regarding the kinship between human beings and other creatures, and its implications for our treatment of animals, was already being conducted in an early chapter of Greek philosophy.

In this respect the moral universe of the Pythagoreans extended further than that of most subsequent Greek philosophers. Individually, too, Pythagoras himself is held to have entertained a radically different relationship with wild animals. He is said to have moved an eagle to fly down, allow itself to be caressed and then fly off again.[69] In another episode he is said to have 're-educated' a dangerous female bear by stroking it and feeding it with barley bread and acorns and moving it to swear never again to harm a living being.[70] This kind of authority over wild animals is a characteristic element and motif of the stories told of saints.[71] There are unmistakably moral overtones here, for the behaviour of animals is often judged and even condemned on implicitly ethical grounds. "Even on the unintelligent beasts his words had a corrective and exhortative effect," writes Iamblichus. Among the Greek philosophers there are few others who adopted a similar attitude towards nature. Such a stance was regarded as unphilosophical, moreover. Burkert, who describes this 'dissident' viewpoint at length, considers it to be at the

[67] It should be remarked that Dierauer (1977, 21-22) states that this "Homogenität aller Lebewesen" does not preclude man from occupying a special niche at the apex of the hierarchy. Only in the case of human beings can the soul undergo an "Erlösung aus dem irdisch-kreatürlichen Dasein".

[68] Iamblichus *Vita Pyth.*, 168.

[69] Iamblichus *Vita Pyth.*, 62.

[70] Porphyry *Vita Pyth.*, 21.

[71] A. Nitschke 1966. As with Francis of Assisi, for example; *cf.* H. Nolthenius 1988, R.D. Sorrell 1988 and J.J. Boersema 2002.

very heart of the philosopher's (historical) significance. It is perhaps on the wrong grounds that Pythagoras has his place in our collective psyche.

Summarizing, Burkert writes: "The tradition of Pythagoras as a philosopher and scientist is, from the historical point of view, a mistake. But the fascination that surrounded, and still surrounds, the name of Pythagoras does not come, basically, from the specific scientific connotations or from the rational method of mathematics, and certainly not from the success of mathematical physics. More important is the feeling that there is a kind of knowing which penetrates to the very core of the universe, which offers truth as something at once beatific and comforting, and presents the human being as cradled in a universal harmony. In the figure of Pythagoras an element of pre-scientific cosmic unity lives on into an age in which the Greeks were beginning, with their newly acquired method of rational thought, to make themselves masters of the world, to call tradition into question, and to abandon long-cherished beliefs. The price of the new knowledge and freedom was a loss in inner security; the paths of rational thought lead further and further in different directions, and into the Boundless. There the figure of the ancient Sage, who seemed still to possess the secret of unity, seemed more and more refulgent. Thus after all, there lived on, in the image of Pythagoras, the great Wizard whom even an advanced age, though it be unwilling to admit the fact, cannot entirely dismiss".[72]

This figure of the sage, blessed moreover with a special gift in his dealings with nature, was never again to emerge in this form in Greek philosophy.[73]

With the passing of Pythagoras, dissenting viewpoints on the subject of humanity's relationship with the animals did not vanish entirely, however. Plutarch (*c.* 40-*c.* 120 AD), for example, in the *Moralia*, says of the shepherd poet Bion (*c.* 100 BC) that he stood up in defence of animals: "Boys stone a frog in sport, but the frog dies in earnest".[74] In a footnote to this passage the translators Cherniss and Helmbold write: "Bion and Xenocrates were almost alone among the Greeks in expressing pity for animals". Both have unfortunately fallen into obscurity as defenders of 'animal rights'. Whether this

[72] W. Burkert 1972, 482.

[73] This also applies to the beliefs and habits described above, which only reemerge in the ascetic tradition of the Essenes and early Christians.

[74] Plutarch *Moralia* 965 B.

is regrettable is hard to judge. We know too little of Bion to form any real impression of his beliefs and influence. Although we know more about Xenocrates (396—314 BC), who was prominent as Plato's second successor at the Academy, nothing of his (anomalous) views regarding animal life is reported in major contemporary philosophical reference works.[75] De Vogel however, cites Aelianus (*Var. Hist.* XIII 31) who calls Xenocrates friendly (φιλάνθϱοπος) to the the animals.[76]

Rather surprisingly, the translators do not mention Plutarch himself, who should certainly be included in this short list of 'animal-friendly' philosophers. Plutarch is also the only dissident whose beliefs have had a demonstrable influence in later ages, on the French writer Montaigne among others. This influence relates specifically to the question of humanity's relationship with the animals. Plutarch's main work, the *Moralia,* contains three essays on the topic. The first, and longest, deals with the shrewdness of terrestrial and marine animals (*De Sollertia Animalium*). The second is concerned with rationality among the animals (*Bruta animalia ratione uti*) and the third with the consumption of meat (*De Esu Carnium*). In the first two essays Plutarch argues against the prevailing Stoic belief that all animals are irrational. He also strikes out perceptively at those fools who hold that animals know no feelings of joy or fear and are merely 'feigning'.[77] The moral dimension of what is defensible in our dealings with animals, and what not, is certainly an issue for Plutarch. At the same time, though, his reasoning remains firmly within the conceptual framework of the Stoics and the decisive question is thus whether animals are endowed with logos, the sole basis for any moral obligations humankind might have towards them. Plutarch endeavours to convince the Stoics that a great deal of rationality can indeed be discerned in animals and in animal behaviour. Recognition of this fact would bring the animals within the moral compact and make them a subject of moral concern.[78] In principle, Plutarch does not distance himself from the belief that moral concern should be based on common possession of rational faculties. He is certainly not a

[75] In the authoritative *Der Kleine Pauly* and in Guthrie 1965, for example.

[76] C.J. de Vogel II, fragment 769 p282.

[77] Plutarch *Moralia* 961 E,F. Although Plutarch is here addressing the Stoic Chrysippus (SVF, ii p. 240), among others, Descartes might also have benefited from his remarks.

[78] A good example of this line of reasoning is to be found in Plutarch *Moralia* 976 C.

'biocentrist' in the sense that he held non-human life in undivided respect. "There is surely no injustice," he writes, "in punishing and striking down animals that are anti-social and can inflict injury on man, nor in taming animals that are amiable to man, certainly not if we can make them our helpers in the purpose for which nature has equipped them".[79] In the latter phrase the spirit of Aristotle shines through.[80] With Plutarch the difference between domesticated and wild (and harmful) gains a moral significance. In *On Eating Meat* he writes: "We do not eat lions or wolves out of self-protection; no, we leave them in peace, preferring to slaughter innocent creatures lacking harmful spine or fang, creatures which, I vow, have been brought forth by nature for their beauty and their grace".[81] This latter, aesthetic argument clearly demonstrates that, even though Plutarch's views are essentially utilitarian, they are evidently open to very broad interpretation. Plutarch, too, holds that nature creates nothing without a purpose, but for some elements of nature beauty, and implicitly humanity's enjoyment of that beauty, is the very purpose of their existence and consequently their utilitarian value. It is still evidently an anthropocentric viewpoint. For Plutarch there are at least three (moral) categories of animals: harmful pests that may be treated as one pleases, harmless creatures that should be left alone, and harmless creatures that may be used by human beings, but certainly not consumed as food. In his eyes, human beings are not carnivorous. The consumption of meat is not based in any measure on considerations regarding our 'true nature'. If that were so, nature would have equipped us to kill other animals. Plutarch challenges his antagonists to use their teeth to rend a lamb asunder and consume it raw, as true carnivores do. With this line of reasoning, he follows the Stoic tradition of taking nature as the norm. In debating the issue with the Stoics, in this diatribe he again employs their own arguments to oppose their consumption of meat. "Of course human beings have no compact of justice (δίκαιον) with the irrational animals (ἄλογα).[82] But with perfume such is also absent. Refrain,

[79] Plutarch *Moralia* 964 F. Similar views are also expressed by Plato *Republic* 352 E.

[80] Aristotle *Politics* I, 1255 a 1-5.

[81] Plutarch *Moralia (De Esu Carnium)* 994 B.

[82] The translators here give the word οἰκεῖον as a textual variant on δίκαιον, rendering even more explicit the link with the Stoics.

therefore, from animals too.."[83] To a point Plutarch follows their reasoning, but his conclusions are entirely different. Based on his beliefs regarding (the nature of) humanity, in nature he sees greater scope for moral nuance than the Stoics.[84] Plutarch's beliefs can in fact be justifiably termed animal- and nature-friendly and at the same time anthropocentric. His writings exude a clear emotional love of animals, but never at the expense of the human race.[85]

In the debate on meat consumption, the neo-Platonist Porphyry (*c.* 250 AD) also deserves mention. In his work *De Abstinentia* (On Abstinence) he also seeks to establish moral grounds for vegetarianism. Perhaps we may kill harmful animals, he holds, but with most animals we have a binding relationship. In contrast with Plutarch, then, in Porphyry we have an explicit rejection of the Stoic doctrine of appropriateness.[86] Porphyry is concerned not so much with 'animal welfare' as with how humanity should lead a virtuous life. He has no wish to impose vegetarianism on the masses, but rather on philosophers. As a "man of virtue", the philosopher should aspire to higher moral standards.[87] For Porphyry, the crux of the matter is human affinity with animals and he is thus pursuing the same motif as that with which Aristotle and Theophrastus had already grappled and which the Stoic philosophers had developed into a pivotal dogma. Despite his extensive writings, Porphyry gained only limited recognition in later eras.

---

[83] Plutarch *Moralia (De Esu Carnium)* 999 B.

[84] In this regard Plutarch certainly does not stand alone. The distinction between 'domesticated', 'wild' and 'harmful pest' has moral significance in many cultures (*cf.* Chapters 2 and 3 and also, *inter alia*, R. Bulmer 1967; C. Houtman 1982; S. Atran, 1990; S. R. Kellert, 1984, 1989, 1996 and J.J. Boersema, 1994, 2002).

[85] His love of animals is also reflected in the following quotation, ascribed by Plutarch to the Roman statesman Cato and translating freely as: "One should not treat living beings as one does shoes or tools, which one discards when they are broken or worn out" (Cato the Elder 5, 339c, translated from citation in Dierauer 1977, 292). At the time of writing (2001) here in the Netherlands and western Europe , where an epidemic of foot and mouth disease has given us the tableau of lethally-injected cows, pigs, sheep being 'processed' in an industrial incinerator or a pyre, this would appear as pertinent as ever. At the time I wrote my thesis in Dutch (1997) the same sad story for swine fever had been told.

[86] Porphyry *De Abstinentia* III, 19-25, who takes this passage explicitly from Theophrastus.

[87] Porphyry *De Abstinentia* IV, 18.

4.6 Summary

The ideas discussed can be summarized as follows. The principal Greek philosophers saw the cosmos as a logical, ordered and immaculate whole. With the logos with which they has been endowed, human beings are able to fathom that order: the world can be grasped and understood. The order is a linear hierarchy, the *scala naturae*, and the human race stands at the apex above all other living creatures. Within the natural order, 'inferior' beings derive their value from their usefulness to those higher up on the ladder, but 'usefulness' is to be interpreted in a broad sense. Nature does not have any intrinsic value. Possession of logos is taken as the yardstick for affinity and moral obligation. There is a crucial difference between the value assigned to tame, domesticated nature and 'harmless' nature on the one hand, and wild, 'harmful' nature on the other. Among a handful of Greek philosophers, this difference gains moral significance. These 'dissidents' distinguish themselves precisely in their beliefs regarding human attitudes and deeds towards the category of 'harmless' nature. In this context, Pythagoras adopts the most anomalous position and in certain respects conforms more to the tradition of the (early Christian) saints than that of the Greek philosophers.[88]

4.7 General Motifs in Humanity's Relationship with Nature

To conclude this chapter, let us attempt to put these specific philosophical beliefs into a broader perspective by tracing several general motifs in humanity's relationship with the natural world. We shall then return to these motifs in the following, concluding chapter.

In the cultures of antiquity, humanity's outlook on nature was determined largely by whether or not nature was perceived as benevolent to human beings. Wild, untamed nature was, in principle, unpredictable and often threatening. Besides inspiring awe and wonder, such chaotic elements of nature that were inimical to human beings, and thus 'anti-social', were also to be combated. Not only was such action sanctioned; it was indeed mandatory. This

[88] A similar conclusion was also drawn by Burkert, on different grounds, in an article in 1960.

pivotal conviction can be tied in with a number of more general, cultural-historical motifs.

The most general of these is the civilization motif. In the active variant, conquering the (forces of) nature is regarded as a sign of civilization; in the passive, more descriptive variant, it is held that where (the influence of) untamed nature is driven back, civilization can take hold. In an article on the history and background of deforestation, one writer has summarized this outlook as follows: "Everywhere where cultures take shape, the forests vanish".[89] With many Greek philosophers, the civilization motif also gains a moral dimension: it is humanity's moral duty to actively combat nature when it encroaches on the human domain.

In this process of civilization, there was a special role for 'culture heroes' and for kings.[90] We see the culture hero motif perhaps most clearly expressed in the Greek hero Heracles. His labours against well-nigh invincible mythical monsters symbolically protects the motherland from threat and in later eras sets an example for others who do battle with 'real' wild animals.[91] Greek kings like Alexander and particularly the later Roman emperors take the lead in 'conquering' nature. With the Mesopotamian, Egyptian and Greek rulers, this regal motif does not yet have particularly dramatic consequences for nature. Animals are kept in captivity, they take part in victory parades and certain species are hunted. Lion-hunting, in particular, is a regal affair. The Roman emperors go significantly further, however. From the regions their armies conquer they not only seize slaves and wild animals, as a sign of their absolute command over both human beings and beasts; they also intensify their hunting of certain animals. For some species, the consequences are disastrous.

In the Mediterranean basin the extinction of wild mammals gave the Pax Romana an extra dimension. The 'games' (*ludi*) in the amphitheatres were a major factor in this respect. Here, in the *venationes*, the fights with wild animals, thousands of animals died a

[89] From an article in the September 1989 science supplement of the Dutch newspaper NRC/Handelsblad. In the words of K. Thomas (1983, 25): "Human civilization indeed was virtually synonymous with the conquest of nature".

[90] On the heroes see W. Burkert 1985, 203-211.

[91] In an account of his travels, Pausanias (*c.* 150 AD), for example, relates the story of a boy named Poulydamas who fought and killed a lion to imitate Heracles and safeguard his village from danger (Pausanias VI, 5, 5).

bloody death.[92] For centuries, it was not recognized that such numbers were sufficient to entail the extinction of certain species, particularly the more vulnerable larger mammals.[93] Such *ludi* were also held at many Roman garrisons, but nowhere were such enormous numbers of animals employed as in Rome itself. Surprisingly, scholars have devoted little effort to uncovering the logistics used by the Romans to transport such large numbers of unmanageable animals.[94]

Weeber considers it likely that the *ludi* had the indirect effect of facilitating the extension of agriculture in the Mediterranean region. Wild animal populations were reduced substantially as a result, and farmers could thus leave their herds to graze and crops to ripen without fear of attack or damage.

Even in the present century the civilization motif can still be traced. In a work translated into Dutch in 1955, for example, the French classicist and ex-minister (in the Vichy regime) Jerôme Carcopino wrote: "The killing of wild animals during the *venationes* in the circuses may perhaps fill us with abhorrence, but it served an important purpose. The emperors liberated their lands from wild animals. With the games in the amphitheatres, the Roman Empire passed on the glorious works of Heracles to western civilization".[95] In order to attain an acceptable quality of life, an extensive degree of control over nature was deemed not only necessary but also entirely justified. To no small degree this justification was provided by the Greek philosophical heritage. As the above discussion demonstrates, this heritage was not some blend of contempt for animals and exploitation of nature, but was rooted in a clearly perceived ideal of civilization, supported by positive, classical-humanist, anthropocentric arguments.

---

[92] At the opening of the Colosseum (70 AD) some 5000 animals were 'used up' in the course of just a few days. The arena, where today thousands of tourists gape at the marvels of ancient Roman culture, must have been stained blood-red. It is only very recently that any consideration has been given to the ethical questions this raises (R. Auguet 1994).

[93] *Cf.* K.W. Weeber 1990/1993, *Smog über Attika. Umweltverhalten im Altertum.* Artemis Verlag München, Reinbek bei Hamburg, 131-150.

[94] In a recent sourcebook this question is even passed over entirely, so that no consideration at all is given to the transportation of the animals (F. Meijer & O. van Nijf 1992).

[95] J. Carcopino 1955, vol. II, 126 (translated from the Dutch).

CHAPTER FIVE

DISCUSSION AND CONCLUSIONS

5.1. Introduction

We begin this chapter by comparing the ideas expressed in the Hebrew Bible with the Greek philosophical tradition. We then go on to consider the subsequent influence of those ideas. What were the perceptions and interpretations of the early Christian church? How were these passed down within the western tradition, and who were the main contributors? What elements of this heritage have come to predominate in the western worldview? What does this information tell us about our relationship with the natural world today? The aim of the first part of this chapter is to demonstrate that certain beliefs crucial to the development of West European culture can be traced back to the very sources of that culture. I shall argue that these beliefs are, or have been, of major influence on human dealings with the natural environment. This brief excursion into history is unpretentious in its aims. I shall outline the principal ideas and beliefs and consider some general historical trends and motifs in humanity's relationship with nature.

In the final part of this concluding chapter, three developments in the twentieth century are identified that have given grounds for reappraising the dominant (western) worldview, or 'cosmology'. Subsequently, three currents of thought to have emerged from such a reappraisal are discussed. The most promising of these is then elaborated in greater detail in the context of humanity's relationship with nature, with reference to the historical gathered together in the present study. In doing so, I shall weave in the notions of sustainability and quality.

5.2 Two Cosmologies Compared

In the previous chapter we considered the ideas of the Greek philosophers. Let us now return to the cosmology of the Hebrew Bible and attempt a comparison. The Israelites can be said to have perceived the creation as an ordered whole, but they held that order to have been violated by humanity, at least in part. The created world exists ultimately not for humankind, but for the Creator. In the triangular relationship between God, humanity and the world (the 'trangle of meaning' of Zweers and Glacken; *cf.* Section 1.6) there is a clear hierarchy, but it is a three-sided affair that does not merely revolve around humankind. Humans certainly occupy a very special place, however, and an appeal is made to the responsibility that comes with that status. Certain persons such as kings and prophets may stand in a different relationship to nature. Besides this regal motif, the saint motif is also encountered in the HB (*cf.* Section 5.3). The order of creation experienced by humankind is not something immanent and static; it also involves an assignment. It points to a better order, in which God's creation has once again been restored. In the present world the forces of chaos are still operative. The importance of these forces in the creation order is also indicated in another section of the HB: in the wisdom literature and particularly in the Book of Job. A major difference from mainstream Greek philosophy emerges when it comes to the question of what mortal human beings can know about the world. The Greek philosophers held that the world can be rationally understood, but the conclusion of the Book of Job points to there being limits to such knowledge.

In the HB, and especially in the Torah, a clear distinction is made between domesticated and 'wild' nature. This distinction has implications for human actions. That part of nature that has been domesticated becomes absorbed, as it were, into religiously inspired ethics. This is reflected, for example, in the rules pertaining to the sabbath and their formulation in the fourth Commandment. Among the domesticated animals a further subdivision is made, as set out by the priestly authors in the dietary laws of Leviticus and Deuteronomy. The slaughter of animals is strictly regulated, by rules essentially reflecting the idea that life is part of the creation order, while death is not. Certain 'clean' animals and plants may be offered as a sacrifice and eaten, provided they satisfy certain additional criteria. While untamed nature is admired for its magnificence and beauty, it is also feared as dangerous and as God's punishment of humankind.[1]

---

[1] In this context C. Houtman 1982 refers to nature being 'Janus-faced'.

Figure 1 identifies some of the main characteristics of the dominant Greek and ancient Israelite cosmologies.

Fig. 1 Dominant ancient Israelite and Greek cosmologies compared.

Ancient Israelite cosmology	**Dominant Greek cosmology**
	* GODS
GOD	DEMIURGE
	UNMOVED MOVER
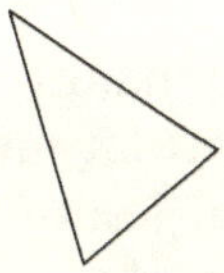 PROPHET/SAINT/	LOGOS
KING/	
MEN/WOMEN	* HUMANS
	PHILOSOPHERS
	MEN
NON-HUMAN CREATURES	WOMEN
* CLEAN/DOMESTICATED	BARBARIANS
- SACRIFICE	SLAVES
- FOOD	
- CARE	* ANIMALS
* UNCLEAN/WILD	DOMESTICATED
- RESPECT	USEFUL
- AWE	WILD
- ABOMINATION	VERMIN
- SANCTION	
	* PLANTS
	USEFUL
	WILD
	POISONOUS
	* NON-LIVING WORLD
	HEAVENLY BODIES
	EARTHLY MATTER
* Hierarchy: triangular	* Hierarchy: linear
* Theocentric	* Anthropocentric
* Order: assignment and referential	* Order: immanent and normative

5.3 The evolution of West European Cosmology

5.3.1 *Early Christian cosmologies*

Christianity evolved from Judaism, but did so within a Greek setting. In this chapter we shall be concerned mainly with Greek influences, but the evidently Jewish origins of Christianity should obviously not be forgotten.[2] One of Christianity's primary sources

[2] On the relation between the Jewish community and the Hellenistic rulers from

is the New Testament (NT), which together with the Hebrew Bible (in this chapter also termed OT) forms the canon, the holy scripture, of the Christian church.[3] In this section we shall first consider certain of the motifs elaborated in the NT, drawing attention to the different emphasis of the gospels and the epistles of Paul. We shall then examine the early Christian interpretation of the HB, as evidenced in the writings of the Church Fathers.

The New Testament

In the NT we read relatively little about nature and humanity's relationship with nature.[4] This is not particularly surprising, for the NT is concerned with the life and teachings of one man, Jesus of Nazareth, and not, like the HB, with the history of the world and of a people journeying towards and living in the Promised Land. The period of time over which the books of the NT were written is also significantly shorter than in the case of the HB: several decades at most. In our present context, the following motifs are of relevance.

Over and against the precise legalistic thinking of the Jewish priesthood of the era, and more specifically that of the Pharisees and the scribes, Jesus posits the original designs of the Law, the Torah and the Prophets.[5] His intention is not to do away with the law, but rather to fulfil it. It is through Jesus that the law finds its incarnation, for it is He who reconciles the fallen world with God the Father. It is through Him that the creation can be restored. As related in the gospels, the actions of Jesus stand in stark contrast to the strict interpretation of the law that had come to hold sway among

---

the conquest of Palestine by Alexander the Great in 332 BC to the revolt of the Maccabees see E.J. Bickerman 1988; On Jews and Judaism and the classical world, G. Alon 1977; On Christianity and classical culture J. Pelikan 1993; On the jewish origens of Christianity, D. Flusser 1964. An extremely useful collection of texts from Greek and Latin authors on Jews and Judaism is edited by M. Stern 1980.

[3] The canon evolved into its ultimate form and was adopted by the church between the second and fourth centuries of the common era, a status enjoyed by the HB since the end of the first century. *Cf.* S.Z. Leiman 1976, E. Earle Ellis 199, A.S. van der Woude 1993 and J.C. Trebolle Barrera 1998. Although agreement on what was to constitute the canon was reached at the Councils of Hippo (393 AD) and Carthage (419 AD), no formal decision was taken on this issue at the time. On the NT, *cf.* A.J.F. Klijn 1995.

[4] One of the few books addressing the relevance of the New Testament in environmental discourse (even in the title) is the one edited by Calvin de Witt. In this collection Gordon Zerbe writes "Nowhere does the New Testament directly call for the care of the earth. At the same time, the New Testament is by no means silent on the matter of humanity's relationship to creation" (1991, 73).

[5] *Cf.* E. Earle Ellis (1991, 138) for a similar, and extensively argued, viewpoint.

the spiritual leaders of the people of Israel. This is exemplified by Jesus' conduct on the sabbath and his opinions regarding the seventh day. In his eyes, plucking an ear of corn to appease one's hunger, saving an animal's life and curing the sick are activities that are more in harmony with the sabbath than the measured conduct of the Pharisees.[6] I use the phrase 'in harmony with' rather than 'permitted on' because Jesus' activities were close to the original designs of the sabbath.[7] "The sabbath was made for man, not man for the sabbath," Jesus declares in the Gospel of Mark.[8] It sounds very human-oriented, and so it is, to the extent that Jesus sides with humanity rather than with legalism. It should not be taken for anthropocentrism, though, with humanity standing over and against the rest of creation to its exclusion. The creation order resonates through Jesus' teachings and works, as it were, with every creature counted in—even the sparrow on the rooftop and all the more so, therefore, human beings.

At the beginning of the Gospel of Mark it is related how Jesus, having been baptised by John, is tempted for forty days in the wilderness by Satan.[9] "He was with the wild beasts; and the angels ministered to him," thus ends Mark's pericope.[10] This preparation for His earthly work links Jesus to Moses and Elijah, the two great prophets of the OT. Both these prophets sojourned in the desert and it was there that they were instructed to lead (Moses) or continue to lead (Elijah) the people of Israel.[11] In a sense, the great leaders of the people of Israel, with Moses representing the Torah and Elijah the Prophets and the two consequently the OT as a whole, prove themselves more powerful than the wilderness. They defy the desert, that symbol of the forces in God's creation that are hostile to humankind and the dwelling-place of devils and demons. It is for this reason that they do not die in the manner of ordinary mortals. At

---

[6] Matt 12:1-8, Luke 6:6-11, Mark 3:1-6.

[7] *Cf.* Chapter 2, particularly the conclusion of Section 2.4.

[8] Mark 2:23-28.

[9] Mark 1:9-13.

[10] The Greek for 'with the wild beasts' in Mark 1:13 implies that Jesus was 'with the beasts' as one is 'with friends' and not merely in an area that was also inhabited by wild beasts. *Cf.* W. Foerster 1938, W.A. Schulze 1955 and E. Fascher 1965.

[11] Moses: Exod 3:1-22; Elijah: 1 Kgs 19. In both narratives we encounter Horeb, 'the mount of God', and in the story of Elijah the period of forty days and forty nights is the same as in the Gospel of Mark. In his sojourn in the desert, Moses also remains on Mount Horeb for forty days and forty nights (Exod 24:18).

the time of his death, the power of Moses remains unbroken. His grave is never found and he is buried by God himself. Elijah is 'taken up' after his death, and Jesus is resurrected. In terms of the NT, Jesus conquers death once and for all. For this and for other reasons, He is seen as completing the unfinished work of his two predecessors.

To my mind this parallel is imbued with a deeper symbolism, which I would refer to as the saint motif, for it recurs regularly with the later Christian saints.[12] These saints frequently choose to live in the desert and appear to have restored their bond with wild nature, in part at least. Exchanging a worldly for a ascetic life, the most common motive for living in solitude, is thus not necessarily congruent with renouncing nature, let alone with contempt of nature. In the descriptions that have come down to us, not only is the menacing aspect of wild nature absent or at least tempered; there is also some reinstatement of humanity's moral authority over nature.[13] We read of animals being blessed by saints, but also reprimanded or even cursed if they commit an outrage against human beings or other creatures.[14] Among the Greek philosophers, it is only with Pythagoras that we encounter this characteristic, though not everyday, saint motif. Other Greek thinkers were a long way off this mark, and we in turn stand far-removed from Stoic thought.[15]

Moving from the Gospels and the life of Jesus to the rest of the NT, we find rather more similarity with the Greek tradition. This

[12] See, for example H. Waddell 1934, C.G. Loomis 1948, *Historia Monachorum in Aegypto: The Lives of the Desert Fathers* 1980, Butler's *Lives of the Saints* 1985, H. Nolthenius 1988, R.D. Sorrel 1988, S. Power Bratton (1988, 1993), *The Desert Fathers* (*Vitae Patrum* translation H Waddell 1998) and J.J. Boersema 2002.

[13] In this way the saints set themselves apart from ordinary human beings. This is also testified to in the NT, for example in the description of events on the island of Malta, where the apostle Paul was bitten in the hand by an adder (Acts 28:1-6). When nothing untoward occurred (to be interpreted as the adder bite proving not to be life-threatening) Paul was taken to be a god.

[14] The most comprehensive collections of accounts of the Christian saints in connection with the animal world are given by W.E.H. Lecky 1877, H. Waddell 1934, C.G. Loomis and E.A. Armstrong 1973. For a briefer review, see W.J. Short 1983, L.G. Regenstein (1991, 57-69) and, on St. Francis of Assisi, H. Nolthenius (1988, 139, 167 *ff.*), D. Sorrel 1988 and J.J. Boersema 2002.

[15] On Greek religion see W. Burkert 1985. The Christian tradition has always embraced the idea that humankind can live in harmony with the animal kingdom, and this also held for the Church Fathers. See e.g. Irenaeus *Adv. Haer.* V, 33, 3-4, cited by Papias in his description of the restored Kingdom. This belief is also to be found in Luther's writings; see, for example, J. Pelikan 1961.

is indeed understandable, for the apostles brought the evangelical teachings to a largely Hellenized world. Although military power was held by the Romans, the spiritual climate was dominated by Greek thought, and above all by the philosophers of the Stoic, (neo)Platonic and, to a lesser degree, Epicurean schools. The apostles defended the new teachings against both Jewish (Judaic) and Greek beliefs, but the rift with the Jews appears to have been sharper as well as socially more divisive. Peter, for example, is given to understand in a vision that the distinction between clean and unclean animals and the associated dietary practices (and sacrificial cult) no longer pertain.[16] While opening the faith to uncircumcized Gentiles, in everyday life it created a divide between Christians and Jews, for the sharing of meals became problematical. This issue played a major role in the conflicts confronting the first Christian communities.[17]

Paul's epistles contain only meagre reference to nature, although his attitude towards the natural world belies the saint motif (*cf.* note 13). Without holding earthly life in contempt, Paul is a very spiritually oriented man. On the subject of marriage he holds very different views from those expressed in such Old Testament writings as the wisdom literature and the Song of Songs. The relationship between the law and the gospel is an important theme of Paul's works. Although he advocates a reappraisal of Jewish law, Paul does not hold it to be invalid: "By no means! On the contrary, we uphold the law".[18] He stands more opposed to the 'works of the law' than to the Law itself, which retains authority as the 'Torah of faith'.[19] Paul is all too familiar with Greek philosophical thought. In Athens he enters into open debate with a number of Stoic and Epicurean philosophers, arguing in defence of a personal God who created the world, who maintains it and who will judge it by way of one man, the resurrected Lord.[20] This eschatological expectation, fully embraced by Paul and described elsewhere in his letters as encompassing "all of creation in all its parts", places him in an authentic Judeo-Christian tradition.[21] Passmore, in contrast, thereby following other

---

[16] Acts 11: 1-18.

[17] *Cf.* Acts 15:1-34 and Gal 2:11-14.

[18] Rom 3:31. On Paul's relationship with the law, see e.g. H. Küng, 1991, Ch. B III*: Um des Menschen willen.*

[19] Rom 3:27.

[20] Acts 17:15-34.

[21] Rom 8:18-30.

authors, concludes from certain passages in Paul's letters that here (and thus in the NT) we see an incipient Hellenization of early Christian thought.[22] Passmore's conclusions have recently received strong additional support from Van Kooten in his study on the Graeco-Roman cosmology and the Jewish eschatology in Paul and in the pseudo-Pauline letters.[23]

The Church Fathers

From the second century onwards a good deal of evidence for such a Hellenization can indeed be found in the Church Fathers' interpretation of the HB.[24] In response to the Gnostics and the Manicheans, religious movements which rejected the to their mind immoral God of the HB, the Church Fathers embark on a process of theological reflection on the HB in general and the Creation in particular. In my view a number of important choices were made in the process, with a Greco-Christian view of humanity's relationship with nature emerging which deviated in a number of clear respects from the beliefs expressed in the Hebrew Bible. Virtually all the views and beliefs discussed earlier, in Chapter 4, are encountered in the writings of the Church Fathers, but seldom in unadulterated form. Many views became intertwined, with ignorance sometimes begetting confusion. In some cases the original texts were also lacking. Greek conceptions thus coalesced with Christian beliefs or with other Greek ideas. Marcia Colish, who has made an extensive study of the pervasive influence of Greek and particularly Stoic philosophy in the world of antiquity, consequently warns against drawing over-hasty conclusions on that influence.[25] Similarities of phrasing and ideas and use of specific terms may be due to a wide variety of factors and, in the absence of additional evidence, can never constitute conclusive proof of the direction of influence. There is, nonetheless, a reasonable degree of consensus on how early Chris-

---

[22] As reflected, for example, in Paul's reference to 'nature as the norm' in his arguments against men wearing long hair (1 Cor 11:14 "does not nature itself teach you.."). *Cf.* J.N. Sevenster 1961, J. Passmore (1974, 16 and note 44) and H. Küng 1992, 22.

[23] G.H. van Kooten 2001.

[24] An excellent overview of this development is given by J. Pelikan in his impressive 5 volume work on the Christian Tradition (1971-1989). In this chapter I restrict myself to the western branch of Christianity. For the influence of Greek philosophy on the eastern tradition, see also J. Pelikan 1993.

[25] M.L. Colish 1979, 1985.

tian views on humanity's relationship with nature were influenced by Hellenistic thought and where the new emphasis came to lie.[26]

Neoplatonic views on nature and the super- or supranatural, on body and mind, come to play a significant role in explaining earthly reality and assigning a value to it. Overemphasis on the spiritual aspect leads to allegorical interpretations of the Bible and particularly of the 'nature passages'. We have already encountered such interpretations in Chapter 3, in the explanations of the dietary laws.[27] It is in this climate that the belief emerged that creation was to be read and interpreted as a book, and with it developed the associated hermeneutic tradition. In this respect, Colish regards St. Augustine (354-430) and St. Ambrose (339-397) as bridges between Plato and later Christian thought. In contrast to Sorabji, she deems Ambrose more important than Augustine in this respect.

A remarkable role in the Hellenization of the Jewish heritage during the early Christian era was played by Philo of Alexandria (*c.* 25 BC-*c.* AD 50).[28] The works of this Hellenized Jew exerted a major influence on Church Fathers like Origen and Clement. The influence of Plato's the *Timaeus* is very much present in Philo's work and particularly in his commentary on the Pentateuch. Philo adopts the main elements of Platonic cosmology (rational structure, perfection, etc.).[29] In *De Animalibus*, Philo is particularly concerned with the hierarchy of the cosmos and the status of animals relative to humankind. The key issue for Philo in the relationship between human beings and animals is possession of reasoning faculties, the logos. The classic question in this context is: Do animals possess such faculties? In a debate with his nephew Alexander, Philo defends the position that only human beings are blessed with the logos, and that this sets humanity above the beasts.[30] That human beings were created 'in God's image' implies that they possess a logos that is an image of

[26] *Cf.* R. Sorabji 1993, 195-205.

[27] *Cf.* S. Stein 1957.

[28] On Philo's intermediate status between the Grecan and Judaic worlds, see Y. Amir 1983; on Philo and Poimandres, Kl. Steur 1935; on Philo and Plato, D.T. Runia 1983 and on Philo and the Church Fathers D.T. Runia 1993.

[29] D.T. Runia 1983.

[30] Runia (1983, 300-307) considers this "in line with Biblical and Judaic views", but holds that Philo reasons his point using Stoic argumentation. I personally regard it as very debatable whether the connection between possession of logos and the exercising of dominion is as 'biblical' as is so widely assumed. The link between the logos and the image of God is perhaps even more dubitable (see § 2.3.1). To my mind, Philo falls back on the 'real source': the Stoics.

the divine Logos. In his commentary on Genesis, which takes the form of question and answer, Philo writes (in loose translation): "Humankind is a copy of the original seal. And that is the Logos of God, the first principle, the archetypal idea."[31] Lacking logos, animals can neither possess nor lack virtue; ethical categories are simply not applicable. Here we see the unmistakable influence of the Stoic philosophers. That influence is also apparent in Philo's discussion of the creation hierarchy, the *scala naturae*. He defends the position that the irrational exists for the sake of the rational and that everything ultimately exists for the benefit of humankind.[32]

This view is held by many of the Church Fathers.[33] Here we encounter what is to my mind an important difference between the original substance of the HB and its interpretation by the Church Fathers under the influence of Greek philosophical thought. God created the world and all it comprises for the sake of humanity; human beings are the purpose as well as the pivotal focus of creation. Thus the Church Fathers perceived and interpreted the writings of the HB. In doing so, they leaned more towards dominant Greek cosmological thought than they would perhaps have liked, not only as regards their position but also in the line of reasoning they took. This is well illustrated by the Church Father Origen when he defends the writings of the 'heretic' Celsus.[34] Celsus had raised the following question: If the earth was created for humankind rather than for the animals, why then "are all things produced for them without their sowing and ploughing?" We already encountered this question (in Section 4.5) with Lucretius, who used it to oppose the arguments of the Stoics. For Lucretius and the Atomists, the 'harshness' of nature was an argument for rejecting providence. In his response Origen takes a very different line, thereby concurring seamlessly with the teleological anthropocentrism of Aristotle and the Stoics. He holds that these circumstances were foreseen, in order that humanity might use its intellect to develop agricultural and metal-working technologies to satisfy his essential needs. Nature is thus conceived as a challenge to the human intellect. Providence has ensured that nature throws up sufficient impediments for that intellect to thrive.

---

[31] Philo *Questiones in Genesin* I, 4.

[32] A. Terian (1981, 45) and SVF II 1152-1167 for the Stoic sources.

[33] For a similar conclusion, see T.J. van Bavel 1990, 25.

[34] Origen *Contra Celsum* IV 76; Celsus may well have adopted the question from Cicero, who has it posed by Velleius (*De Natura Deorum* I, 9, 23).

In contrast to the Greek philosophers, however, the Church Fathers did not hold God's creation to be geared entirely to human purpose. Thus, Didymus considered that human influence is limited to his immediately adjacent environment, and should remain so.[35] Nowhere in the writings of the Church Fathers is there explicit reference to our ruling over or exploiting untamed nature. Nature was not their prime concern.[36] Many were convinced that the original harmony between human beings and other creatures had been disrupted. Humanity's relationship with nature had originally been untainted by the unpleasant phenomena of poisonous snakes, thorns and spines.[37] Ultimately, however, they all shared the belief that harmony can be restored, certainly in the end time.[38]

A major issue on which the Church Fathers differed was whether nature and the natural order could serve as a yardstick for judging human deeds, and how this related to the divine order. On this point the Stoic philosophers were unambiguous: nature was the norm. On the whole, the Church Fathers accepted nature as the norm, while differing when it came to the implications. One of the areas in which such differences were apparent was in what is referred to as the theology of cosmetics, the treatment of such questions as: May a man wear his hair long; is such behaviour natural? May clothing be dyed? If God had meant us to wear blue clothing, He would have created us with a blue coat of hair, and so on and so forth. As we have seen, similar questions are also raised in the NT (1 Cor 11:14). This debate still in fact continues: the notion of nature as the norm recurs time and time again, as we shall see below.[39]

Let us conclude this section with a few final remarks on St. Augustine, the most influential Church Father of the early Christian era. Originally an upholder of the Manichaean doctrine, in his conversion to Christianity Augustine was also influenced by the

---

[35] Didymus *In Gen.* I, 28-31.

[36] See, for example, the study by J. Cohen 1989 on the 'career' of the *dominium terrae* text of Gen 1:26-28.

[37] Basil *Hom. Hex.* 5, 1 and 6. With Philo, too, we find the view that before 'the Fall' animals were not set against man, but were his helpers and allies (*Questiones in Genesin* I, 18).

[38] "When man ceases doing evil, the animals will regain their benign nature" (Theophilus *Ad Autolicum* II, 17). The Church Fathers also make reference to Daniel 6 and Acts 28:3-6. On the end time, see Irenaeus, who quotes Papias in his description of the restored kingdom (*Adv. Haer.* V, 33, 3-4).

[39] Lately dealt with by S.R.L. Clark in his *Biology and Christian Ethics* 2001.

Neoplatonist Plotinus.[40] Although in Augustine's writings we find the notion of humanity as the pivot of, and ruler over, creation, that 'ruling' is tempered. Epistemologically, Augustine adopts a very different stance from the Stoics, because for him 'there can be no knowledge without love'.[41] On the subject of agriculture, Augustine entertains similar ideas to Plato, holding that it is best for us to enter into a dialogue with nature.[42] In reference to animals and our relationship towards them, he states that they deserve our respect and that domesticated animals are particularly deserving, for they live under our care. Over and against the Manichaean belief that the world is permeated with evil, he maintains that there is great beauty and that humankind may stand in awe of the animals.[43] Animals are not included in the moral compact (*societas legis)*, however. On this point, Augustine fully embraces the Stoic line of reasoning that animals are excluded from this compact because they lack a logos, a rational soul.[44] In discussing the commandment "Thou shall not kill", Augustine also considers the slaughter of animals. This he deems permissible, because unlike us they have no faculties of reason.[45] In life as in death, the creatures of the earth are subject to human will. At the same time, though, Augustine was familiar with human nature and was realistic enough to write "Sometimes we are in ignorance of the place animals hold in nature, and sometimes though we know it, we sacrifice it to our own convenience."[46] In assessing Augustine's role in moulding human attitudes towards animals, Sorabji draws a rather negative conclusion: "In the Latin West it was Augustine above all who stamped the test of rationality into Christian discussions of how to treat animals".[47]

Summarizing, then, with the early Church Fathers we see a gradual shifting of thought towards the Greek philosophical mainstream in a number of areas. With regard to humanity's relationship with

---

[40] Augustine, St. *Confessions* VII, c.9. In *The Christian Doctrine* (II, c.40) Augustine writes that (even) heathen philosophy comprises knowledge to which Christians have a right.

[41] An argument that re-emerges last century with Max Scheler, for example; on this point, *cf.* P. Kockelkoren 1994.

[42] Augustine, St. *De Gen. Ad Litt.* III, 18, 27-28.

[43] Augustine, St. *City of God* XI, 16.

[44] Augustine, St. *On the Morals of the Manichaeans* II, 17, 59.

[45] Augustine, St. *City of God* I, 20.

[46] Augustine, St. *City of God* XI, 16 and XII, 4.

[47] R. Sorabji 1993, 201.

nature, particularly, their worldview is more in line with Greek than with ancient Israelite cosmology. Table 1 provides a summary review of the themes discussed in the previous few paragraphs, indicating nuances and historical shifts of accent.

Table 1 Status of key themes in the respective cosmologies.

Idea	HB	Plato	Arist.	Stoics	NT	Church Fathers	Western cosmology
Cosmos ordered and good	-	++	++	++	-	±	±
Order immanent	-	++	+	+	-	±	±
Order cognizable and logical	±	++	+	++	±	±	+
Order hierarchical; Humankind at apex	+	+	++	+	+	++	+
Human beings as purpose of cosmos	-	+	++	++	-	+	+
Cosmos/nature the norm	-	+	++	++	-	±	±
Distinction between natural and supernatural	-	++	±	±	±	±	±
Allegorical explanation of nature	±	+	-	-	±	+	±
Moral bond between human beings and animals	+	-	-	- -	+	±	±
Domesticated animals in different moral category	++	-	±	-	+	+	±
Linear conception of history	+	-	-	-	+	+	++

++ theme plays an important role
- - idea explicitly disputed
+ theme encountered
- theme absent or idea rejected
± considerable nuance and/or no unanimity of opinion

5.3.2 *From the Middle Ages to Modernity*

Historical threads

In any historically oriented study there is a danger of projecting contemporary concerns onto ancient texts and other materials, thereby twisting the latter into our own particular conceptual framework.

This becomes even more problematical where issues and notions are concerned that were perceived differently in an earlier era, or were altogether absent.[48] At the same time, though, it is perhaps even more hazardous to take an entirely non-historic approach to the topic at hand. This is the attitude essentially adopted both by those who hold the historical record to hold no significance for today's environmental problems and by those who maintain that we must make a clean break with the western tradition if the earth is to have a future. Both positions underrate the importance of continuity in human culture. It is therefore useful to start out with a few remarks regarding the status of the cosmologies discussed and their role in the history of our (western) culture, before moving on to consider their relevance to the contemporary environmental debate.

By merely identifying certain key Greek philosophical concepts, elaborating the main thrust of ancient Israelite thought and outlining the beliefs that the early Christian church may possibly have adopted from Greeks we obviously do insufficient justice to the rich tapestry of views and beliefs and the wealth of nuance to be found within these, the roots of our civilization. With regard to the Hebrew Bible, for example, mention should at least be made of the wisdom literature (the Book of Job, as well as Ecclesiastes), which differ in emphasis from the Torah. In this respect, Job is particularly interesting.

Job is struggling with the question of why he should be suffering, a fate he deems to have overcome him without good reason. His friends all have their own measured explanation, but Job remains unconvinced. Eventually God appears on the scene and, speaking from a whirlwind, proceeds to describe the creation in the marvellously, and uncharacteristically, poetic imagery of chapters 38-41. The entire natural order is reviewed, starting with the non-living world, and each time Job is asked the same question: Can you do that? Did you make that? Do you understand how it works? And

[48] Statements like Plato being a 'green dictator' (Koolschijn 1990) 'The Bible is full of ecology' and 'The Bible is a green book' (examples taken from two Dutch environmental writers, Hans Bouma and Roel van Duyn) are therefore immediately suspect. Neither has Saint Francis of Assisi proved entirely immune. Following Lynn White's proposal in 1967 that he be made the patron saint of ecology and Pope John Paul II's fulfilment of that wish on 13 November, 1979, he has been so 'ecologized' that precious little remains of the medieval thinker, saint and mystic. This picture is in need of revision; on this point, *cf.* H. Nolthenius 1988, R.D. Sorrel 1988 and, especially J.J. Boersema 2002.

the answer is implicitly negative: no, this is far beyond the comprehension and capacities of Job, in all his wisdom. Only the Creator himself can comprehend and perform these works, and at times even He appears to find it difficult. The problem of human suffering proves to be related to the question of who rules over the universe, but not in the way his friends think. Their logic was unrelenting: Job must have erred, for otherwise God would not have allowed him to suffer so, since He was obviously capable of preventing such suffering. Or is God not all-powerful, after all? And the answer to this human dilemma is ultimately posited as being embodied in the natural order, of which humankind evidently has only very deficient knowledge and in which there are forces at work that he scarcely comprehends. There is the beauty of the creation, its harmony and coherence, but there are also the enigmas of primeval *bᵉhēmôt* and the overwhelming, well-nigh unrestrainable nature of *liwyātān*. Together, this appears to answer Job's questions, where his friends had failed. The theological determinism of his friends, the question of what can be known about the Creation, and the relationship between human suffering, the power of God and the order and chaos of that Creation give the Book of Job a theme that today is still as relevant as ever. Within the present study, we shall not consider the matter any further. Despite these reservations, however, to my mind the above schematic review retains its general validity.

Among the Greeks we find an enormous diversity of views, for theirs was a civilization exploring the limits of the intellectual universe.[49] In terms of influence, however, certain philosophers stood out, with Aristotelian cosmology gaining dominance. By comparison, writers 'sympathetic' towards animals such as Pythagoras and Plutarch exerted far less influence, and Porphyry still less, to say nothing of Xenocrates and the virtually unknown poet Bion alluded to earlier.[50] For our present purposes, it is important to note that on the issues of concern there was a fair degree of convergence between the principal schools of Greek thought, as argued in Chapter 4.

---

[49] A Russian author once gave the following advice: 'Start every lecture with the Greeks, then you can keep it short.'

[50] Of these, Plutarch was the most influential, particularly through his attacks on the Stoics in his *Morals*, for example in the books on the shrewdness of animals and the consumption of meat. The French thinker Montaigne was strongly influenced by Plutarch (see, for example, M. de Montaigne *Essays*, Book II, Ch. 11, *On Cruelty*, 491-507).

Among the founding fathers of the early Christian church, too, there was a greater nuance of belief than has generally been held, as van Bavel has recently shown.[51] In this respect, John Chrysostomos and St. Basil may be mentioned.[52] The great importance of this kind of study is that it also shows why certain theological positions were adopted and considers the context in which certain statements came to be committed to writing. Unlike certain Greek philosophers, the Church Fathers were not primarily concerned with animals, nor with the relationship between human beings and the animal kingdom, still less with employing biblical texts to justify human exploitation of nature.[53] They were struggling with the issue of the goodness of creation and the status of evil, with humanity's freedom of will in relation to predestination and, more generally, with our position relative to God and the cosmos.[54] The extinction of the lion, say, in the Middle East, while constituting the historical background to their words, both written and spoken, did not serve as their inspiration. This was certainly due in part to the fact that early Christianity was associated far more closely with an urban environment, in which it indeed came to flourish.[55]

It is undeniable, nevertheless, that a particular perspective on

---

[51] T.J. van Bavel 1990, see also J. Pelikan 1971-89.

[52] In the same vein: R. Sorabji 1993, 203. With Chrysostomos, who spent ten years in the desert, the saint motif is particularly evident. He is part of a tradition that runs from Pythagoras, by way of the desert fathers, to the later Celtic and Christian saints see C.G. Loomis 1948, W.J. Short 1983 and A. Jones 1985.

[53] This is also demonstrated in the studies of D.K. Jobling 1972 and J. Cohen 1989 on the reception of Gen 1:26-28 down the ages. J. Cohen (1989, 271-305) draws the interesting conclusion:

> Rarely, if ever, did premodern Jews and Christians construe this verse as a license for the selfish exploitation of the environment. Although most readers of Genesis casually assumed that God had fashioned the physical world for the benefit of human beings, Gen 1:28 evoked relatively little concern with the issue of dominion over nature.

For a review and discussion of the historical reception of this crucially important text, *cf.* H.-J. van Soest 1996, 22-26.

[54] And in this they were not alone, as is reflected, among other places, in the *Corpus Hermeticum* (1991, 13-27), a collection of gnostic writings probably written in Alexandria in the first century AD and in which Judaic, Greek and Egyptian influences are to be discerned.

[55] For a historical outline of the environment into which Christianity was born, see C. Clemen 1908/1912, J. Pelikan 1971-89, S. Safari & M. Stern (eds.) 1974, M.L. Colish 1985, M.J. Mulder (ed.) 1988, R.Lane Fox 1989 and E.P. Sanders 1990.

humanity's status in nature and in the *scala naturae* and a particular interpretation of the Old Testament texts have been absorbed by Christian thought by way of the Church Fathers. This perspective has been inappropriately termed 'Judeo-Christian' by Lynn White, for 'Greco-Christian' is a more fitting epithet.

Hellenization

West European culture has been shaped to a major extent by Christianity and that is certainly true of the cosmologies predominating in that culture today. On further consideration, 'Christianity' proves to be an umbrella term that also takes in the heritage of classical antiquity and the Middle East.[56] As we have seen, Christianity was influenced by Greek thought at a very early stage of its development. In the western tradition, these influences have come down to us as 'Christian'.[57] In this first wave of Hellenization, specifically Jewish elements that had been inherited from the ancient Israelites fell away into the background. In later eras, too, the Jewish tradition continued to exert very little influence on Christianity. Indeed, the emancipation of Christianity from Judaism was attended by strongly anti-Jewish sentiment. That the Jewish tradition lost its influence on the mainstream is reflected in the minimal influence of the Talmud and Mishnah on western thought. This divergence is to be regretted in a number of areas, but perhaps nowhere more so than in the evolution of human attitudes towards nature. Greater exchange, in both directions, might have led to a more balanced interpretation of the law and the prophets. In certain respects, it may be added, the Jewish tradition has itself also moved considerably beyond the basic tenets of the HB, towards a doctrinal stance that is almost entirely anthropocentric in orientation.[58] Modern Jewish writers have explained this in terms of the Jewish loss of national territory and forced abandonment of an agrarian lifestyle. In exile, Jews soon began to focus on other affairs, having no option but to gear their lives to survival.[59]

---

[56] Among many other writers, in this context reference may be made to W. Banning & J.D. Bierens de Haan 1939/1961; E.J. Dijksterhuis 1950, D. van Arkel 1974, C.J. Glacken 1967, K. Thomas 1983, J.H. van der Pot 1988, D.C. Lindberg & R.L. Numbers 1986, D.C. Lindberg 1992, P. Singer (1975, 1981, 1990) and J. Goody 1995.

[57] It was from Augustine, for example, that Thomas Aquinas inherited the typically Stoic notion of the logos as the criterion for man's moral bond with the animals, which he used to defend their slaughter. (*Summa Theologica* 2.2 q.64, a.1).

[58] On this point, *cf.* S.S. Schwarzschild 1984 and H. Eilberg-Schwarz 1990.

[59] *Cf.* E.G. Freudenstein (1972), who considers it time for a "deeper under-

Still, the Jewish community is also currently engaged in a reappraisal of its past.[60]

During the late Middle Ages and the Renaissance there was renewed interest in the classical writers, particularly the Greek philosophers. Their works were translated into Latin from Greek or Arabic and studied in the western centres of learning. In its impact, this second wave of Hellenization was certainly as far-reaching as the first, for links were forged with the natural sciences, which were now evolving in an experimental, technological direction.[61] As a result, intellectual debate soon took on practical implication. Towards the end of the Middle Ages new interpretations of Genesis 1:26-28 begin to emerge that seek to justify active human intervention in nature. By means of cropping and cultivation, nature is to be 'perfected'.[62] Keith Thomas has studied human attitudes towards nature in 16th to 18th century England, with reference to numerous everyday sources. Contrary to prevailing opinion, he demonstrates convincingly that humanity's relationship with nature, as reflected particularly in the status of animals did not in fact deteriorate after the Middle Ages but gradually improved.[63] During this period of scientific and social upheaval, Aristotelian and Stoic thought long continued to serve as the Christian standard. Although this is not the place to consider the matter in any detail, let me cite one episode that may be taken as exemplary: the celebrated debate between Las Casas and Sepúlveda, in Valladolid in 1550, regarding the status of Indians. Although both took it as axiomatic that there was a hierarchy within the human race, the debate was concerned with exactly how the Indians fitted into this (Christian-)Aristotelian

standing and acceptance of the concern for the environment evinced by the Jewish tradition" (274).

[60] See M. Landmann 1959, N.J. Cohen 1959, M.R. Konvitz (ed.) 1972, A. Rose (ed.) 1972, G. Stein 1980, A. Schaffer 1982, D. Ehrenfeld & Ph.J. Bentley 1985, H. van Praag & A. Oliverio 1986 and especially M. Gerstenfeld 1999 and numerous articles in journals, including the theme issue of the Melton Journal (1991, no. 4), *Judaism and Ecology. Our Earth and our Tradition.*

[61] For a general review of these developments, see E.J. Dijksterhuis 1950, on the role of Judaism, L. Finkenstein (ed.) 1974 and on the relationship between science and Christianity, D.C. Lindberg 1992, M. Ruse 2001, R. Pollack 2001 and on science and religion in the Western tradition G.B. Ferngren (ed.) 2000 .

[62] See, for example, U. Krolzik (1979, 1989), who has examined the interpretation of these verses from Genesis from the Middle Ages to the Enlightenment.

[63] K. Thomas (1983, 25) qualifies the situation thus: "But at the start of our period exploitation, not stewardship was the dominant theme".

scheme, with a view to deciding how these natives might and indeed should be treated. Sepúlveda considered them barbarians, and slaves by nature, and argued for their subjugation. While not denying the Aristotelian scheme *per se*, Las Casas took issue with his opponent's interpretation thereof. He himself deemed Indians intelligent, good-natured and eminently eligible for conversion to Christianity. Although no clear-cut conclusion was in fact reached in this historic debate, the position taken by Las Casas was not to gain widespread credence. The subsequent historical fate of the American Indians and their treatment by European settlers is sadly familiar.[64]

Despite all the important variations and numerous fascinating undercurrents, since the late Middle Ages a worldview has undeniably evolved that can be deemed characteristic of our western culture. This process has been described by E.J. Dijksterhuis in his poignantly entitled book *The mechanization of our worldview*, a powerful metaphor that permeated his work.[65] The tenets of that 'mechanized' worldview, or cosmology, can be summarized succinctly as follows:

- The world is an ordered whole.
- That order is hierarchical.
- The natural order operates mechanically and is intelligible to the mind.[66]
- Human beings stand at the apex of the *scala naturae*.
- The value of nature derives from its usefulness to humankind.

In this reading of reality, science and the natural sciences in particular come to play an increasingly important role. In pursuit of the deeper meaning of the universe, though, science can as yet be no more than a useful tool. The vast majority of natural scientists still pursue their study of nature and the laws of universe in the convic-

[64] For an extensive treatment of this debate and its historical context, *cf.* T. Lemaire (1986, 77 *ff.*). For the views of Las Casas himself, *cf.* Bartolemé de Las Casas (1552/1992); the cited debate is also treated in the introduction of this edition (28 *ff.*) For the influence of the colonists on the cultures of the Indians and the Indians' relationship with animals, see C. Martin (1974, 1978).

[65] *Cf.* van Berkel 1996.

[66] Use of the term 'mechanical' reflects the incorporation of the atomists' ideas regarding the nature of the cosmos. The seventeenth century saw a revival of this current of thought (D.J. Furley 1989, 226). That the cosmos is intelligible is not an atomist viewpoint.

tion that this will augment our knowledge of the Creator. Until the early nineteenth century, this remained the generally accepted motive for the scientific project.[67] More generally, the responsibility towards a superior power still felt in many circles, led to a certain prudence in how nature, the Creation, was treated. Despite the ubiquitous degradation of the natural world (and wild nature in particular), this did place an implicit ideological limit on 'ruthless' exploitation. Nonetheless, a 'nature-unfriendly' and exploitative attitude was certainly the rule rather than the exception.[68] Keith Thomas provides the following sobering account:

> In drawing a firm line between human beings and other creatures, the main purpose of early modern theorists was to justify hunting, domestication, meat-eating, vivisection (which was quite common in the late seventeenth century) and the wholesale extermination of vermin and predators.[69]

As yet, dissident views were rarely heard, but voiced they were and in growing numbers. Passmore demonstrates that it was in this intellectual climate that the notion of humanity as the 'steward' of nature resurfaces, the term being explicitly introduced in the late seventeenth century.[70] 'Resurfaces', because the notion of 'humankind as the keeper of the garden' had always remained part of the western tradition, handed down as part of the second story of creation, known as the story of the Garden of Eden for that very reason.[71]

In the contemporary Dutch environmental debate, the term stew-

---

[67] This led, *inter alia*, to the very interesting school of 'naturalist' theology, represented in England by John Ray (late 17th/early 18th c.) and his major work *The Wisdom of God Manifested in the Works of Creation*, in France by Comte de Buffon (late 18th c.) and in the Netherlands by the Zutphen preacher Martinet (late 18th c.) and Groningen professor J.A. Uilkens (early 19th c.). *Cf.* J. Bots (1972) and IJ. Botke 1984.

[68] This largely exploitative stance has been extensively documented. In addition to the literature already cited, see also, among many others, B. Noske 1988, R.H. Grove 1995, W. Beinart & P. Coates 1995, as well as C. Merchant 1980, who has elaborated a convincing connection between environmental exploitation, scientific attitudes and gender relations.

[69] K. Thomas 1983, 41.

[70] For the first time probably in Sir Matthew Hale's *The Primitive Originations of Mankind* (London, 1677, 370). Passmore refers to a persistent but never potent undercurrent in the western tradition that may go back to the work of the neoplatonist Iamblichus (J. Passmore 1974, 28 *ff.*)

[71] For the influence of the paradise/garden ideal on western culture, *cf.* J. Prest 1981.

ardship (*rentmeesterschap*) is used exclusively by political parties with Christian affiliations, who take it to denote a biblically motivated, more caring attitude towards the natural environment.[72] This is a historical misconception, however, for the term is in fact being used as a latter-day typification of an older ideal. The image of a 'steward' cannot possibly serve as a model for Christian believers in their dealings with nature. In fact the term *rentmeester*, so in favour with political parties, occurs nowhere in the Dutch translation of the HB, and no wonder, for the modern notion is indeed also lacking. In English translations the term 'steward' is used to describe someone who acts as a kind of household caretaker or supervisor.[73] Although the term is employed in Dutch translations of the New Testament, it has no connotations with nature (i.e. as a whole).[74] As we saw in Chapters 2 and 3, the HB does not conceive of 'man' as a 'steward' of nature. Humanity has a moral bond with domesticated nature, rather, and rules are laid down for our dealings with the animals that have been tamed. The rest of nature has not been given to humankind to 'master and manage'. Human beings may admire the world of nature in which they live, seek protection against its varied influence, and indeed stand in awe of it; but at no point is humankind appointed 'steward' over nature.[75]

[72] The specific term was used by all four Christian-inspired parties, in the following respective sections of their 1994 election manifestos: CDA (4, 46), GPV (52), RPF (3, 38) and SGP (7, 68); the term was not used by any other party *Verkiezingsprogramma's 1994*. Sdu. The Hague, 1994.

[73] Gen 43:16,19 and Gen 44:1,4. Three other Hebrew terms used in the sense of 'steward' are found in Is 22:15, Dan 1:11,16 and 1 Chr 27:31, 28:1. In none of these passages does nature serve as the context.

[74] The word most frequently employed for 'steward' (*οἰκονόμος* 'the economist') occurs a total of eight times: four times in the Gospel of Luke (12:42 and 16:1,3,8), twice in 1 Cor 4:1,2, once in Titus 1:7 and once in 1 Pet 4:10. All the passages in Luke are concerned with financial management, Corinthians with "the mysteries of God", Titus with "a bishop, as God's steward" and Peter's first epistle with "good stewards of God's varied grace". The same holds for the other Greek term that can be translated as 'steward' (*ἐπίτροπος*). This term occurs twice. In Luke 8:1 it is used in the sense of a household supervisor, without further details. In the second instance, too, in Matt 20:8, where the reference is to a vineyard owner's chief supervisor, the steward is involved with paying the labourers. Any connection between the term 'steward' and care for nature or any element thereof is simply lacking in the Bible.

[75] Compare the 1994 election manifestos of the aforementioned Dutch parties (in translation): CDA (§ 46): "The Creator has set the earth under mankind's supervision."; GPV (§ 7): "mankind is to manage His creation.."; RPF (§ 38): "Proper management of God's creation is an assignment deriving directly from

With the Enlightenment and the process of secularization that gained momentum in the nineteenth century, the situation described by Passmore changed. Western people began to lose their faith in God, and what remained was faith in human potential.[76] Ideologically, this paved the way for a stance towards nature whereby (western) humanity came to assume dominance over the whole of creation, from so-called 'savages' all the way 'down the ladder' to other creatures and plants and to 'dead' nature.[77] Darwin's new theory of evolution did not essentially alter this hierarchical conception of the earth's lifeforms. On the contrary: despite its emphasis on the ultimate kinship of human beings and other organisms, it lent a certain scientific credence to the 'top-down' view. Although the new evolutionary science marked a break with the Christian-Platonic image of eternal, immutable species, then, the Christian-Aristotelian conceptualization of a linear hierarchy remained unshaken.

Peter Singer, who discusses the western tradition in his book *Animal Liberation*, argues that it was only in the last century that the dominant worldview started to be seriously questioned. Even Francis of Assisi remained within the mainstream, he holds.[78] With the nineteenth century, though, Singer sees the beginnings of a breakthrough in human attitudes towards animals. At a theoretical level, he interprets this change as "an extension of moral rights to animals". Animals came to be included in the moral compact, from which they had for centuries been excluded in accordance with Stoic views on the matter.[79] On this point Passmore interprets the change differently, as "a restriction of human rights with regard to animals".[80] What took place, he argues, is that it gradually dawned that humanity's presumed unbridled rights vis-à-vis nature must needs be limited. This was not through any questioning of the *scala naturae* itself, however. Rather, there arose a sensitivity that inhumane treatment

---

the biblical instruction to 'master and manage'"; SGP (§ 7): "The earth has been given to humanity to master and manage".

[76] This may sound negative, but it obviously depends entirely on man's perspective on himself, his anthropology, what such secularization implies for nature and the environment.

[77] In the writings of Karl Marx (1971, 94), for example, we read that subsequent to the 'death of God' nature "simply becomes an object for mankind, purely a matter of utility."

[78] P. Singer 1975 (revised edition 1990, 198).

[79] For a description of this development, *cf.* E.C. Hargrove 1987, 3-31.

[80] J. Passmore 1974, 217.

of slaves and certain animals was no longer the 'done thing' for civilized people. It was thus more a question of civilization, a shifting of the accepted image of 'civilized human beings', than of the moral compact becoming an 'extended circle'.[81] Passmore's assessment of the change is nearer the truth, I feel. From numerous studies (in the United Kingdom and British colonies) he concludes that the more enlightened attitude to emerge during the last century was not towards animals in general, but merely a particular subset thereof. In seeking to explain this selective focus, Keith Thomas points to the Industrial Revolution, which reduced our dependence on animals, and on horses and dogs specifically. This opened the way for other affections, for sentiments for which in an earlier era there had been neither opportunity nor cause.[82]

The anthropocentric, economic and selective elements of the emerging 'love of animals' have been remarked upon by a number of authors. This love was directed very much towards the horse and the dog, and was eminently compatible (at the personal level, too) with fox hunting at home and big-game hunting in the colonies. In the ensuing shooting match, some regions were literally emptied of their fauna.[83] Reasonable treatment of certain (farm) animals is regarded as a sign of civilization.[84] At the same time, though, 'controlling' the wild animals in the colonies—even to the point of extinction—is also regarded as humane.[85] Consequently, cultural de-

---

[81] *Cf.* J. Wynne-Tyson's anthology *The Extended Circle* 1980.

[82] We here have a familiar paradox. Today, too, 'love of nature' often transpires to flourish apace once nature has all but disappeared and no longer counts as a serious factor in the economic system.

[83] Among others: D. Harwood 1928, N.H. Pollock 1968, G. Carson 1972, B. Harrison 1973, H. Ritvo 1987/1990, J.M. MacKenzie (1989, 1990) and R.D. Ryder 1989. On the destruction of the American bison see Ted Frank Belue 1996 and A.C. Isenberg 2000. Belue and Isenberg lay bare strong similarities between the attitudes towards the Indians ('wild human beings, hard to tame') and the bison ('wild bovine species, hard to tame').

[84] Charles Darwin also wrote to this effect, stating in *The Descent of Man* that "humanity to the lower animals seems to be one of the latest moral acquisitions. It is apparently unfelt by savages excepts towards their pets" (2nd ed., 1989, Part I, Chapter IV, 123). He is here referring, *inter alia*, to North and South American Indians, to whom such a harmonious relationship with nature is ascribed in contemporary debates. On the allegedly nature-friendly attitudes of American Indians see Shephard Krech III 1999.

The celebrated anti-vivisectionist Frances Power Cobbe held similar views. He posed the question: "Why is a landseer 'a thing' though we confer the title of person on a savage that consumes its grandmother?" F. Power Cobbe 1895, 504.

[85] In contrast with the Saint motif, discussed earlier, here we see a modern

velopments in the nineteenth century lead merely to a partial improvement or change. While humanity's ability to impose his will on nature is growing by leaps and bounds, and there is held to be ample justification for subjugation, certainly of 'wild' nature, attitudes towards specific elements of domesticated nature are at the same time becoming gradually more humane.[86] Whatever changes may be in the air, though, they occur entirely within the framework of the dominant worldview, which still remains fundamentally unshaken.

5.4 A New Cosmology?

Three arguments

The twentieth century brought on a series of dramatic events and developments, however, which have come to be interpreted as a compelling rationale for thoroughly reappraising our worldview, which even today is still predomantly mechanistic. In this context three principal arguments are frequently encountered.

In the first place, humanity's belief in progress and optimism regarding the innate goodness of humankind have been dealt a serious blow by two world wars and particularly by the role of the west therein.

Second, modern physics has undermined classical conceptions of 'facts' existing independent of the observer and of what can ultimately be known about reality.[87] At the same time, chaos theory has shown us that (even deterministic) complex systems may, in principle, be unpredictable.[88]

---

variant of the Culture Hero or Heracles motif. The Greek hero Heracles had ten well-nigh impossible tasks to complete. In some of these he defeated (mythical) creatures always held to have been invincible. Although the original narrative is highly symbolic, in later ages it is cited in defence of the extermination of predators deemed dangerous. According to J. Carcopino 1955, the carnage of the ancient Roman ludi ('games') may disgust us today, but at least they served a practical purpose: the emperors purged their territories of wild beasts. The lion was banished from Mesopotamia, the tiger from Hyrcania and the elephant from North Africa. Carcopino concludes (in translation): "Through the *venationes* of the amphitheatre, the Roman Empire has handed down to civilization the benefactions of Heracles' labours" (Part 2, 126). *Cf.* Chapter 4.

[86] In this connection Brian Harrison (1973, 786-821) reports that the British animal welfare society RSPCA "was founded largely by evangelical humanitarians". The Anglican priest Arthur Broome was the RSPCA's founder and first secretary. On this point, *cf.* L.G. Regenstein 1991, 93-94.

[87] I. Prigogine & I. Stengers 1985, 1988/1989.

[88] J. Gleick 1988. *Cf.* H. Tennekes 1994, H.W. Broer & F. verhulst (eds.) 1990

Third, the current scale and intensity of the environmental problematique is focusing our minds once more on the (unbalanced) relationship with nature and on the imperative of adopting a new position in this realm, too.

Although it is universally recognized that these three developments signal the intellectual and moral failure of the dominant western worldview, their implications for a new cosmology can scarcely yet be grasped. Essentially, we see ourselves confronted with two pivotal questions: how novel can and should that cosmology be and, in tandem, how radical the break with the old?

Three currents of thought

In the contemporary debate on our relationship with nature within the terms of a new cosmology, three currents of thought can be broadly distinguished.[89] Moreover as Andrew Giddens has noticed, "in the current phase of post-modernity, we can observe the mutual emergence of movements of ecological resistance and a new quest for spiritual meaning and the transcendent."[90]

The first, and most radical, holds that a new cosmology is to be sought entirely outside the western tradition, one direction of inquiry being the Far East.[91] As a comprehensive strategy for society, this proposal seems rather rash, not for lack of appealing elements in eastern, or non-western modes of thought—on the contrary—but more because it is surely an impossibility, certainly within the time frame available, to simply 'adopt' a cosmology from traditions that have in many cases developed separately from our culture for many hundreds of years.[92] A worldview is not, after all, something that can slipped on and off at will like a coat, nor for sale in parts. It takes

and J. Polkinghorne 1989. There are many misconceptions regarding these developments in modern physics; *cf.* N.G. van Kampen 1988 and Ph.B. Smith 1989.

89 Much of the (ever more voluminous) literature in this area is 'environmental philosophy'. For a useful review, see D.E. Davis (1989). For a more recent and broader overview, see W. Zweers in *Agriculture and Spirituality* 1995, 117-140.

90 A. Giddens 1991, 151.

91 One advocate being the Dutch botanist V. Westhoff, who has become a Buddhist himself. *Cf.* V. Westhoff 1989.

92 For a review of Hindu beliefs, see O.P. Dwivedi & B.N. Tiwari 1987, for a comparison between Europe and China, Yi-Fu Tuan 1968, for an Islamic viewpoint, Seyyed Hossein Nasr 1968 and A.R. Agwan (ed.) 1997, for Buddhism, Ch. Kabilsingh 1998, for a broad and varied anthology, R.S. Cottlieb (ed.) 1996 and Peter Gerlitz' *Mensch und Natur in den Weltreligionen* 1998.

time to grow and change. The problem with the modern supermarket of religions is that it purveys transcendence or spirituality as one more commodity rather than offering a genuinely alternative way of living based on a coherent worldview.[93]

The second school of thought is represented by a variegated tradition of radical eco-philosophers, of which the Land Ethic and subsequent Deep Ecology movements are the major exponents.[94] Although less oriented towards other cultures, or inspired explicitly by the past, these thinkers are certainly crafting a new kind of worldview. They deem it imperative for society to embark on a fundamental change of course.[95] Here is an environmental philosophy that indeed seems to be making new inroads. The most striking cultural break with the past perhaps concerns the status of humankind. Deep Ecologists argue for the equality of all life forms, with the favouring of one species above all others being termed 'speciesism', on the analogy of other discreditable -isms. For our culture, this is certainly a new perspective. In the western tradition there have always been those who have regarded human beings as an insignificant speck, for whom modesty should be the rule, as well as those who, on whatever grounds, consider humans to be near-divine. However, the notion that humans are not essentially any different from any other entities, living or non-living, is fundamentally new and indeed runs against the very grain of our culture.[96] Deep Ecologists also argue for a different view of science, thereby seeking a synthesis between physics and metaphysics. Such a synthesis is to be found in ecology, it is held, which is thereby more or less transformed into an all-encompassing worldview. In the process, the laws and concepts of ecology in the stricter sense gain moral significance: they are deemed of relevance for human society.[97] Decisions involving

---

93 M.S. Northcott 1996, 318; See also Wouter Hanegraaff's impressive study *New Age Religion and Western Culture* 1996

94 With the Deep Ecologists being the more influential of the two, certainly in the Anglo-Saxon countries. See A. Naess (1973, 1988); B. Devall & G. Sessions 1985 and particularly E. Goldsmith 1988, 1992. The term 'Land Ethic' derives from Aldo Loepold's classic *A Sand County Almanac*, first published in 1949.

95 P.W. Taylor (1984, 160), for example, speaks of "nothing less than a revolution in our ordinary ethical vision".

96 Based on historical analysis, E.C. Hargrove 1987 arrives at a similar conclusion in his article on the 'Foundations of Wildlife Protection Attitudes'.

97 On this point, *cf.* Rolston Holmes III (1975, 1988) and P.W. Taylor 1986.

difficult choices are resolved by appealing to notions of completeness and diversity. To some extent, we see here a repetition and an extension of the 1970s debate, in the Netherlands and elsewhere, regarding the presumed 'crypto-normative' nature of ecological concepts.[98] In that era, too, the main criticism concerned the sometimes subconscious blending of factual description and normative judgement.[99] The maxim that human activity should be 'ecologically compatible' is a vestige of that era. In Dutch policy, today this has made way for the pursuit of 'sustainable development'. As I have argued in Chapter 1, a subjective choice to pursue such an aim does not automatically follow from a factual analysis of our biophysical environment.[100]

This said, though, the break with the past envisaged by Deep Ecologist is in fact not quite as fundamental as is sometimes supposed.[101] Although rarely explicitly, and at times perhaps even unawares, the radical ecologists also weave various classical themes into their philosophy. The Aristotelian concept of entelechy (imbued with innate purpose) resurfaces as 'self-realization', for example.[102] While for Aristotle this concept served primarily to describe what he saw as a vital force in nature and only partly as a yardstick for human action, Deep Ecologists are concerned above all with its normative elaboration. In their adopting 'nature as the norm', the influence of the Stoics can also be discerned, of course, but again in modified form. While for the Stoic philosophers living 'according to nature' was still living 'according to the human view of how nature works', for today's Deep Ecologists the bottom line is: 'Nature knows best,

---

Although Taylor is not a 'Deep Ecologist' as such, in his environmental philosophy he steers by the compass of ecology. The same holds for the Dutch environmental philosopher W. Zweers, whose views are similar to those of the Deep Ecologists; see, for example, Zweers 2000.

[98] *Inter alia*, in a discussion paper by H. Derks 1975.

[99] G. Manenschijn (1988, 104, my translation), in discussing the views of environmental philosophers such as Taylor, writes: "We cannot simply proceed from facts to conclusions on values and standards".

[100] The Dutch environmental economist Opschoor expresses a similar view: "If people want their conduct to be ecologically compatible, they must make a conscious choice" (J.B. Opschoor, 1987, my translation).

[101] This 'two-track' strategy is in fact consciously adopted by W. Zweers 1994, who seeks continuity within the tradition, but in such a way as to yet constitute a radical break with contemporary thought.

[102] Self-realization is a pivotal concept in Naess' philosophy (*cf.* A. Naess 1976/1989, 1979, and a discussion of his work by W. Fox 1990. Taylor speaks of a "teleologically ordered hub of life" P.W. Taylor 1986, 119-122.

and ecology reveals what nature wants of us'. I use the term 'reveal' intentionally, to signal the fact that in this sense ecology fulfils a religious role.[103]

As defined and used in the writings of radical ecologists, the term 'intrinsic value' also echoes a religious attitude towards nature.[104] It represents a renaissance of animism (which has never died out entirely), with intrinsic value becoming akin to a divine presence concealed in nature and in natural entities. In the contemporary environmental debate we would do well to appreciate this religious flavour; it may well indeed be imperative. In our cosmology, the 'world-ordering triad' that imbues our lives with meaning, a void is now apparently felt where God once stood.

With this radical school of environmental philosophers we see a cosmology emerging that is a remarkable blend of animism with Aristotelian and Stoic thought. Remarkable, because while proceeding from the Greek premise of nature as the norm, these thinkers go on to fuse this with a very non-Greek, and indeed non-western, view of humanity. With these traditions thus combined, humankind and its particular responsibilities, for this world in particular, do not always fare particularly well.[105]

Over the past few decades, eco-philosophers have made a major contribution to analysing our relationship with the natural world, in both the historical and the contemporary context. They have also generated a wealth of sound and practicable ideas for building a new cosmology. This said, though, I do not feel that the more extreme variants of this school of thought we have here briefly considered, including Deep Ecology, offer a viable perspective.[106] It is also very

[103] *Cf* the title of E. Goldsmith' work *The Way: An ecological Worldview* 1992, which to the Christian reader clearly alludes to the use of the word 'way' in the New Testament (e.g. John 14:6 where Jesus said to his disciples: "I am the way, and the truth and the life").

[104] The crucial point being the definition of intrinsic value as existing separate from the subject.

[105] Hargrove points out that an ecological philosophy *pur sang* in fact leaves little or no scope for morality of action. Nature does not operate in terms of good and evil. He also poses the question whether many conservation activities are not in conflict with the 'rules of ecology' (E.C. Hargrove 1987).

[106] It is significant that extreme articulations of this vision have in North America met with accusations of 'eco-fascism' (*cf.* R.F. Nash 1989, 155 *ff.* for an evaluation of this controversy). Those who make such a charge see clear tensions between a strictly ecocentric vision and the upholding and defence of certain human values. In the Netherlands the term 'green dictatorship' is occasionally heard in this connection (Koolschijn). Although there are theoretical grounds for such

doubtful whether a cosmology so firmly grounded in an animistic and rather static view of the natural order is sufficiently dynamic and capable of providing an adequate answer to the questions confronting us as we endeavour to elaborate our notions of 'quality', in the widest sense.

There is a striking difference between this philosophical ecology, or 'deep ecology', and the traditional Christian outlook on nature. The former takes nature as the norm ('nature knows best'), with scientific ecology providing valuable support by analysing nature in such a manner as to 'show us the way'. And even if there is communication with nature, the main purpose is to listen to what nature has to tell us or, at most, to 'exchange feelings'. For deep ecologists too, though, the notion of nature being immoral or of a wild boar, wolf or killer whale being addressed authoritatively and corrected by particular people (as is done by saints) is beyond the pale.[107] This point is indeed also made by Lynn White.

In the Christian tradition ethical standards are derived from 'above', that is to say from concrete revelations of how God intended creation. Epistemologically, we are on thin ice here of course, for the existence of an 'above' and of divine revelation may also be regarded as earthly, human projections. In normative terms, modern 'eco-faiths' and religions of nature are not in fact dissimilar to classical religions of revelation, even though the former derive their ethics from the natural world and the latter from the revelations of the Bible. In a practical sense, though, there remains a world of difference. For many Christians, this earthly life is not the ultimate measure of all things, for it is their conviction that this world has been corrupted by the Fall and must also be redeemed.

Within the third current of thought I would identify, the ongoing process of reappraisal is based at least in part on a quest explic-

---

tensions, opposition to such 'dictatorship' is in part artificial and is certainly also motivated by human unwillingness to take ecological limits seriously at all. The terms fascism and dictatorship are entirely misplaced, though, and are unnecessarily insulting to those on the receiving end of such a charge, generally friendly conservationists and only very rarely haters of democracy.

[107] It may be queried, though, how much longer the characterisation of nature as wholly amoral will hold up. In California 'animal jurisdiction' was recently reintroduced in response to the (evidently morally and legally reprehensible) conduct of aggressive dogs. Previously, the dog's owner was held legally responsible. In scientific circles, too, there is a debate emerging on (the) morality of and in nature. See, for example, Frans de Waal's *Good Natured. The Origins of Right and Wrong in Humans and other Animals* 1996

itly focused within our own (western) historical heritage. A number of interesting efforts have been made in this direction.[108] In their work, social scientists of the Norbert Elias school have given new substance to the civilization motif. The Dutch sociologists Goudsblom and Tellegen, for example, advocate a "civilizing offensive" in the environmental field.[109] A responsible attitude towards the environment they hold to be a token of civilization and an improved relationship with the natural world a measure of the quality of a society, analogous to decent housing or the abolition of child labour last century.[110] In debating the pursuit of a sustainable society, it is indeed absolutely essential that there be a reappraisal of our notions of quality.[111] Once the deep-rooted historical linkage between human conceptions of the quality of life and human domination of nature is recognized, though, the conclusion is all but inevitable that such an offensive will prove almost as difficult as it is essential. The British historian Toynbee has written—thereby following Hippocrates—that overcoming (the resistance of) nature may well indeed feed the civilization process.[112] In certain areas, then, it is nothing less than a complete U-turn that must be made. Matters which for centuries

[108] Among others by G. Liedke (1972, 1979), J. Passmore 1974, B. Goudzwaard 1976, A. Linzey (1976, 1986, 1987, 1995), A. Linzey & D. Yamamoto (eds.) 1998, R. Attfield (1983, 1983[a]), L. Reijnders 1984, Ch. 10, W. Achterberg 1986, G. Manenschijn 1988, E.C. Hargrove 1989, J.B. Opschoor 1989, M. Schlitt 1992, Colin A. Russell 1994, Herman E. Daly 1996, Part VIII. and Stephan Bede Scharper 1997.

[109] J. Goudsblom, for example, in a 1992 lecture for a major Dutch environmental organization (published 1993). Tellegen in: E Tellegen & M. Wolsink 1992, Ch. 10 and E. Tellegen 1995.

[110] It is to be queried, though, whether the advances referred to have not been precisely at the expense of 'the environment' and whether nature and the environment can therefore simply be slotted into the process as the next issue. In the words of K. Thomas: "Human civilization indeed was virtually synonymous with the conquest of nature" K. Thomas, 1983, 25. On the same dilemma, *cf.* J.J. Boersema 1995.

[111] See, for example, CLTM (1990, 1993) and J.J. Boersema 1995.

[112] A. J. Toynbee (1934, Vol. III, 2) speaks of the "fallacious view that civilizations grew out of easy environments". He points to Egypt and the "stupendous human effort involved, not only in transforming the prehistoric jungle-swamp of the Lower Nile Valley into the historical land of Egypt, but also in perpetually preventing this magnificent but precarious work of men's hands from reverting to its primeval state of Nature". D. Arnold (1996, 37-38), who also discusses this connection, expresses this classical viewpoint as follows: "lands that were by nature blessed with fertile soils and natural abundance produced indolence, not civilization". He holds that "our view of the balance between nature and culture has shifted significantly since then".

have been regarded as a sign of civilization—the protection of citizens from attack by wild animals, for example, through extermination of such creatures—are now under attack for being uncivilized.

The present study can be placed within the third of these schools of thought, with an appeal being made to our own history in undertaking the required reappraisal of our cosmology. To my mind, such a reappraisal might usefully take advantage of several valuable elements of the worldviews of the ancient Israelites and Greeks, as described in the present study. In the closing pages, let me therefore outline the most important of these, with reference to the questions posed in the initial chapter of this study.[113]

Sustainability with quality

The challenge facing society today is to achieve and maintain a suitable quality of life, while reducing to a sustainable level the environmental burden to which our activities give rise. This will probably also require a different kind of economy, rooted less in material throughput (amount of material circulating in the economy per unit of time or place). In theory such an economy is feasible. By making wise use of direct and indirect solar energy, an 'ecological economy' can also grow. A number of interesting initiatives have indeed been taken in this direction, in both the technological and organisational realm.[114] Whatever the case, a change is also required at the deeper level of our *Weltanschauung*. Reduced to its essence, the 'quality of life' we are after should be sought more in activities requiring less energy, nature and raw materials: not so much in pursuance of an ideal of poverty, but ultimately because we sincerely experience that this leads to greater quality. That quality must be derived less from matters (goods and services) embodying high environmental pressure and more from activities having little impact on the environment and nature. Such activities do not necessarily have to be purely 'spiritual'; there are numerous more homely alternatives. Nature should be valued more as a defining factor of our well-being. More music and less Formula 1, an inspiring work of art

[113] The present emphasis is on conceptual ideas. For my position on several other aspects of the environmental problem, see Boersema 1998. I also largely endorse the approach taken by Opschoor 1987 and by Stephen R.L. Clark 2001.

[114] For elaboration on this route see Umweltbundesambt 1997, E.U. von Weizsäcker, A.B. Lovins & L. Hunter Lovins 1997, P. Hawken, A.B. Lovins & L. Hunter Lovins 1999.

in the garden instead of a new kitchen suite. Rather than going to the tropical swimming pool, have a conjuror come to the children's party, things like that. Instead of speeding down a dry ski slope with a group of friends, join the amateur theatre group or enjoy a good glass of wine.

As stressed in the early pages of this book, this necessitates a fundamental reappraisal of our economy, geared as it is to material throughput, as well as of the cultural norms and values associated with that economy. For this reason, it is imperative that the environmental problem be studied in its historical and cultural context. Every culture grapples with questions relating to human dealings with the natural world. Such questions, and the ensuing answers, are part of a greater whole that I have referred to as cosmology (Section 1.6). The sources of contemporary western cosmology can be traced back to the cultures of Greek antiquity and the ancient Near East, as I have demonstrated in earlier chapters with reference to a range of written sources. The texts considered in the present study still prove to be all too relevant in this context.

In conclusion, then, I would offer the following considerations as a contribution to the contemporary debate:

1. The earth does not exist for the sole benefit of humankind, and our species does not constitute the purpose of creation. The biblical creation narrative does not culminate in the creation of the human race, but in the seventh day, the sabbath. In the cosmology of the ancient Israelites, nature has a 'value of its own', a value that does not derive from nature's significance to human society. The fact that the God of the Hebrew Bible makes a covenant with the animals and blesses them is unique. This is a conception encountered neither in the contemporaneous Near East, nor among the Greek philosophers. From the creation and deluge narratives it may be concluded that God's love extends to life in all its myriad forms, a view that is echoed elsewhere in the Hebrew Bible.[115] This can serve as a cornerstone for a general ethics of nature and the environment. In justifying such an ethics, no appeal need be made to some innate attribute of animal or plants: their possession of logos, for example, or a capacity to suffer. An ethics thus grounded grants to non-human species a basic right to existence and deems any form of extinction at human hands entirely unacceptable.

[115] In Ps 24:1, for example: "The earth is the Lord's and the fullness thereof".

2. Nature is not the norm. The Hebrew Bible describes for us the non-immanent, referential character of the manifest order of creation, an order which may be admired, feared and studied, but which may never have the last word. There is thus a clear place for human responsibility and for hope. Behind the reality we perceive and in which we live, there is the hidden reality of how the world was intended. This means, on the one hand, that we must uphold a reality in which to viably live, by means of rules scaled to the human level, and, on the other, that those rules be imbued with reference to shared ideals. Such rules acknowledge imperfection, both human and worldly, while providing a rudder by which to steer, by their reference to the ideal. The dietary laws of the Hebrew Bible are a good example of such rules. The fact that such rules have vanished from the Christian tradition without anything taking their place is to be regretted.[116] There is much to be said for formulating modern-day (variants of the) dietary laws. In such an exercise, Christians might find inspiration in both the Old and the New Testament.

3. The Hebrew Bible takes a differentiated view of the natural world, with a distinction being made between domesticated and 'wild' nature. This might serve as a point of departure for further elaboration of a general ethics.[117] One of the aims of such a project would be to elaborate the terms of our dealings with nature and the associated responsibilities. Humanity has a moral obligation towards the natural world to the extent that it is under our influence and control. Many biblical passages cited in the present-day environmental debate as embodying 'benign' attitudes towards animal life and the environment were written in reference to nature-as-domesticated.[118] In today's world that category of nature evidently encompasses far more than it did two thousand years ago. The rules we adopt in deciding how we treat other life-forms, both plant and animal, hence have far greater relevance than ever before. In Chapters 2 and 3 the special status of this 'acculturated' nature was described at length. The sabbath rhythm provides a certain breathing space, in which

[116] Although in the modern world too, there is more to food than eating see Sack 2001.

[117] Such an ethics approximates the Moral Pluralism of C.D. Stone 1988 and also dovetails well with the work of Kellert *et al.* on human attitudes to animals, see A.S. Gunn 1980, S.R. Kellert (1984, 1989, 1996), S.R. Kellert & E.O. Wilson (eds.) 1993. For a practical application see N. Myers 1983.

[118] In Prov 12:10, for example: "A righteous man has regard for the life of his beast" The Hebrew word for beast (*bᵉhēmāh*) meaning domesticated animals or cattle.

human beings, the land and the animals can recover from the pace of everyday existence. The importance of this weekly rhythm for the economy as well as our social lives is enormous and is not sufficiently appreciated. Animals may never be treated simply as 'production factors'. The taking of an individual life should never be self-evident. The quality of animal and plant life is a value in its own right.[119] In this context the words of Cato the Elder (Chapter 4 note 83) are as relevant as ever. The manner in which plants and animals are exploited in the modern world as part of intensive production processes is in blatant contradiction with these tenets.

4. Humanity should exercise restraint vis-à-vis 'wild' nature. The HB provides absolutely no grounds for the conviction that humankind has been appointed as some kind of 'viceroy' over the earth. While it is obviously still necessary, as well as acceptable, for human society to protect itself against the threats and forces of nature, on large portions of the earth today this is scarcely the situation, with nature in far greater need of protection from human impact. We have only limited knowledge of complex systems, however, and there may indeed be fundamental limits to human knowledge. On this point, too, we would do well to admit that things are not quite as new as they seem: long before our own era, these fundamental limits were already appreciated by Job. And he was by no means the only one: the writings of the ancient Greek philosophers also belie a recognition of this basic truth.[120] In many instances it is therefore wiser to adopt 'the precautionary principle'.[121] Protecting nature from adverse anthropogenic impact, and simply leaving it undisturbed where possible, are extremely intelligent strategies.

5. In the way we use nature's resources, there should be a prudent balance between means and ends. The rate at which human society is presently harnessing nature for its own ends, and the encroachments being made on the natural world, are thoroughly excessive. The intensity of the present onslaught on the living environment can no longer be justified or explained in terms of the imperative to meet the basic needs of a growing world population. Human beings do not have an automatic prerogative to utilize the

[119] Even though in this case, too, it is a humanly bestowed value.

[120] As with the Sceptics and the Roman humanist Seneca: "We do not know everything, for the greater part of the universe, God, remains concealed from us" *Naturales Questiones* VII, 30.4.

[121] Timothy O'Riordan & James Cameron 1994.

whole of nature and its fruits as they see fit.[122] In striving for a balance between means and ends, human values should not predominate simply by virtue of their being human. This means that human values should not be allowed to automatically prevail over the intrinsic value of non-human life. At the same time, though, this is not to say that the two kinds of values are entirely equal; they must be weighed up, one against one another. Under threat, human life may and should take precedence over animal life.[123] In the day to day practice of nature conservation and environmental protection this is not generally the issue, however. In the vast majority of cases it is not human survival that is in any way involved, but a miscellany of affairs which, while important for the quality of our lives, can in no way be construed as life-threatening. Across the world, animal habitats are being destroyed for the sake of industrial development, dam construction, logging, fish farming and similar large-scale projects. In such cases far greater weight should be given to the intrinsic rights of the life forms at threat than is presently the case.

6. The quality of human existence will, increasingly, have to be derived from the quality of non-human creation rather than from the extent to which the latter can furnish us with material goods and services. What this entails, in a word, is a 'greening' of our system of values and norms, not as a cosmetic exercise, but from a deep-felt conviction. There must be a sea-change in our conception of nature: from serving merely as a source of prosperity, nature must also become a source of knowledge and inspiration, as well as a touchstone for the level of civilization embodied in our and any culture.[124] Ultimately, nature knows no moral standards, nor can it serve as a moral reference for our dealings with the natural world.

---

[122] Historically, our terminology has been highly coloured by a utilitarian philosophy and reflects our system of values. We thus speak of 'natural resources', 'pests', 'waste land', and so on.

[123] Peter Singer defends the position that in balancing human and animal interests due consideration should be given to the quality of both human and animal life. He writes: "The more intellectually sophisticated non-human animals have a mental and emotional life that in every significant respect equals or surpasses that of some of the most profoundly intellectually disabled human beings. This is not my subjective value-judgement. It is a statement of fact that can be tested and verified over and over again" P. Singer 2000, 221. In my view this is by no means a 'fact', but indeed a value-judgement—and one that can lead to a very dubious brand of ethics.

[124] UNEP 1999

How we treat the natural environment, though, is certainly determined by our morals and duly reflects those morals. In this respect our track record in the Christian, humanist west is nothing to be proud of, as this study has shown. We must therefore work towards a change that can serve as a basis for forging a sustainable relationship with nature. This change, this 'greening of our values', will have a far greater chance of succeeding, I suggest, if it is grounded within our familiar western tradition, while achieving a radical reappraisal on certain fundamental issues. In such an undertaking, we can find support in a variety of valuable undercurrents in the western tradition, inspired at the same time by other cultures as to how we are to shape the world of tomorrow.

7. Perhaps the most important conclusion to be drawn from this historical inquiry is that modesty is befitting: a modesty that we shall sorely need in endeavouring to resolve the environmental problems confronting us today. In this endeavour, let us hope that familiarity with the main wellsprings of our culture may allow us to avoid both the scylla of hubris and the charybdis of cynicism.[125]

[125] Such hubris, arrogance, is reflected in the title of a feature issue of *Scientific American* (September 1989) on environmental protection entitled *Managing Planet Earth*. Cynicism, scepticism and pessimism, often parading as realism, have one thing in common: they give us an alibi not to do what we as human beings must surely do: make the best of things and do more than others (Matt 5:47).

ABBREVIATIONS

ABD	*Anchor Bible Dictionary*
ACEBT	*Amsterdamse Cahiers voor exegese en bijbelse theologie*
AHW	*Akkadisches Handwörterbuch*
AJSL	*American Journal of Semitic Languages*
ANET	*Ancient Near Eastern Texts*
AS	*Asiatischen Studien*
ASOR	American Schools of Oriental Research
AUSS	*Andrews University Seminary Studies*
BAR	*Biblical Archaeologist Reader*
BASOR	*Bulletin of the American Schools of Oriental Research*
BN	*Biblische Notizen*
BZAW	*Beihefte zur Zeitschrift für die alttestamentliche Wissenschaft*
CB	*Coniectanea Biblica*
CBQ	*Catholic Biblical Quaterly*
CLTM	Commissie voor Lange Termijn Milieubeleid (Dutch Committee for Long-Term Environmental Policy)
CML-RUL	Centre for Environmental Science – Leiden University
CPB	Centraal Plan Bureau (Central Planning Agency)
DDD	*Dictionary of Deities and Demons in the Bible*
DTT	Distance-to-Target method
EKC	Environmental Kuznets Curve
ERE	*Encyclopaedia of Religion and Ethics*
ESB	*Economisch Statistische Berichten*
EUS	Environmental Utilization Space
FAO	Food and Agriculture Organisation
FRLANT	Forschungen zur Religion und Literatur des Alten und Neuen Testaments
GDP	Gross Domestic Product
GNP	Gross National Product
GTT	*Gereformeerd Theologisch Tijdschrift*
HAT	*Handbuch zum Alten Testament*
HB	Hebrew Bible
HUCA	*Hebrew Union College Annual*
I.I.S.D.	International Institute for Sustainable Development (Winnipeg)
IUCN	International Union for the Conservation of Nature (Geneva)
IVEM	Centre for Energy and Environmental Science- State University Groningen
JAOS	*Journal of the American Oriental Society*
JBL	*Journal of Biblical Literature*
JBTh	*Jahrbuch für Biblische Theologie*
JJS	*Journal of Jewish Studies*
JNSL	*Journal of Northwest Semitic Languages*
JPS	The Jewish Publication Society (Philadelphia)
JSOT	*Journal for the Study of the Old Testament*
JTS	*Journal of Theological Studies*
K-B	L. Koehler & W. Baumgartner, *Hebräisches und Aramäisches Lexikon zum Alten Testament*
KNNV	Koninklijke Nederlandse Natuurhistorische Vereniging

LCA	Life Cycle Analysis / Life Cycle Assessment
LXX	Septuagint
MT	Masoretic Text
N.S.	New Series (*MAN*)
NBG	Nederlands Bijbel Genootschap
NEPP	(Dutch) *National Environmental Policy Plan* (1989)
NICOT	New International Commentary on the Old Testament
NRLO	Nationale Raad voor Landbouwkundig Onderzoek (National Council for Agricultural Research)
NT	New Testament
NTT	*Nederlands Theologisch Tijdschrift*
NWO	Nederlandse Organisatie voor Wetenschappelijk Onderzoek (Netherlands Organization for Scientific Research)
OBO	*Orbis Biblicus et Orientalis*
OT	Old Testament
OuTS	*Oudtestamentische Studiën*
PEQ	*Palestine Exploration Quarterly*
R.L.A.C.	*Reallexikon für Antike und Christentum*
R.S.P.C.A.	Royal Society for the Protection of Cruelty against Animals
RIVM	Rijks Instituut voor Volksgezondheid en Milieu (Netherlands Institute of Public Health and the Environment)
RMB	Raad voor het Miliebeheer (Council for the Environment)
RMNO	Raad voor Milieu- en Natuur Onderzoek (Advisory Council for Research on Nature and Environment)
RVS	Revised Standard Version
SBB	*Stuttgarter Biblische Beiträge*
SCP	Sociaal en Cultureel Planburau (Social and Cultural Planning Office)
SER	Sociaal Economische Raad (Social Economic Council)
SHCANE	*Studies in the History and Culture of the Ancient Near East*
SPCK	*Society for Promoting Christian Knowledge*
SVF	*Stoicorum Veterum Fragmenta*
THAT	*Theologisches Handwörterbuch zum Alten Testament*
TUAT	*Texte aus der Umwelt des Alten Testaments*
TWAT	*Theologisches Wörterbuch zum Alten Testaments*
TWNT	*Theologisches Wörterbuch zum Neuen Testaments*
UFAW	Universities Federation for Animal Welfare
UNCED	United Nations Conference on Environment and Development
UNEP	United Nations Environmetal Program
UNESCO	United Nations Educational, Scientific and Cultural Organization
VAT	Value Added Tax
VROM	Ministerie van Volkshuisvesting, Ruimtelijke ordening en Milieubeheer (Ministry of Housing, Spatial Planning and the Environment)
VT	*Vetus Testamentum*
WCC	World Council of Churches
WCED	World Commission on Environment and Development
WCMC	World Conservation Monitoring Centre
WH	W. Houston 1993
WRR	Wetenschappelijke Raad voor het Regeringsbeleid (Netherlands Scientific Council for Government Policy)
WWF	World Wildlife Fund
ZAW	*Zeitschrift für die alttestamentliche Wissenschaft*
ZNW	*Zeitschrift für die neutestamentische Wissenschaft*

BIBLIOGRAPHY

Unless otherwise stated, the classical writers and Church Fathers have been read in the following edition: *Loeb Classical Library* W. Heinemann, London / Harvard University Press, Cambridge, Massachusetts.

Aalders, G.CH. (1949), *Het boek Genesis.* Retranslation from the original Hebrew and commentary by G.CH. Aalders, J.H. Kok, Kampen.

Achterberg, W. (1986), *Partners in de natuur. Een onderzoek naar de aard en de fundamenten van een ecologische ethiek.* Jan van Arkel, Utrecht.

Achterberg, W. (1994), *Samenleving, natuur en duurzaamheid: een inleiding in de milieufilosofie.* Van Gorcum, Assen.

Achterberg, W. & W. Zweers (eds.) (1984), *Milieucrisis & Filosofie. Westers bewustzijn en vervreemde natuur.* Ekologische Uitgeverij, Amsterdam.

Achterhuis, H.J. (1992), *De illusie van groen. Over milieucrisis en de fixatie op techniek.* De Balie, Amsterdam.

Achterhuis, H.J. (1992[a]), *Technologie en filosofie: beeld en werkelijkheid.* Inaugural speech, 24 September 1992, Twente University, Enschede.

Achterhuis, H.J. (1995), *Natuur tussen mythe en techniek.* Ambo, Baarn.

Achterhuis, H.J. (1998), *De erfenis van de utopie.* Ambo, Amsterdam.

Achtner, W. u.a. (1998), *Dimensionen der Zeit: Die Zeitstrukturen Gottes, der Welt und des menschen.* Primus, Darmstadt

Ackerman, S. (1992), *Under Every Green Tree. Popular Religion in Sixth-Century Judah.* Harvard Semitic Monographs 46, Scholars Press, Atlanta, Georgia.

Agenda 21. cf. UNCED.

Agwan, A.R. (ed.) (1997), *Islam and the Environment.* Synergy Book International, Kuala Lumpur, Malaysia.

Aharoni, J. (1926), Ueber das Vorkommen und Aussterben Palästinische Tierarten. *Zeitschrift des Deutschen Palästina Vereins,* XLIX, 247-262.

Aharoni, Y. (1979), *The Land of the Bible: A Historical Geography.* Second revised Edition, translated from the Hebrew and edited by A.F. Rainey, Burns & Oates, London.

Ahlström, G.W. (1986), *Who were the Israelites?* Eisenbrauns, Winona Lake, Indiana.

Albertz, R. (1992), *Religionsgeschichte Israels in alttestamentlicher Zeit.* 2 vols., Vanden-Hoeck & Ruprecht, Göttingen.

Alblas, W. & H. Sips (1995), Voortgaande keuzes. *Tijdschrift voor Politieke Economie,* 4, 94-95.

Albrektson, B. (1967), *History and the Gods. An Essay on the Idea of Historical Events as Divine Manifestations in the Ancient Near East and in Israel.* Coniectanea Biblica, Old Testament Series 1, CWK Gleerup, Lund, Sweden.

Albrektson, B. (1981), Difficilior Lectio Probabilior. In: A.S. van der Woude (ed.), *Remembering all the Way...A Collection of Old Testament Studies Published on the Occasion of the Fortieth Anniversary of the Oudtestamentisch Werkgezelschap in Nederland.* Oudtestamentische Studiën, vol. XXI, E.J. Brill, Leiden, 5-18.

Albright, W.F. (1950), *The Biblical Period.* The Biblical Colloquium, Pittsburg.

Algra, K. (1988), The Early Stoics on the Immobility and Coherence of the Cosmos. *Phronesis,* XXXIII, 2, 155-181.

Alon, A. (1969), *The Natural History of the Land of the Bible.* Hamlyn, London.

Alon, Gedalyahu (1977), *Jews, Judaism and the Classical World*. Translated from the Hebrew by Israël Abrahams, The Magnes Press, Hebrew University, Jerusalem.

Alsford, S.E. (1991), Evil in the Non-Human World. *Science and Christian Belief*, 3, 119-130.

Alster, B. (1999), lemma *Tiamat* in *DDD*, 867-869.

Alter, R. (1981), *The Art of Biblical Narrative*. George Allen & Unwin, London/Sydney.

Altner, G. (1985), Aktuelle und historische Tendenzen im Gespräch zwischen Theologie und Naturwissenschaften. *Evang. Theologie*, 45, 248-61.

Altner, G. (ed.) (1989), *Ökologische Theologie: Perspektiven zur Orientierung*. Kreuz Verlag, Stuttgart.

Altner, G. (1991), *Naturvergessenheit. Grundlagen einer umfassenden Bioethik*. Wissenschaftliche Buchgesellschaft, Darmstadt.

Altner, G., G. Liedke *et al.* (1984), *Manifest zur Versöhnung mit der Natur. Die Pflicht der Kirchen in der Umweltkrise*. Neukirchener Verlag.

Altpeter, G. (1976), *Hebräisch-Deutsche Präparation zum Deuteronomium*. Oekumenischer Verlag Dr. R-F. Edel, Marburg.

Amery, C. (1971), *Das Ende der Vorsehung. Die gnadenlosen Folgen des Christentums*. Verlag Reinbek (bei Hamburg).

Amery, C. (1985), *Die ökologische Chance. Das Ende der Vorsehung & Natur als Politik. Mit einem 'Nachwort 1985'*. Gesammelte Werke in Eizelausgaben, Süddeutscher Verlag, München.

Amir, Y. (1983), *Die hellenistische Gestalt des Judentums bei Philon von Alexandrien*. Neukirchener Verlag, Neukirchen-Vluyn.

Anderson, J.K. (1985), *Hunting in the Ancient World*. University of California Press, Berkeley, California.

Animal Rights (1979), Special Issue of: *Inquiry*, 22, 1/2, Summer.

Anon. (1963), *Het dier als motief in de antieke kunst*. exhibition catalogue no. 6, Rijksmuseum van Oudheden, Leiden, 2 March-4 June 1963.

Anon., *Das Tier in der Antike*. 400 Werke Ägyptischer, Griechischer, Etruskischer und Römischer Kunst aus privatem und öffentlichem Besitz. Archäologisches Institut der Universität Zürich, Rämistrasse 73, Ch-8006 Zürich.

Aquinas, Thomas (1942-1960), *Summa contra Gentiles*. Translated from the Latin by H. Fahsel, 6 vols., Zürich.

Aristotle (1989), *Problemata Physica Arabica*. (*see:* Filius, 1989).

Arkel, D. van (1974), Society and Technology: 30,000 Years in Shorthand. In: H.G. van Raay & A.E. Lugo (eds.) (1974), *Man and Environment Ltd., Natural Imbalances and social Justice*. Rotterdam University Press, Rotterdam, 19-33.

Armstrong, A.H. (1947/1981), *An Introduction to Ancient Philosophy*. University Paperback Reprint, 1981, Methuen & Co. Ltd., London.

Armstrong, A.H. (1972), Neoplatonic Valuations of Nature, Body and Intellect. *August. Stud.*, 3, 35-54.

Armstrong, A.H. (1973), Man in the Cosmos: A Study of Some Differences between Pagan Neoplatonism and Christianity. In: *Romanitas et Christianitas Studia J.H. Waszink oblata*, Amsterdam, London.

Armstrong, E.A. (1973), *St. Francis: Nature Mystic. The Derivation and Significance of the Nature Stories in the Franciscan Legend*. University of California Press, Los Angelos, Ca.

Armstrong, G.T. (1962), *Die Genesis in der Alten Kirche. Die drei Kirchenväter*. J.C.B. Mohr (Paul Siebeck), Tübingen.

Arnim, Ioannes Ab. (1903), *Stoicorum Veterum Fragmenta (SVF)*. 4 volumes, Lipsiae in Aedibus B.G. Teubneri.

Arnold, D. (1996), *The Problem of Nature. Environment, Culture and European Expansion.* Blackell Publishers, Oxford.

Arnold, E.V. (1911), *Roman Stoicism.* Cambridge University Press, Cambridge.

Aronsson, T., P.-O. Johansson & K.-G. Löfgren (1997), *Welfare Measurement, Sustainability and Green National Accounting. A Growth Theoretical Approach.* Edward Elgar, Cheltenham UK / Brookfield Vermont.

Arrow, K.J. (1951), *Social Choice and Individual Values.* Yale University Press, New Haven, Connecticut.

Arrow, K.J. *et al.* (1995), Economic Growth, Carrying Capacity and the Environment. *Science*, 268, 520-521.

Asselin, D.T. (1954), The Notion of Dominion in Genesis 1-3. *CBQ*, 16, 277-294.

Atran, S. (1990), *Cognitive Foundations of Natural History: towards an Anthropology of Science.* Cambridge University Press, Cambridge.

Attfield, R. (1983), Christian Attitudes to Nature. *Journal of the History of Ideas* 44(3), 369-386.

Attfield, R. (1983[a]), *The Ethics of Environmental Concern.* Basil Blackwell, Oxford.

Attfield, R. (1991), Has the History of Philosophy ruined the Environment? *Environmental Ethics*, 13, 2, 127-137.

Attfield, R. (1999), *The Ethics of the Global Environment.* Edinburgh Studies in World Ethics, Edinburgh University Press, Edinburgh.

Attfield, R. & B. Wilkins (eds.) (1992), *International Justice and The third World. Studies in the Philosophy of Development.* Routledge, London/New York.

Auguet, R. (1994), *Cruelty and Civilization. The Roman Games.* Routledge, London/ New York.

Baarschers, W.H. (1996), *Eco-facts & Eco-fiction. Understanding the Environmental Debate.* Routledge, London/New York.

Baarsma, B. (2000), *Monetary Valuation of Environmental Goods. Alternatives to Contingent Valuation.* Dissertation Amsterdam University, Amsterdam.

Bacon, Fr. *The Works of Francis Bacon.* Spedding, Ellis and Heath, London.

Bacon, Fr. (1978), *Essays.* Translated and and with an introduction by Willem Dijkhuis, Boom, Meppel, Amsterdam.

Bahn, P. & J. Flenley (1992), *Easter Island, Earth Island.* Thames and Hudson, London

Bailey, R. (ed.) (1995), *The True State of the Planet.* The Free Press, New York.

Banton, M. (ed.) (1966), *Anthropological Approaches to the Study of Religion.* SCM Press, London

Baranzke, H. (1991), Ökologie—Natur—Schöpfung. Zur Funktion einer Schöpfungstheologie im Rahmen der Umweltproblematiek. *Stimmen der Zeit*, 209, 695-706.

Baranzke, H. & H. Lamberty-Zielinski (1995), Lynn White und das Dominium Terrae (Gen. 1, 28[b]). Ein Beitrag zu einer doppelten Wirkungsgeschichte. *BN*, 76, 32-61.

Barbour, I.G. (ed.) (1972), *Earth Might be Fair.* Prentice Hall, Englewood Cliffs, New Jersey.

Barbour, I.G. (ed.) (1973), *Western Man and Environmental Ethics. Attitudes towards Nature and Technology.* Addison-Wesley, Reading, Massachusetts.

Barbour, I.G. (1998), *Religion and Science: Historical and Contemporary Issues.* Revised and expanded edition, SCM Press, London.

Barbour, J. (1999), *The End of Time.* Weidenfeld & Nicolson, London/New York.

Barr, J. (1972), Man and Nature: The Ecological Controversy and the Old Testament. *Bulletin of the John Rylands Library*, 55, 1, 9-32.

Barrow, C.J. (1995), *Developing the Environment. Problems & Management.* Longman Scientific & Technical, Essex.

Bartels, G.C. (1994), *Saillante resultaten uit het milieucommunicatie-onderzoek.* VROM, The Hague.

Bartels, Kl. & L. Huber (eds.) (1965), *Lexikon der Alten Welt.* Artemis Verlag, Zürich/ Stuttgart.

Barth, K. (1945/1970), *Die Kirchliche Dogmatik 3/1. Die Lehre von der Schöpfung.* Zürich.

Bateson, G. (1979), *Mind and Nature, a necessary Unity.* E.P. Dutton, New York.

Bauer, J.B. (1985), Hase. *R.L.A.C.*, XIII/99, 662-678.

Baumann, H. (1982), *Die Griechische Pflanzenwelt in Mythos, Kunst und Literatur.* Hirmer Verlag, München.

Bavel, T.J. van (1990), De Kerkvaders over de Schepping. *Tijdschrift voor Theologie*, 30, 1, 18-34.

Beauchamp, P. (1969), *Création et Séperation. étude exégétique du Chapitre premier de la Genèse.* Biliothèque de Sciences religieuses. Aubier Montaigne—éditions du Cerf. Delachaux & Nestlé—Desclée de Brouwer.

Beavis, I.C. (1988), *Insects and Other Invertebrates in Classical Antiquity.* Exeter University Publications, Exeter.

Beck, H.L., M. de Jonge, P.S. van Koningsveld, K. van der Toorn & T.E. Vetter (1997), *Grondleggers van het geloof.* Prometheus, Amsterdam.

Beck, U. (1992), *Risk Society. Towards a New Modernity.* Sage, London.

Beckerman, W. (1992), *World Development Report.* 20, World Bank, Washington DC.

Beckerman, W. (1995), *Growth, the Environment and the Distribution of Incomes: Essays by a Sceptical Optimist.* Edgar Elgar, Aldershot.

Beckerman, W. (1995[a]), *Small is Stupid: Blowing the Whistle on the Greens.* Duckworth, London.

Beek, A. van de (1996), *Schepping. De wereld als voorspel voor de eeuwigheid.* G.F. Callenbach, Baarn.

Beekman, V. (2001), *A Green Third Way? Philosophical Reflections on Government Intervention in non-sustainable Lifestyles.* Thesis, Wageningen University, Wageningen.

Beinart, W. & P. Coates (1995), *Environment and History. The taming of nature in the USA and South Africa.* Routledge, London/New York.

Belue, Ted Franklin (1996), *The Long Hunt. Death of the Buffalo East of the Mississippi.* Stackpole Books, Mechanicsburg, PA.

Ben-Amos, Paula (1976), Men and Animals in Benin Art. *Man* (N.S.), 11, 243-253.

Bentham, J. (1948), *Introduction to the Principles of Morals and Legislation.* Reprint, Hafner, New York.

Bereishis (1977), *Genesis / A New Translation with a Commentary anthologized from Talmudic, Midrashic and Rabbinic Sources.* Translation and Commentary by Rabbi Meir Zlotowitz, Overviews by Rabbi Nosson Scherman, Mesorah Publications Ltd.

Berger, A. (1953), Encyclopedic Dictionary of Roman Law, *Transactions of the American Philosophical Society* (N.S.), 43, part 2, 1953, 333-808.

Berkhof, H. (1968), God in Nature and History.in: *New Directions in Faith and Order,* Bristol 1967, 7-31 World Councel of Churches, Geneva.

Berkhof, H. (1973), *Christelijk geloof. Een inleiding tot de geloofsleer.* Callenbach, Nijkerk.

Bertels, J. *et al.* (1995), *Ruimte als schaars goed.* CML-RUL, Leiden.

Berry, R.J. (1991), Christianity and the Environment. *Science and Christian Belief,* 3, 3-14.

Bickerman, E.J. (1988), *The Jews in the Greek Age.* Harvard University Press, Cambridge, Massachusetts.

Biesiot, W. & A.J.M. Schoot Uiterkamp (eds.) (1993), *Transitie naar duurzaamheid en kwaliteit.* Report no. 63, IVEM, State University of Groningen, Groningen.

Biesiot, W. (1995), Personal communication.

Bilsky, L.J. (ed.) (1980), *Historical Ecology Essays on Environment and Social Change.* National University Publications, Kennikat Press, Port Washington, New York/London.

Biodiversity. See WCMC, World Conservation Monitoring Centre.

Birch, L.C. (1965), *Nature and God.* SCM Press Ltd., Bloomsbury Street, London.

Birch, Ch. & J.B. Cobb Jr. (1981/1992), *The Liberation of Life.* Cambridge University Press, Cambridge, (paperback edition, Environmental Ethics Books 1992).

Birch, Ch., W. Eakin & J.B. McDaniel (eds.) (1991), *Liberating Life. Contemporary approaches to ecological theology.* (report to the WCC) Orbis Books, Maryknoll, New York.

Birch, Ch. & L. Vischer (1997), *Living with the Animals. The Community of God's Creatures. Rosk Book Series,* WCC (World Council of Churches) Publications, Geneva.

Black, J. (1970), *The Dominion of Man, the Search for Ecological Responsibility.* University Press, Edinburgh.

Blackstone, William (ed.) (1974), *Philosophy and Environmental Crisis.* University of Georgia Press, Athens, Georgia.

Blavatsky, H.P. (1886/1887), Have Animals Souls. *Collected Writings,* vol. VII, The Theosophical Publishing House, Wheaton, Illinois/Madras, India, 12-49.

Bleich, J.D. (1986), Judaism and Animal Experimentation. In: T. Regan (ed.) *Animal Sacrifices. Religious Perspectives on the Use of Animals in Science.* Temple University Press, Philadelphia, 61-115.

Blidstein, G.J. (1966), Man and Nature in the Sabbatical Year. *Tradition,* 9, 48-55.

Bodenheimer, F.S. (1935), *Animal Life in Palestine.* Jerusalem (transl. of *Die Tierwelt Palästinas,* Leipzig, 1920).

Bodenheimer, F.S. (1960), *Animal and Man in Bible Lands.* E.J.Brill, Leiden.

Bodson, L. (1983), Attitudes toward Animals in Greco-Roman Antiquity. *International Journal for the Study of Animal problems,* 4, 4, 312-320.

Boer, Margaretha de (1995), *The Environment, Space and Living Quality. Time for Sustainability.* VROM, The Hague.

Boersema, J.J. (1985), *Overzicht van theorieën die de milieuproblematiek proberen te verklaren.* Syllabus cursus Milieukunde, IVEM-RuG, Groningen.

Boersema, J.J. (1986), Arend onder de liederen. Het raadsel van de vale gier en zijn jongen. *Trouw,* 28-11-1986.

Boersema, J.J. (1994), First the Jew but also the Greek: In search of the Roots of the Environmental Problem in Western Civilization. In: W. Zweers & J.J. Boersema (eds.) *Ecology, Technology and Culture.* The White Horse Press, Knapwell, Cambridge, 20-56.

Boersema, J.J. (1995), Environmental Quality and the Quality of our Way of Life. *Environmental Values,* 4, 97-108.

Boersema, J.J. (1998), Milieu en energie. In: J.W. van Deth & J.C.P.M. Vis (eds.) *Actuele politieke problemen.* Instituut voor Publiek en Politiek, Amsterdam, 105-135.

Boersema, J.J. (2001), How to Prepair for the Unknown? On the Significance of Future Generations and Future Studies in Environmental policy. *Environmental Values,* 10, 35-58.

Boersema, J.J. (2002), Why is Francis of Assisi the patron saint of ecologists? *Science and Christian Belief,* 14/1 (in press).

Boersema, J.J. *et al.* (eds.) (1994), *Basisboek Milieukunde.* 5e druk, Boom, Meppel, Amsterdam.

Boersema, J.J. & J. Bertels (2000), Sustainable Development in the Developed Countries: Will Theory and Practice Meet? In: Keekok Lee, Alan Holland & Desmond McNeil (eds.), *Global Sustainable Development in the Twenty-first Century*, Edinburgh University Press, Edinburgh, 77-97.

Boessneck, J. (1988), *Die Tierwelt des Alten Ägypten. Untersucht anhand kulturgeschichtlicher und zoologischer Quellen.* C.H. Beck, München.

Boessneck, J. & A von der Driesch (1978), "Preliminary Analysis of the Animal Bones from Tell Hesbân". *AUSS*, 16, 259-287.

Böhl, F.M.Th. de Liagre (1923), *Genesis.* vol. 1, Groningen/The Hague.

Boholm, A. (1996), Risk Perception and Social Anthropology: Critique of Cultural Theory. *Ethnos*, 61, 1-2, 64-84.

Boman, Thorleif (1952/1968), *Das hebräische Denken im Vergleich mit dem griechischen.* Vandenhoeck & Ruprecht, Göttingen.

Boon, D. (1983), *Nederlands Dierenrecht.* Gouda Quint B.V., Arnhem.

Boon, D. (1986) De eigenwaarde van dieren en het dierenrecht. *Filosofie & Praktijk*, 7, 4, 198-204.

Bos, A.P. (1989), *Aristoteles Over de kosmos.* Introduction, translation and provided with explanatory notes by A.P. Bos, Boom, Meppel

Bos, A.P. (1991), *In de greep van de Titanen. Inleiding tot een hoofdstroming van de Griekse filosofie.* Buijten & Schipperheijn, Amsterdam.

Boserup, E. (1981), *Population and Technology.* Basil Blackwell, Oxford.

Boserup, E. (1993), *The Conditions of Agricultural Growth.* Earthscan Publications Ltd., London.

Bossel, H. (1998), *Earth at the Crossroads. Paths to a Sustainable Future.* Cambridge University Press, Cambridge.

Bossel, H. (1999), *Indicators for Sustainable Development: Theory, Method, Applications.* A Report to the Balaton Group, International Institute for Sustainable Development (I.I.S.D.), Winnipeg.

Botke, IJ. (1984), *Jacobus Albertus Uilkens (1772-1825). Predikant te Lellens en Eenrum. Hoogleraar in de Landhuishoudkunde te Groningen.* Grongingen University Museum.

Botkin, D.B. (1996), *Our Natural History. The Lessons of Lewis and Clark.* A Perigee Book, New York.

Bots, J. (1972), *Tussen Descartes en Darwin. Geloof en natuurwetenschap in de achtiende eeuw in Nederland.* Van Gorcum & Comp. N.V., Assen.

Bottema, S. (1993), *East is East and West is West.* Inaugural speech, 18 June 1993, Leiden State University.

Bottéro, J. (1994), *Babylone et la Bible. Entretiens avec Hélène Monsacré.* 2nd edition, Les Belles Lettres, Paris.

Bouma, Hans (1974), *De Aarde is er ook nog. Op zoek naar een ethiek van het milieubeheer.* Zomer & Keuning, Wageningen.

Bowler, P.J. (1992), *The Fontana History of The Environmental Sciences.* Fontana Press, London.

Boxel, P.W. van (1987), *Sjabbatskind; Vroegjoodse tradities over leven en dood.* Gooi en Sticht, Hilversum.

Bramwell, A. (1994), *The Fading of the Greens. The Decline of Environmental Politics in the West.* Yale University Press, New Haven.

Brands, J. (1935), *Griekse diernamen.* Dissertation, Catholic University of Nijmegen.

Brandt, W. (1910), Jüdische Reinheitslehre und ihre Beschreibung in den Evangelien. *Beihefte zur ZAW (Zeitschrift für die Alttestamentliche Wissenschaft)*, XIX. Verlag Alfred Töpelmann, Gießen.

Bratsiotis, N.P. (1973), Lemma *'îš/'iššāh* in *TWAT*, vol. I, 238-252.

Bratton, S.P. (1984), Christian Ecotheology and the Old Testament. *Environmental Ethics*, 6, 195-209.

Bratton, S.P. (1988), The Original Desert Solitaire: Early Christian Monasticism and Wilderness, *Environmental Ethics*, 10, 1, 31-53.

Bratton, S.P. (1993), *Christianity, Wilderness and Wildlife. The Original Desert Solitaire.* Associated University Presses, Inc.

Bream H.N., R.D. Heim & C.A. Moore (eds.) *A Light unto My Path: Old Testament Studies in Honour of Jacob M. Myers.* Temple University Press, Philadelphia.

Brehm, A.E (1890-), *Brehms Tierleben: allgemeine Kunde des Tierreichs.* 10 vols., Bibliographical Institute, Vienna.

Bremmer, J.N. (1983), *The Early Greek Concept of the Soul.* Princeton University Press, Princeton, New Jersey.

Bremmer, J.N. (1992), Symbols of Marginality from Early Pythagoreans to Late Antique Monks. *Greece & Rome,* 24, 2, 205-214.

Bremmer, J.N. (1994), *Greek Religion.* Oxford University Press, Oxford.

Bremmer, J.N. (1997), Near Eastern and Native Traditions in Apollodorus' Account of the Flood. In: F. García Martínez & G. Luttikhuizen (eds.), *Interpretations of the Flood.* Kok Pharos, Kampen, 45-60.

Bremmer, J.N. & F. García Martínez (eds.) (1992), *Sacred Texts in early Judaism.* A Symposium in Honour of A.S. van der Woude. Kok Pharos, Kampen.

Brennan, A. (1988), *Thinking About Nature. An Investigation of Nature, Value and Ecology.* Routledge, London.

Breukelman, F.H. (1992), *De theologie van het boek Genesis.* Bijbelse Theologie I, 2, Kok, Kampen.

Brink, C.O. (1956), Oikeiosis and oikeiotes: Theophrastus and Zeno on Nature in Moral Theology. *Phronesis*, 1, 123-45.

Brimblecombe, P. (1975), Industrial air pollution in thirteenth-century Britain. *Weather*, XXX, 388-96.

Brimblecombe, P. (1987), *The Big Smoke.* London/New York

Broer, H.W. & F. Verhulst (ed.) (1990), *Dynamische systemen en chaos; revolutie vanuit de wiskunde.* Epsilon, Utrecht.

Brooke, J. & G. Cantor (1998), *Reconstructing Nature: The engagement of Science and Religion.* Gifford Lectures, T&T Clark, Edinburgh.

Brown, J.P. (1983), Men of the Land and the God of Justice in Greece and Israel. *ZAW,* 95, 376-402.

Brown, L.R. (1989-), *State of the World.* W.W. Norton & Company, New York.

Brown, L., H. Kane & E. Ayres (1993-), *Vital Signs 1993-1994. The Trends that are shaping our Future.* Earthscan Publications Ltd, London.

Brown, P. (1967), *Augustine of Hippo.* University of California Press, Berkeley and Los Angeles.

Brown, W.N. (1974), The Unity of Life in Indian Religion. In: Klaits, J. and Klaits, B.G. (eds.) *Animals and Man in Historical Perspective,* Harper & Row, New York, 51-59.

Brueggemann, W. (1993), *Using God's Resources Wisely. Isaiah and Urban Possibility.* Westminster/John Knox Press, Louisville, Kentucky.

Brueggemann, W. (1997), *Theology of the Old testament. Testimony, Dispute, Advocacy.* Fortress Press, Minneapolis.

Bruijel, F.J. (1939), *Bijbel en Natuur, studies over planten en dieren uit de Heilige Schrift.* J.H. Kok, Kampen.

Bruijel, F.J. (1948), Roofvogels in den Bijbel. *Gereformeerd Theologisch Tijdschrift*, I: 112-124, II: 186-191, III: 236-245.

Bruyn, S.M. de (1997), De ontkoppeling ontmaskerd. *ESB*, 82, 515-517.

Bruyn, S.M. de (1999), *Economic Growth and the Environment. An Empirical Analysis.* Tinbergen Institute Research Series 216, Free University, Amsterdam.

Brümmer, V. (ed.) (1991), *Interpreting the Universe as Creation. A Dialog of Science and Religion.* Kok Pharos, Kampen.

Buber, M. (1964), *Werke II. Schriften zur Bibel.* München.

Buffon, Comte de, Georges-Louis Leclerc (1749-1804), *Histoire Naturelle, Générale et Particulière.* 44 vols., Impremerie Royale, puis Plassan, Paris.

Buijs, L. & M. Storm (1994), *Tora. De onderwijzing van Mosje.* Em. Querido's Uitgeverij, Amsterdam.

Bulmer, R. (1967), Why is the Cassowary not a Bird? A Problem of Zoological Taxonomy among the Karam of the New Guinea Highlands. *Man* (N.S.), 2, 5-25.

Bulletin of the Smithsonian Institution, CLXVII (1937), 302.

Butler's *Lives of the Saints.* (1756-59/1985), Concise Edition, Michael Walsh (Ed.), Harper & Row, San Francisco/New York.

Burkert, W. (1960), *Hermes,* 88, 159-177.

Burkert, W. (1972), *Lore and Science in Ancient Pythagoreanism.* Translation with revisions by Edw. L. Minar, Harvard University Press, Cambridge, Massachusetts.

Burkert, W. (1983), *Homo Necans.* Translation P. Bing, University of California Press, Berkeley, California.

Burkert, W. (1985), *Greek Religion.* Translation J. Raffin, Harvard University Press, Cambridge, Massachusetts.

Burkert, W. (1987), The Problem of Ritual Killing. in: Hamerton-Kelly, R.G. (ed.) (1987), *Violent Origins.* Burkert, W., R. Girard & J.Z. Smith on Ritual Killing and Cultural Formation. With an introduction by Burton Mack and a commentary by Renato Rosaldo, Stanford University Press, Stanford, California, 149-188.

Burkert, W. (1996), *Creation of the Sacred. Tracks of Biology in Early Religions.* Harvard University Press, Cambridge, Massachusetts.

Caduff, G.A. (1986), *Antike Sintflutsagen.* Hypomnemata Heft 82, Vandenhoeck & Ruprecht, Göttingen (Dissertation, Zürich, 1978).

Callicott, J. Baird, (1985), Intrinsic Value, Quantum theory and Environmental Ethics. *Environmental Ethics*, 7, 257-275.

Callicott, J. Baird (1986), The Metaphysical Implications of Ecology. *Environmental Ethics,* 8/3, 301-316.

Callicott, J. Baird, (1989), *In Defense of the Land Ethic. Essays in Environmental Philosophy.* State University of New York Press, Albany, New York.

Callicott, J. Baird, (1990), Genesis Revisited: Murian Musing on the Lynn White, Jr. Debate, *Environmental History Review,* 14, 65-90.

Callicott, J. Baird (1990), The Case against Moral Pluralism. *Environmental Ethics,* 12, 2, 99-124.

Callicott, J. Baird (1992), Can a Theory of Moral Sentiments support a Genuinely normative environmental Ethic? *Inquiry,* 35, 2, 183-198.

Calvin, J. (1554/1965), *Commentary on Genesis.* Translation 1847, reprinted in 1965, Banner of Thruth publishers, London.

Campbell, E.F. Jr. & D.N. Freedman (eds.) (1983), *The Biblical Archaeologist Reader.* Vol. 4 The Almond Press / ASOR, Sheffield.

Cansdale, G. (1970), *All the Animals of the Bible Lands*. Zondervan Publishing House, Grand Rapids, Michigan.
Capponi, Filippo (1979), *Ornithologica Latina*. Inst. van Filologia Classica e Medievale, no. 58, University of Genua.
Caras, R.A. (1998/1999), *A Perfect Harmony. The Intertwining Lives of Animals and Human Thought*. Simon & Schuster, New York.
Carcopino, Jérôme, *Het dagelijks leven in het oude Rome*. 2 vols. Prisma boeks no. 360, 361, Spectrum Utrecht/Antwerp.
Carey, H.C. (1848), *The Past, the Present and the Future*. Philadelphia.
Carmody, John (1983), *Ecology and Religion: Toward a Christian Theology of Nature*. Paulist Press, New York.
Carpenter, B. (1991), A Panther by another Name. *U.S. News & World Report*, June 17, 1991.
Carrier, L.S. (1980), Perception and Animal Belief. *Philosophy*, 55, 193-210.
Carson, Gerard (1972), *Man, Beasts and Gods, A History of Cruelty and Kindness to Animals*. Charles Scribner's Sons, New York.
Carson, R. (1962), *Silent Spring*. Crest Book, Fawcett Publications Inc., Greenwich, Connecticut.
Casetti, P. (1982), Gibt es ein Leben vor dem Tod? *OBO*, 44, Freiburg/Göttingen.
Casas, Bartholomé de Las (1992), *De verwoesting van de West-indische landen*. Translation and introduction by M.J. van Nieuwstadt, SUN, Nijmegen.
Cassuto, U. (1944/1961/1989), *A Commentary on the Book of Genesis*. 2 vols. Translated from the Hebrew by Israel Abrahams, The Magnes Press, The Hebrew University, Jerusalem.
Castiglioni, A. (1974), The Contribution of the Jews to Medicine. In: L. Finkelstein (ed.), *The Jews, Vol. III, their Rôle in Civilization*. Schocken Books, New York.
Causey, A.S. (1989), On the Morality of Hunting. *Environmental Ethics*, 11, 4, 327-343.
Chadwick, Henry (1966), *Early Christian Thought and the Classical Tradition. Studies in Justin, Clement and Origen*. Oxford, 1966.
Chantraine, P. (1968-), *Dictionnaire étymologique de la langue Grècque*, Paris.
Charlesworth, J.H. (ed.) (1985), *The Old Testament Pseudepigrapha*. 2 vols., Doubleday, Garden City, New York.
Cheney, J. (1989). The Neo-Stoicism of Radical Environmentalism. *Environmental Ethics*, 11, 4, 293-325.
Choosing a Sustainable Future (1993) The Report of the National Commision on the Environment, Island Press, Washington D.C.
Clagett, Marshall (1955/1988), *Greek Science in Antiquity*. Ayer Publishing Salem, N.H. (1955), Scholars Bookshelf Paperback Printing (1988), Princeton Junction, New Jersey.
Clark, G. (1992/1994), *Space, Time and Man. A Prehistorian's View*. Canto Cambridge University Press, Cambridge.
Clark, Stephen R.L. (1979), The Rights of Wild Things. *Inquiry*, 22, 171-88.
Clark, Stephen R.L. (1977/1984), *The Moral Status of Animals*, Clarendon Press, Oxford. Oxford University Press, Oxford (paperback 1984).
Clark, Stephen R.L. (1982/1984), *The Nature of the Beast. Are Animals Moral?* Oxford University Press, Oxford (paperback 1984).
Clark, Stephen R.L. (1990), Notes on the Underground. *Inquiry*, 33, 1, 27-37.
Clark, Stephen R.L. (2001), *Biology and Christian Ethics*. New Studies in Christian Ethics 17, University of Cambridge Press, Cambridge.

Clark, W.M. (1968), The Animal Series in the Primeval History. *Vetus Testamentum,* XVIII, 433-449.
Clarke, F., B.J. Wright, G.H.J. Leonard, R. Roemes, Hicks & Hicks, P.& P. Clark & P. Brook (1967), Notes on Griffon Vultures, *14, Bird Rep. Cyprus Orn. Soc,* 41-42.
Clarke, W.K. Lowther (1953), *Concise Bible Commentary.* MacMillan Company, New York.
Clason, A.T. (ed.) (1975), *Archaeozoological Studies.* North Holland Publishing Company, Amsterdam/Oxford.
Clemen, C. (1908/1912), *Primitive Christianity and its non-Jewish Sources.* Translated by R.G.Nisbet, T.& T. Clark, Edinburgh.
Clifford, R.J. (1994), *Creation Accounts in the Ancient Near East and the Bible.* The Catholic Biblical Quarterly Monograph Series 26, The Catholic Biblical Association of America, Washington DC.
CLTM (Netherlands Committee for Long-Term Environmental Policy) (1990), *Het Milieu: denkbeelden voor de 21ste eeuw.* Kerckebosch, Zeist.
CLTM (1993), *The Environment: Towards a Sustainable Future.* Kluwer Academic Press, Dordrecht/Boston/London.
Clutton-Brock, J. (1987), *A Natural History of Domesticated Animals.* Cambridge University Press, Cambridge.
Clutton-Brock, J. & C. Grigson (eds.) (1984), Animals and Archeology. III Early Herders and their Flocks. *BAR International Series,* 202, Oxford.
Cobb, John B., Jr. (1972). *Is it too late? A Theology of Ecology.* Bruce, Beverly Hills, California.
Cobb, John B. Jr. (1992), *Sustainability. Economics, Ecology and Justice.* Orbis Books, Maryknoll, New York.
Cobb, John B. Jr. (1994), *Sustaining the Common Good. A Christian Perspective on the Global Economy.* The Pilgrim Press, Cleveland, Ohio
Cobbe, Fr. P. (1895). The Ethics of Zoophily. *Contemporary Rev.*, 68.
Cohen, J. (1989), *"Be fertile and increase, fill the earth and master it" : The Ancient and Medieval Career of a Biblical Text.* Cornell University Press, Ithaca.
Cohen, N.J. (1959). *Tsa'ar Ba'ale Ḥayim—The Prevention of Cruelty to Animals: Its Bases, Development and Legislation in Hebrew Literature.* PhD theses, Cath. University of America, The Catholic University of America Press, Washington D.C.
Cohn, Leopold (ed.) (1967), *Philonis Alexandrini libellus De Opificio Mundi.* Georg Olms Verlag, Hildesheim.
Cohn, N. (1996), *Noah's Flood: The Genesis Story in Western Thought.* Yale University Press, New Haven/London.
Colborn, T. et al. (1996), *Our stolen Future.* E.P. Dutton, New York.
Coleman, W. (1976), Providence, Capitalism and Envrionmental Degradation: English Apologetics in an Era of Economic Revolution. *Journal of the History of Ideas,* 37, 27-45.
Colish, M.L. (1979), Pauline Theology and Stoic Philosophy: a Historical Study. *Journal of the American Academy of Religion,* 47, 1, supplement B, 1-21.
Colish, M.L. (1985), *The Stoic Tradition from Antiquity to the Early Middle Ages.* 2 vols., E.J. Brill, Leiden.
Collingwood, R.G., (1965), *The Idea of Nature.* paperback no. 97, Clarendon Press, Oxford.
Commoner, B. (1971), *The Closing Circle.* A.A. Knopf, New York.
Concordance, The Eerdmans Analytical. (1990), Compiled by R.E. Whitaker to the Revised Standard Version of the Bible, W.B. Eerdmans Publishing Company, Grand Rapids, Michigan.

Cook, R. (1974/1988), *The Tree of Life. Image for the Cosmos.* Thames and Hudson, Reprint 1988, London/New York.

Cooper, J. (1993), *Eat and be satisfied. A Social History of Jewish Food.* Jason Aronson, Northvale, NJ.

Corbey, R. & P. van der Grijp (eds.) (1990), *Natuur en Cultuur. Beschouwingen op het raakvlak van antropologie en filosofie.* Ambo, Baarn.

Cornfeld, Gaalyah (1976), *Archaelogy of the Bible: Book by Book.* Harper & Row, New York/London.

Corpus Hermeticum (1991), Introduced, translated and with comments by R. van den Broek & G. Quispel, 3rd revised edition, In de Pelikaan, Amsterdam, Ambo, Baarn.

Costanza, R. & B.C. Patten (1995), Defining and Predicting Sustainability. *Ecological Economics* 15, 193-196.

Costanza, R. a.o. (1997), The Value of the World's Ecosystems, Services and Natural Capital. *Nature*, 387, 253-260.

Cottlieb, R.S. (ed.) (1996), *This Sacred Earth. Religion, Nature, Environment.* Routledge, New York/London.

Cottingham, J. (1978), A Brute to the Brutes?: Descartes' Treatment of Animals. *Philosophy*, 53, 551-559.

CPB (1996), *Economie en Milieu: op zoek naar duurzaamheid.* Sdu, The Hague.

Cramer, J. (1989), *De groene golf.* Uitgeverij Jan van Arkel, Utrecht.

Crosby, A.W. (1986/1993), *Ecological Imperialism. The Biological Expansion of Europe, 900-1900.* Canto / Cambridge University Press, Cambridge.

Crutchfield, J.P., J.D. Farmer, N.H. Packhard North & R.S. Shaw (1986), 'Chaos'. *Scientific American*, 255, 38-49.

Cunningham, W.P. & B.W. Saigo (2001), *Environmental Science: a Global Concern.* Sixth Edition, McGrawHill, New York.

Currier Bell, Barbara (1981), Humanity in Nature: Toward a Fresh Approach. *Environmental Ethics*, 3, 3, 245 et seq.

Curry-Roper, J.M. (1990), Contemporary Christian Eschatologies and their Relation to Environmental Stewardship. *Professional Geographer*, 42, 2, 157-169.

Dake, K. (1992), Myths of Nature. Culture and the Social Construction of Risk. *Journal of Social Issues*, 48, 21-37.

Dalley, S. (1991), *Myths from Mesopotamia. Creation, the Flood, Gilgamesh and others.* Oxford Paperbacks, Oxford University Press, Oxford.

Daly, H.E. (1972/1991), *Steady State Economics.* W.H. Freeman and Co, San Francisco.

Daly, H.E. (1996), *Beyond Growth.* Beacon Press, Boston, Massachusetts.

Daly, H.E. (1999), *Ecological Economics and the Ecology of Economics. Essays in Criticism.* Edward Elgar, Cheltenham.

Daly, H.E. & J.B. Cobb Jr. (1989/1990), *For the Common Good. Redirecting the Economy toward Community, the Environment and a Sustainable Future.* Beacon Press, Boston (1989), Green Print, London.

Daly, H.E. & K.N. Townsend (eds.) (1993), *Valuing the Earth. Economics, Ecology, Ethics.* The MIT Press, Cambridge, Massachusetts, London, England.

Dam, R.J. (1949), *Stoa en litteratuur in het licht der Schrift.* Oosterbaan & Le Cointre, Goes.

Darling, F.F. (1956), Man's Ecological Dominance through Domesticated Animals on Wild Lands, in: Thompson, W.L. (ed.). *Man's Rôle in Changing the Face of the Earth.* University Chicago Press, Chicago, Ill, 778-787.

Darlington, C.D. (1969), *The Evolution of Man and Society.* George Allen and Unwin, London.

Darwin, Ch. (1889), *The Descent of Man and Selection in relation to Sex.* Second edition, revised and augmented, John Murray, London.
Davids, C.A. (1989), *Dieren en nederlanders. Zeven eeuwen lief en leed.* Matrijs, Utrecht.
Davidson, B. (1970), *The Analytical Hebrew and Chaldee Lexicon.* Zondervan Publishing House, Grand Rapids, Michigan.
Davidson, J. (2000), Sustainable Development: Business as Usual or a New Way of Living? *Environmental Ethics*, 22 25-43.
Davies, P.C.W. (1989), *The cosmic Blueprint. New discoveries in Nature's Creative ability to order the Universe.* A Touchstone Book, Simon & Schuster, New York.
Davies, W.D., (1974) *The Gospel and the Land: Early Christianity and Jewish Territorial Doctrine.* University of California Press. Berkeley, California.
Davies, W.D. (1983), *Jewish and Pauline Studies.* Fortress Press, Philadelphia.
Davis, D.E. (1989), *Ecophilosophy A Field Guide to the Literature.* R. & E. Miles, San Pedro, California.
DDD (Dictionary of Deities and Demons in the Bible). Toorn, K. van der, B. Becking & P.W. van der Horst (eds.) (1999), Second revised edition, E.J. Brill, Leiden.
De Bardeleben, J. (1986), *The Environment and Marxism-Leninism: the Soviet and East German Experience.* Westview Press, Boulder, Colorado.
Debus, A.G. (1978), *Man and Nature in the Renaissance.* Cambridge University Press, Cambridge.
Dembowski (1985), Natürliche Theologie—Theologie der Natur. Erwägungen in einem weiten Feld. *Evang. Theologie*, 45, 224-48.
Derks, Hans (1975). *De Armoede van het ekologisme.* Werkgroep Milieukunde, Utrecht State University.
Descartes, R. (1637), *Discours de la méthode.* Jean Maire, Leiden.
Descartes, R. (1987), *Over de Methode.* Translation and introduction by Th. Verbeek, 3rd ed., Boom, Meppel/Amsterdam.
Desert Fathers cf *The Desert Fathers*
Deth, J.W. van & J.C.P.M. Vis (eds.) (1994), *Actuele politieke problemen.* Stichting Burgerschapskunde/Instituut voor Publiek en Politiek, Amsterdam.
Deurloo, K.A. (1986), Naar Gods gelijkenis, manlijk en vrouwlijk (Gen. 5:2,3; 1:27). *ACEBT* 7, 25-34.
Deurloo, K.A. (1988), *De mens als raadsel en geheim. Verhalende antropologie in Genesis 2-4.* Ten Have, Baarn.
Deval, Bill & G. Sessions (eds.) (1985), *Deep Ecology.* Peregrine Smith Books, Salt Lake City, Utah.
Dever, W.G. (1995), Archaeology and the Religions of Israel. *BASOR*, 301, 83-90.
DeWitt, Calvin B. (ed.) (1991), The Environment & the Christian. What can we learn from the New Testament? Bake Book House, Grand Rapids, Michigan.
Diamond, J. (1997), *Guns, Germs and Steel. A short History of Everybody for the last 13,000 Years.* Chatto & Windus, London.
Diamond, J.M. (1966), Zoological Classification System of a Primitive People. *Science*, 151, 1102-1104.
Dickens, P. (1992), *Society and Nature. Towards a Green Social Theory.* Temple University Press, Philadelphia, PA.
Dierauer, U. (1977), *Tier und Mensch im Denken der Antike: Studien zur Tierpsychologie, Anthropologie und Ethiek.* Verlag B.R. Grüner, Amsterdam (Dissertation, Zürich, 1972).
Dijk, J. van (1983), *LUGAL UD ME-LÁM-bi NIR-GÁL. Le récit épique et didactique des Travaux de Ninurta, du Déluge et de la Nouvelle Création.* E.J. Brill, Leiden

Dimont, M.I. (1962/1994), *Jews, God, and History*. Revised and updated Edition, A Mentor Book published by the Penguin Group, London.

Diogenes Laertius (1989), *Leven en leer van beroemde filosofen*. Translation by R. Ferwerda & J. Eykman, Ambo, Baarn.

Doherty, B. & M. de Geus (eds.) (1996), *Democracy & Green Political Thought. Sustainability, Rights and Citizenship*. Routledge, London.

Döller, J. (1917), Die Reinheits und Speisegesetze des A.T. in Religionsgeschichtlicher beleuchtung *Alt. Abhandlungen*, Band VII, Heft 2-3, Münster.

Donner, H. (1984), *Geschichte des Volkes Israel und seiner Nachbarn in Grundzügen*, Göttingen.

Doron, Pinchas (2000), *Rashi's Torah Commentary. Religious, Philosophical, Ethical and Educational Insights*. Jason Aronson, Northvale, NJ.

Doughty, Robin W. (1981), Environmental Theology: Trends and Prospects in Christian Thought. *Progress in Human Geography*, 5, 234-248.

Douglas, M. (1966). *Purity and Danger: An Analysis of Concepts of Pollution and Taboo*, Routledge & Kegan Paul, London.

Douglas, M. (1970), *Natural Symbols: Explorations in Cosmology*. Pantheon Books, Random House Inc., New York.

Douglas, M. (1972), Deciphering a Meal. *Daedalus*, 94, 61-82.

Douglas, M. (1975), *Implicit Meanings: Essays in Anthropology*. Routledge & Kegan Paul, London.

Douglas, M. (1978), *Cultural Bias*. Royal Anthropological Institute of Great-Britain and Ireland, Occasional Paper no. 35.

Douglas, M. (ed.) (1982), *Essays in the Sociology of Perception*. Routledge & Kegan Paul, London.

Douglas, M. (1985), *Risk Acceptability According to the Social Sciences*, Occasional Reports on Current Topics no. 11, Russell Sage Foundation, New York.

Douglas, M. (1986), *How Institutions think*. Syracuse, New York.

Douglas, M. (1993), *In the Wilderness. The Doctrine of Defilement in the Book of Numbers*. *JSOT* Supplement Series 158, Sheffield Academic Press, Sheffield.

Douglas, M. (1993a), Balaam's Place in the Book of Numbers. *Man (N.S.)*, 28, 3, 411-430.

Douglas, M. (1993b), The Forbidden Animals in Leviticus. *JSOT*, 59, 3-23.

Douglas, M. (ed.) (1973), *Rules and Meanings. The Anthropology of Everyday Knowledge*. Penguin Education, Harmondsworth.

Douglas, M. & B. Isherwood (1996/1979), *The World of Goods: Towards an Anthropology of Consumption*. With a new Introduction, Routledge, London.

Douglas, M. & A. Wildavsky (1982), *Risk and Culture. An Essay on the Selection of Technological and Environmental Dangers*. University of California Press, Berkeley, California.

Downs, A. (1972), Up and Down with Ecology—The Issue Attention Cycle. *The Public Interest*, 2, 38-50.

Drees, W.B. (1990), *Beyond the Big Bang. Quantum Cosmologies and God*. Open Court, La Salle, Ill.

Drees, W.B. (1991), Potential Tension between Cosmology and Theology. In: V. Brümmer (ed.), *Interpreting the Universe as Creation. A Dialog of Science and Religion*. Kok Pharos, Kampen, 65-89.

Dresner, S.H., S. Siegel & D. Pollock (1982), *The Jewish Dietary Laws. Their Meaning for our Time, A Guide to Observance*. Revised edition, The Rabbinical Assembly, United Synagogue Commission on Jewish Education, New York.

Drewermann, E. (1981/1990), *Der tödliche Fortschritt. Von der Zerstörung der Erde und des Menschen im Erbe des Christentums*. 6th revised and expanded edition (1990), Verlag Friedrich Pustet, Regensburg.

Driver, G.R. (1955), Birds in the Old Testament. I Birds in Law; II Birds in Life, *Palestine Exploration Quarterly (PEQ)*, I: 5-20, II: 129-140

Driver, G.R. (1956), Mythical Monsters in the OT. in: *Studia Orientalistici.* in Honore Giorgio Levi Della Vida I, Roma, 234-249.

Driver, G.R. (1958), Once Again: Birds in the Bible. *Palestine Exploration Quarterly (PEQ)*, 56-58.

Driver, G.R. (1972), Job 39:27-28: The KY-Bird. *Palestine Exploration Quarterly (PEQ)*, 104, 64-66.

Driver, S.R. (1896), *Deuteronomy. A Critical and Exegetical Commentary on Deuteronomy.* second edition T & T Clark, 38 George Street, Edinburgh.

Duchrow, U. und G. Liedke (1987), *Schalom, Der Schöpfung Befreiung, den Menschen Gerechtigkeit, den Völkern Frieden.* Kreuz Verlag, Stuttgart.

Dumont, L. (1966/1980), *Homo Hierarchicus: The Caste System and Its Implications.* Complete Revised English Edition, translated by Mark Sainsbury, Louis Dumont and Basia Gulati (Editions Gallimard, Paris, 1966), Chicago University Press, Chicago.

Dunlap, R.E. (1991), Trends in public opinion towards environmental issues: 1965-1990. *Society and Natural Resources*, 4, 285-312.

Duns, J. (1863-1868), *Biblical Natural Science, being the Explanation of all References in Holy Scripture to Geology, Botany, Zoology and Physical Geography*, 2 vols., William Mackenzie, 22 Paternoster Row, London.

Düring, Ingemar (1966), *Aristoteles: Darstellung und Interpretation Seines Denkens.* Carl Winter Universitätsverlag, Heidelberg.

Durkheim, É. (1912/1965), *The Elementary Forms of the Religious Life.* translated from the original French edition, *Les Formes élémentaires de la Vie Religieuse. Le système totémique en Australie.* (Félix Alcan, Paris, 1912) by J.W. Swain, 1915, London, reprint 1965, New York.

Durkheim, É. & M. Mauss (1903/1963), *Primitive Classification.* Translated and edited by R. Needham, Cohen & West, London (translation of *De quelques formes primitives de classification.* 1903).

Durning, A.Th. (1992), *How much is enough? The Consumer Society and the Future of the Earth.* Earthscan Publications Ltd., London, Worldwatch Institute, Washington DC.

Dwivedi, O.P. & B.N. Tiwari (1987), *Environmental Crisis and Hindu Religion.* Gitanjali Publishing House, New Delhi 110024.

Easlea, B. (1980), *Witch Hunting, Magic and the New Philosophy: An Introduction to Debates of the Scientific Revolution 1450-1750.* Harvester Press, Brighton.

Easlea, B. (1981), *Science and Sexual Oppression, Patriarchy's Confrontation with Woman and Nature.* Weidenfeld and Nicolson, London.

Easterling, P.E. & B.M.W. Knox (eds.) (1985), *The Cambridge History of Classical Literature.* Cambridge University Press, Cambridge.

Ebach, J. (1989), Schöpfung in der Hebräischen Bibel. In: G. Altner (ed.), *Ökologische Theologie. Perspektiven zur Orientierung*, Kreuz Verlag, Stuttgart, 98-129.

Eberg, J. (1997), *Waste Policy and Learning. Policy Dynamics of Waste Management and Waste Incineration in the Netherlands and Bavaria.* Dissertation, Amsterdam University, Eburon publishers, Delft.

Eckberg, D.L. & T.J. Blocker (1989), Varieties of Religious Involvement and Environmental Concerns: Testing the Lynn White Thesis. *Journal for the Scientific Study of Religion*, 28, 4, 509-517.

Eddington, A. (1927/1958), *The Nature of the Physical World.* Gifford Lectures, Cambridge; University of Michigan Press, Ann Arbor.

Edel, R.F. (1968), *Hebräisch Deutsche Präparation zu Genesis 1-25.* Oekumenischer Verlag Dr. R-F. Edel Marburg/L.

Eerdmans Dictionary of the Bible D.N. Freedman editor in chief (2000), W.B. Eerdmans Publishing Company Grand Rapids, Michican.

Egmond, N.D. van (1996), *Goede raad is duurzaam. Milieukunde en de afweging tussen milieu en economie.* Inaugural speech, 31 October 1996, Utrecht State University.

Egmond, P. van *et al.* (1992), Is een duurzaamheidsindicator (al) een betrouwbare barometer? *Milieu,* 7, 4, 120-123.

Ehrenfeld, D. & Ph.J. Bentley (1985), Judaism and the Practice of Stewardship. *Judaism,* 34, 301-11.

Ehrlich, A.B. (1968), *Randglossen zur Hebräischen Bibel.* Georg Olms Verlagsbuchhandlung, Hildesheim.

Ehrlich, P.R. (1968/1971), *The Population Bomb.* Revised Edition, A Sierra Club/ Ballantine Book, New York.

Ehrlich, P.R. & A.H. Erhrlich (1981), *Extinction. The Causes and Consequences of the Disappearance of Species.* Ballantine Books, New York.

Ehrlich, A.H. & P.R. Ehrlich (1970), *Population, Resources, Environment.* W.H. Freeman, San Fransisco.

Eilberg-Schwarz, H. (1990), *The Savage in Judaism: An Anthropology of Israelite Religion and Ancient Judaism.* Indiana University Press, Bloomington.

Eisenberg, E. (1998), *The Ecology of Eden. Humans, Nature and Human Nature.* A Knopf, New York.

Eisenman, R. & M. Wise (1993), *The Dead Sea Scrolls uncovered.* Penguin Books Ltd., Harmondsworth, Middlesex, England.

Eissfeldt, O. (1940/1963), Das Chaos in der Biblischen und in der Phönizischen Kosmogonie. *Kleine Schriften,* II, 258-262.

Ekins, P. & S. Simon (2001), Estimating sustainability gaps: methods and preliminary applications for the UK and the Netherlands. *Ecological Economics,* 37/1, 5-23.

Eldridge, N. (ed.) (1992), *Systematics, Ecology and the Biodiversity Crisis.* Columbia University Press, New York.

Eliade, Mircea (1959), *The Sacred and the Profane: The Nature of Religion.* New York.

Ellen, R. & Fukui, K. (1996), *Redefining Nature. Ecology, Culture and Domestication.* Berg, Oxford/Washington D.C.

Ellerman, C.P. (1990), New Notes from Underground. *Inquiry,* 33, 1, 3-26.

Elliger, K. (1966), *Leviticus.* Handbuch zum Alten Testament (*HAT*), Erste Reihe, 4, J.C.B. Mohr (Paul Siebeck), Tübingen.

Ellis, E.Earle (1991), *The Old Testament in Early Christianity. Canon and Interpretation in the Light of Modern Research.* J.C.B. Mohr (Paul Siebeck), Tübingen.

Emerton, J.A. (1967), The Meaning of '*nšy qdš* in Lamentations 4:1. *ZAW,* 79, 233-236.

Emerton, J.A. (1987/1988), "An Examination of some Attemps to defend the Unity of the Flood Narratives in Genesis". Vol. I, *VT,* 37, (1987) 401-420; Part II, *VT,* (1988), 38, 1-21.

Emerton, J.A. (ed.) (1990), *Studies in the Pentateuch.* Supplements to Vetus Testamentum, vol. XLI, E.J.Brill, Leiden.

Enckell, P.H., E.S. Königsson & L.K. Königsson (1979), Ecological Instability of a Roman Iron Age human Community. *Oikos,* 33, 328-349.

Encyclopaedia Judaica (1971-), Keter Publishing House Ltd, Jerusalem.

Encyclopedie van het Milieu (1984), Eindredactie J.G. ten Houten & H. Gysels, Oosthoek's Uitgeversmaatschappij BV, Utrecht.

Evers, R. (1986), Jodendom niet afwijzend tegenover vegetarisme. *Nieuw Israëlitisch weekblad*, 29 August, 9.

Fackre, G. (1971), Ecology and Theology. *Religion in Life*, 40, 210-224.

Fairholme, E.G. & W. Pain (1924), *A Century of Work for Animals: The History of the R.S.P.C.A. 1824-1924.* John Murray, Albemarlestreet, London.

Fascher, E. (1965), Jesus und die Tiere. *Theologische Literaturzeitung*, 90, 561-570.

Feliks, Jehuda (1981), *Nature and Man in the Bible. Chapters in Biblical Ecology*. Soncino Press, London.

Fensham, F.C. (1988), Liability of Animals in Biblical and Ancient Near Eastern Law. *Journal of Northwest Semitic Languages (JNSL)*, XIV, 85-90.

Ferngreen, G.B. (ed.) (2000), *The History of Science and Religion in the Western Tradition: an Encyclopedia.* Garland Publishing, New York/London.

Feyerabend, K. (undated), *Langenscheidt's Pocket Hebrew Dictionary to the Old Testament.* 15th Edition, Langenscheidt Publishers Inc., Maspeth, New York.

Filius, L.S. (1989). *Problemata Physica Arabica.* 2 vols, Dissertation, FreeUniversity of Amsterdam.

Finkelstein, I. (1989), The Emergence of the Monarchy in Israel. The Environmental and Socio-Economic Aspects. *JSOT*, 44, 43-74.

Finkelstein, I. (1990), The Emergence of Early Israel: Anthropology, Environment and Archaeology. *JAOS*, 110, 677-686.

Finkelstein, L. (ed.) (1974/1977). *The Jews. Vol. 1: Their History. Vol. 2: Their Religion and Culture. Vol. 3: Their Rôle in Civilization.* Schocken Books, New York.

Firmage, E. (1990), The Biblical Dietary Law and the Concept of Holiness. In: Emerton, J.A. (ed.), *Studies in the Pentateuch.* E.J. Brill, Leiden, 177-209.

Firmage, E. (1992), Zoology. In: *The Anchor Bible Dictionary, VI.* 1109-1167.

Fisher, Jonathan (1834/1972), *Scripture Animals: A Natural History of the living Creatures Named in the Bible.* William Hyde, Portland, Maine, reprint (1972) The Pyne Press, Princeton, New Jersey.

Flandrin, J.-L. & M. Montanari (eds.) (1996), *Histoire de L'alimentation.* Fayard, Paris.

Flusser, D. (1964), *De Joodse oorsprong van het Christendom.* Moussault's Uitgeverij N.V., Amsterdam.

Foerster, W. (1938), lemma θηριον in: G. Kittel *Theologisches Wörterbuch zum Neuen Testament (TWNT),* Dritter band, W.Kohlhammer, Stuttgart, 133-136.

Fokkelman, J.P. (1975), *Narrative Art in Genesis. Specimens of Stylistic and Structural Analysis.* Van Gorcum, Assen.

Folmer, H. & R. Komen (1997), De milieu-Kuznetscurve bestaat! *ESB,* 82, 355

Folmer, H. & R. Komen (1997), De milieu-Kuznetscurve blijft bestaan. *ESB,* 82, 475.

Forst, Rabbi Binyomin (1994), *The Laws of Kashrus. A Comprehensive Exposition of their underlying Concepts and Applications.* Mesorah Publications ltd, New York.

Foster, M.B. (1934), The christian Doctrine of Creation and the Rise of modern natural Science. *Mind*, XLIII, 446-68.

Foster, M.B. (1935), Christian Theology and modern Science of Nature (I). *Mind,* XLIV, 439-66.

Foster, M.B. (1936), Christian Theology and modern Science of Nature (II). *Mind,* XLV, 1-27.

Foster, T.W. (1980), The Amish and the Ethos of Ecology. *The Ecologist,* 10, 10, 331-36.

Fox, M. (1988), *The Coming of the Cosmic Christ*, San Francisco, California.

Fox, R.Lane (1989), *De droom van Constantijn. Heidenen en christenen in het Romeinse Rijk 150 AD—350 AD.* Translation by Anki Klootwijk, Agon BV, Amsterdam.

Fox, R.Lane (1991), *De Bijbel: waarheid en verdichting*. Translation by Tinke Davids, Agon BV, Amsterdam.

France, P. (1986), *An Encyclopedia of Bible Animals*. Photographs by Eric and David Hosking, Croom Helm, London/Sydney

Francis, L.P. & R. Norman, (1978), Some Animals Are More Equal than others. *Philosophy,* 53, 507-527.

Frazer, sir James George (1890/1981), *The Golden Bough. The Roots of Religion and Folklore*. Avenel Books, New York.

Frazer, James George (1919/1923/1988), *Folklore in the Old Testament. Studies in comparative Religion Legend and Law*. 3 vols, MacMillan and Co Ltd., London, Abridged Edition 1923, reprinted in 1988 by Avenel Books, New York.

Freedman, H & M. Simon (eds.) (1981), *Genesis Rabbah*. Soncino Press, London.

French, R. (1994), *Ancient natural History. Histories of Nature*. Routledge, London/ New York.

Freudenstein, E.G. (1972). Ecology and the Jewish Tradition. In: Konvitz, M.R. (ed.) *Judaism and Human Rights*. W.W. Norton & Company Inc., New York, 265-75.

Fritsch, A.J. (1980), Theological Foundations for an Environmental Ethics. In: *Environmental Ethics. Choices for concerned Citizens*, New York, 231-254.

Frossard, J. (1945), Les Aigles du Val de Bagnes II. *Nos Oiseaux, 18*, 21-34.

Furley, D.J. (1987), *The Greek Cosmologists*. Cambridge University Press, Cambridge.

Furley, D.J. (1989), *Cosmic Problems. Essays on Greek and Roman Philosophy of Nature*. Cambridge University Press, Cambridge.

Gambell, R. (1990), Whaling, a Christian Position. *Science and Christian Belief*, 2, 15-24.

García Martínez, F. *et al.* (eds.) (1992), *The Scriptures & the Scrolls*. E.J. Brill, Leiden.

García Martínez, F. (1994), *The Dead Sea Scrolls Translated. The Qumran Texts in English*. E.J. Brill, Leiden.

García Martínez, F. & A.S. van der Woude (1994), *De rollen van de Dode Zee*. Introduction and Dutch translation, 2 vols., Lannoo, Tielt / Kok, Kampen.

García Martínez, F., A. Hilhorst, J.T.A.G.M. van Ruiten & A.S. van der Woude (eds.) (1994), *Studies in Deuteronomy in honour of C.J. Labuschagne on the occasion of his 65the birthday*. Suppl. VT, vol. 53.

García Martínez, F. & G. Luttikhuizen (eds.) (1997), *Interpretations of the Flood*. E.J. Brill, Leiden.

Gensichen, H.P. (1985), Die Sintfluterzählung als Leittext in der Umweltkrise. *Ev. Theol.*, 45, 211-224.

Gerbi, Antonello (1985), *Nature in the New World, from Christopher Columbus to Gonzalo Fernanda de Oviedo*. Translation by Jeremy Moyle, University of Pittsburgh Press, Pittsburgh Pa.

Gerhardt, M.I. (1965), The Antlion, Nature Study and the Interpretation of a Biblical Text, from the Physiologus to Albert the Great. *Vivarium*, III, May, 1-23.

Gerlitz, P. (1998), *Mensch und Natur in den Weltreligionen: Grundlagen einer Religionsökologie*. Primus Verlag, Darmstadt

Gerstenberger, E.S. (1993), *Das dritte Buch Mose, Leviticus*. Das Alte Testament Deutsch, Subvolume 6, Vanderhoeck & Ruprecht, Göttingen.

Gerstenfeld, M. (1998), *Judaism, Environmentalism and the Environment. Mapping and Analysis*. The Jerusalem Institute for Israel Studies, Rubin Mass Ltd., Jerusalem.

Geyer, H.G., J.M. Schmidt, W. Schneider & M. Weinrich (eds.) (1983), *Wenn nicht jetzt, wann dann?* Edition in commemoration of the 65th birthday of Hans-Joachim Kraus, Neukirchener Verlag, Neukirchen-Vluyn.

Giddens, A. (1991), *The Consequences of Modernity.* Polity Press, Cambridge.

Gigon, O. (1959), *Grundprobleme der antiken Philosophie.* A. Francke Verlag, Bern/München.

Ginzberg, Louis (1909/1937), *The Legends of the Jews.* 7 Vols., The Jewish Publication Society of America, Philadelphia.

Girard, René (1987), Generative Scapegoating. In: Hamerton-Kelly, R.G. (ed.) (1987), *Violent Origins.* Burkert, W., R. Girard & J.Z. Smith on Ritual Killing and Cultural Formation. With an introduction by Burton Mack and a commentary by Renato Rosaldo, Stanford University Press, Stanford, California, 73-145.

Gispen, W.H. (1945), *De levietische wet op de melaatschheid.* Inaugural speech, 28 September 1945, Free University of Amsterdam, J.H. Kok, Kampen.

Gispen, W.H. (1966), *Schepping en paradijs verklaring van Genesis 1-3.* J.H.Kok, Kampen.

Glacken, C.J. (1956), Changing ideas of the Habitable World. In: Thomas, W.L. Jr. (ed.) *Man's Rôle in changing the Face of the Earth.* University of Chicago Press, Chicago, 70-93.

Glacken, C.J. (1967), *Traces on the Rhodian Shore, Nature and Culture in Western Thought from ancient Times to the End of the Eighteenth Century.* University of California Press, Berkeley, Los Angeles.

Glasbergen, P. (1995), Bespreking WRR-rapport. *Beleid en Maatschappij,* 4, 254.

Gleick, J. (1988), *Chaos, de derde wetenschappelijke revolutie.* Contact, Amsterdam.

Gloy, K. (1996), *Das Verständnis der Natur. Band I: Geschichte des naturwissenschaftlichen Denkens. Band II: Geschichte des ganzheitlichen Denkens.* Verlag C.H. Beck, München.

Glutz von Blotzheim, Urs N. (ed.) (1971), *Handbuch der Vögel Mitteleuropas,* Band 4, Falconiformes, Akad. Verlags Gesellschaft, Frankfurt am Main.

Goldsmith, E. (1992), The Way: An Ecological World-View. Rider, London.

Goldsmith, E. (1988), The Way: An Ecological World-View. *The Ecologist,* 18, 4, 5, 160-185.

Goldsmith, E. (1990), Evolution, Neo-Darwinism and the Paradigm of Science. *The Ecologist,* 20, 2, 67-73.

Goodin, R.E., B. Headey, R. Muffels & H-J. Dirven (1999), *The Real World of Welfare Capitalism.* Cambridge University Press, Cambridge.

Goodland, R. (1995), The Concepts of Environmental Sustainability. Annual Review of Ecological Systems 26, 1-24.

Goody, J. (1978), *The Domestication of the Savage Mind.* Cambridge University Press, Cambridge.

Goody, J. (1990), *The Oriental, the Ancient and the Primitive. Systems of Marriage and the Family in the pre-industrial Societies of Eurasia.* Cambridge University Press, Cambridge.

Goody, J. (1993), *The Culture of Flowers.* Cambridge University Press, Cambridge.

Goody, J. (1995), Man and the Natural World. *Environment and History,* 1, 3, 255-270.

Goody, J. (1996), *The East in the West.* Cambridge University Press, Cambridge.

Gordis, R. (1974), Love, Marriage, and Business in the Book of Ruth: A Chapter in Hebrew Customary Law. In: Bream H.N., R.D. Heim & C.A. Moore (eds.) *A Light unto My Path: Old Testament Studies in Honour of Jacob M. Myers.* Temple University Press, Philadelphia, 241-264.

Gore, A. (1992), *Earth in the Balance. Forging a New common Purpose.* Earthscan Publications Ltd., London.

Görg, M. (1986), Alles hast Du gelegt unter seine Füße. Beobachtungen zu Ps 8,7b

im Vergleich mit Gen 1,28, Freude an der Weisung des Herrn. In: Haag, E. & F.-L. Hossfeld (eds.), Festschrift für H. Gross, *SBB* 13, Stuttgart, 125-148.
Gottlieb, R. (1996), *This Sacred Earth: Religion, Nature, Environment.* Routledge, London/New York.
Goudie, A. (1981), *The Human Impact: Man's Rôle in Environmental Change.* Basil Blackwell, Oxford.
Goudsblom, J. (1992), *Vuur en beschaving.* Meulenhoff, Amsterdam.
Goudsblom, J. (1993), *De verleiding van het teveel. De milieuproblematiek als beschavingsprobleem.* Amsterdam University Press, Amsterdam.
Goudzwaard, B. (1976), *Kapitalisme en vooruitgang. Een eigentijdse maatschappijkritiek.* Van Gorcum, Assen/Amsterdam.
Goudzwaard, B. (1970), *Ongeprijsde schaarste.* Van Stockum, The Hague.
Gould, J. (1985), On making Sense of Greek Religion. In: P.E. Easterling & J.V. Muir (eds.), *Greek Religion and Society.* Cambridge University Press, Cambridge, 1-33.
Gould, S.J. (1983), *Hen's Teeth and Horse's Toes. Further Reflections in Natural History.* Penguin Books Ltd, Harmondsworth, Middlesex, England.
Gould, S.J. (1989/1991), *Wonderful Life. The Burgess Shale and the Nature of History.* Penguin Books Ltd, London.
Gould, S. J. (1990), The Golden Rule—a Proper Scale for Our Environmental Crisis. *Natural History,* 9, 24-30.
Gould, S.J. (1999), *Rocks of Ages: Science and Religion in the fullness of Life.* Ballantine, New York.
Grant, M. (1973), *The Jews in the Roman World.* Weidenfeld & Nicholson, London.
Green, A. (1991/1992), God, World, Person: A Jewish Theology of Creation. *The Melton Journal,* no. 24 Spring 1991: Part I and no. 25 Spring 1992: Part II, The Melton Research Center of the Jewish Theological Seminary of America, New York.
Green, P. (1970), *Alexander the Great.* Weidenfeld & Nicholson, London.
Green, P. (1990), *Alexander to Actium.* Thames and Hudson, London.
Greenberg, Blu (1983), *How to run a Traditional Jewish Household.* Simon and Schuster, New York.
Greenstein, E.L. (1984), Biblical Law. In: B.W. Holz (ed.), *Back to the Sources. Reading the Classic Jewish Texts.* A Touchstone Book, Simon & Schuster, New York, 83-105.
Gregorios, Paulos M. (1987), *The Human Presence. Ecological Spirituality and the Age of the Spirit.* Amity House, Amity, New York.
Gregorios, Paulos M. (1992), *A Light too bright. The Enlightenment today.* SUNY (State University of New York) Press, Albany.
Groen van Prinsterer stichting (1974), *Bouwen en bewaren.* Groen van Prinsterer stichting (GPV Science Office), Amersfoort.
Groningen, B.A. van (1953), *In the Grip of the Past. Essay on an Aspect of Greek Thought.* E.J. Brill, Leiden.
Groot, R.S. de (1992), *Functions of Nature. Evaluation of Nature in Environmental Planning, Management and Decisionmaking.* Wolters-Noordhoff, Groningen.
Groot, W.T. de (1992), *Environmental Science Theory. Concepts and Methods in a One-World, Problem-Oriented Paradigm.* Elsevier Science Publishers, Amsterdam.
Grove, R.H. (1990), The Origins of Environmentalism. *Nature,* 345, 3 May, 11-14.
Grove, R.H. (1992), Origins of Western Environmentalism. *Scientific American,* 267, 1, 22-27.
Grove, R.H. (1995), *Green Imperialism. Colonial Expansion, Tropical Island Edens and*

the Origins of Environmentalism, 1600-1860. Cambridge University Press, Cambridge.

Grunfeld, I. (1972), *The Jewish Dietary Laws.* 2 vols, The Soncino Press, London/ Jerusalem/New York.

Grunneweg, A.H.J. (1983), A Semantic Revolution. *ZAW,* 95, 437-440.

Gunn, A.S. (1980), Why should we care about rare Species. *Environmental Ethics,* 2, 17-37.

Gunn, A.S. (1991), The Restoration of Species and Natural Environments. *Environmental Ethics,* 13, 291-310.

Guthrie, W.K.C. (1965), *A History of Greek Philosophy.* 4 Vols., Cambridge University Press, Cambridge.

Habel, N.C. & S. Wurst (eds.) (2000), *The Earth Story in Genesis* (*The Earth Bible 2*). Sheffield Academic Press, Sheffield.

Hahm, David E. (1977), *The Origins of Stoic Cosmology.* Ohio State University Press, Columbus, Ohio.

Hajer, M. (1997), *The Politics of Environmental Discourse. Ecological Modernization and the Policy Process.* Oxford University Press, Oxford.

Hakker-Orion, D. (1975), Hunting and Stock-breeding in Israel. in: Clason, A.T. (ed., *Archaeozoological Studies.* North Holland Publ. Company, Amsterdam, 295-301.

Hakker-Orion, D. (1984), The Rôle of the Camel in Israel's early History. In: Clutton-Brock, J. & C. Grigson (eds.), Animals and Archaeology. III Early Herders and their Flocks. *BAR International Series,* 202, Oxford, 207-212.

Hale, Sir Matthew (1677), *The Primitive Origination of Mankind.* Printed by W. Godbid for W. Shrowsbery, London.

Halkes, C.J.M. (1989), *..en alles zal worden herschapen; Gedachten over de heelwording van de schepping in het spanningsveld tussen natuur en cultuur.* Ten Have, Baarn.

Hallo, W.W. (1996), *Origins. The Ancient Near Eastern Background of some Modern Western Institutions.* Studies in the History and Culture of the Ancient Near East (SH-CANE), Vol. 6, E.J. Brill, Leiden.

Hallo, W.W. & K. Lawson Younger Jr. (1997, 2000), *The Context of Scripture.* Vol. I 1997. Vol. II 2000, E.J. Brill, Leiden.

Hallpike, C.R. (1979), *The Foundations of Primitive Thought.* Oxford University Press, Oxford.

Hamerton-Kelly, R.G. (ed.) (1987), *Violent Origins.* Burkert, W., R. Girard & J.Z. Smith on Ritual Killing and Cultural Formation. With an introduction by Burton Mack and a commentary by Renato Rosaldo, Stanford University Press, Stanford, California.

Hamilton, L.S. (ed.) (1993), *Ethics, Religion and Biodiversity. Relations between Conservation and Cultural Values.* The White Horse Press, Knapwell, Cambridge.

Hancock, Graham (1991), *Lords of Poverty.* Mandarin Paperbacks.

Hanegraaff, W.J. (1996), *New Age Religion and Western Culture. Esotericism in the Mirror of Secular Thought.* E.J. Brill, Leiden.

Hanke, L. (1974), *All Mankind is One. A Study of the Disputation between Bartolomé de Las Casas and Juan Ginés de Sepúlveda in 1550 on the Intellectual and Religious Capacity of the American Indians.* DeKalb, Northern Illinois University Press.

Haran, M. (1979), Seething a Kid in its Mother's Milk. *Journal of Jewish Studies (JJS),* 30, 23-35.

Hardin, Garrett J. (1968), The Tragedy of the Commons. *Science,* 162, 1243-1248.

Hardin, Garrett J. (1993), *Living within Limits. Ecology, Economics and Population Taboos.* Oxford University Press, Oxford/New York.

Harding, S. (1986), *The Science Question in Feminism.* Cornell University Press, Ithaca.

Hare, R.M. (1987), Moral Reasoning about the Environment. *Journal of Applied Philosophy,* 4, 1, 3-14.

Hareuveni, N. (1980), *Nature in our Biblical Heritage.* Translated from the Hebrew and adapted by Helen Frenkley, Neot Kedumim, Kiryat Ono, Israel.

Hargrove, Eugene C. (1987), Foundations of Wildlife Protection Attitudes. *Inquiry,* 30, 3-31.

Hargrove, Eugene C. (1989), *Foundations of Environmental Ethics.* Prentice Hall, Englewood Cliffs, N.J.

Hargrove, Eugene C. (ed.) (1986), *Religion and Environmental Crisis.* University of Georgia Press. Athens, Georgia.

Harris, M. (1974), *Cows, Pigs, Wars and Witches.* Random House, New York.

Harris, M. (1977/1978), *Cannibals and Kings. The Origins of Cultures.* Random House/Vintage Books, New York.

Harris, M. (1985), *The Sacred Cow and the Abominable Pig. Riddles of Food and Culture.* A Touchstone Book, Simon and Schuster, New York.

Harris, T.M. (1820), *The Natural History of the Bible; or a Description of all the Quadrupeds, Birds, Fishes, Reptiles and Insects, Trees, Plants, Flowers, Gums and Precious Stones mentioned in the Sacred Scriptures, collected from the best Authorities and Alphabetically arranged.* Boston.

Harrison, Brian, (1973), Animals and the State in nineteenth Century England. *Engl. Hist. Rev.,* 88, 786-821.

Harwood, Dix (1928), *Love for Animals, and how it Developed in Great Britain.* Thesis, Columbia University, New York.

Hasel, G.F. (1992), *Sabbath.* Anchor Bible Dictionary (*ABD*), 5, 849-856.

Hastings, H. (1936/1973), *Man and Beast in French Thought of the eighteenth Century.* The Johns Hopkins Press, Baltimore, Maryland, 1936; Johnson reprint Corporation, New York/London, 1973.

Haussleiter, J. (1935), *Der Vegetarismus in der Antike.* Verlag von Alfred Töpelmann, Berlin.

Havel, Václav (1989/1990), *Do Rùzných Stran.* Scheinfeld-Schwarzenberg, Prague. Translation in Dutch: *Naar alle windstreken.* De Prom, Baarn.

Hawken, P., Amory Lovins & L. Hunter Lovins (1999), *Natural Capitalism. Creating the next Industrial Revolution.* Little, Brown and Company, Boston/New York.

Hawking, S. (1988), *A Brief History of Time.* Penguin, Harmondsworth.

Hawking, S. & R. Penrose (1996), *The Nature of Space and Time.* Princeton University Press, Princeton, NJ.

Hehn, Victor (1911), *Kulturpflanzen und Haustiere in ihrem Uebergang aus Asien nach Griechenland und Italien sowie in das übriche Europa.* Verlag Gebrüder Bornträger, Berlin.

Heidel, A. (1971), *The Gilgamesh Epic and the Old Testament Parallels.* CUP (Chicago University Press), Chicago.

Heidel, A. (1974), *The Babylonian Genesis.* CUP, Chicago.

Heintz, R.J. & H. Verbruggen (1997), Meer groei en toch een schoner milieu? De groene Kuznets-curve. *Milieu,* 2-9.

Hempel, Carl G. (1966/1973), *Filosofie van de Natuurwetenschappen,* Translated by F. van Zetten, 2nd edition, Het Spectrum B.V., Utrecht/Antwerp.

Hempel, J. (1961), Gott, Mensch und Tier im Alten Testament, mit besonderer Berücksichtigung von Gen. 1-3: APOXYSMATA. Festgabe J. Hempel zum 30.7.1961, *BZAW* 81, 198-229.

Hendrickx, L., A. van den Berg & Ch. Vlek (1993), Zorgen over morgen? De factor 'tijd' in de evaluatie van milieurisico's. *Milieu,* 8, 148-152.

Hendry, G. (1980), *Theology of Nature.* Westminster Press, Philadelphia.

Henry, Marie Louise (1993), Das Tier im religiösen Bewußtsein des alttestamentlichen Menschen. In: Janowski, B., U. Neumann-Gorsolke & U. Gleßmer (eds.) (1993), *Gefährten und Feinde des Menschen. Das Tier in der Lebenswelt des alten Israel.* Neukirchener Verlag, Neukirchen-Vluyn, 20-61.

Hengel, M. (1981), *Judaism and Hellenism Studies in their Encounter in Palestine during the Early Hellenistic Period.* Fortress Press, Philadelphia.

Herm, G. (1975/1992), *De Kelten. Voorlopers van een verenigd Europa.* Translated by C.W.A.J.A. Walraven, Tirion, Baarn.

Hermans, C.A.M. (1986), *Morele vorming. Empirisch-theologisch onderzoek naar de effecten van een katechese-curriculum in de morele vorming omtrent de milieucrisis.* Dissertation, Catholic University of Nijmegen, J.H. Kok, Kampen.

Herodotus (1944), *Historiën.* Translation by J.J.E. Hondius en J.A. Schuursma, J.B. Wolters, Groningen/Batavia.

Hess, R.S. (1990), Splitting the Adam: the Usage of *'âdâm* in Genesis 1-V. In: J.A. Emerton (ed.) *Studies in the Pentateuch.* E.J. Brill, Leiden, 1-16

Hetherington, N.S. (ed.) (1993), *Cosmology. Historical, Literary, Philosophical, Religious, and Scientific Perspectives.* Garland Publishing, New York/London

Hiebert, Th. (1996), *The Yahwist's Landscape. Nature and Religion in Early Israel.* Oxford University Press, Oxford/New York.

Hiebert, Th. (1999), *"Air..The First Sacred Thing: The Conception of rûah in the Hebrew Scriptures".* Paper Harvard Seminar on Environmental Values, 20 April 1999. http://divweb.harvard.edu/csvpl/ee/hserv/hiebert-air.ht

Hiers, R. (1984), Ecology, Biblical Theology and Methodology. *Zygon,* 19, 1, 43-60.

Hillel, D. (1994), *Rivers of Eden. The Struggle for Water and the Quest for Peace in the Middle East.* Oxford University Press, Oxford.

Historia Monachorum in Aegypto: The Lives of the Desert Fathers (1980), Translation by Norman Russel, Introduction by Benedicta Ward, Mowbray, London/Oxford.

Hirsch, S.R. (1972), Do not Destroy. In: Konvitz, M.R. (ed.) *Judaism and Human Rights.* W.W. Norton & Company Inc., New York, 259-65.

Hirsch, F. (1977), *Social Limits to Growth.* Routledge & Kegan Paul, London.

Hoage, R.J. (ed.) (1989), *Perceptions of Animals in American Culture.* Smithsonian Institution Press, Washington D.C.

Hoare, F.R. (1991), *Eight Decisive Books of Antiquity.* Dorset Press, New York.

Hobbes, Thomas (1651/1968), *Leviathan.* Edited with an introduction by C.B. Macpherson, Penguin Books, Harmondsworth, Middlesex, 1968.

Hoftijzer, J. & G. van der Kooij (eds.) (1976), *Aramaic Texts from DEIR 'ALLA* E.J. Brill, Leiden.

Holman, J.C.M. (1992), 'A Gentleman's Psalm. Genesis 1,26 naar het beeld van Psalm 8'. In: Verdegaal, C. & W. Weren (ed.), *Stromen uit Eden. Genesis 1-11 in bijbel, joodse exegese en moderne literatuur.* Liber Amicorum for N. Poulssen, Boxtel/Brugge, 75-86.

Holwerda, D. (1955), *Commentatio de vocis quae est fusis vi atque usu praesertim in graecitate Aristotele anteriore.* Dissertation, Groningen State University, Groningen.

Holwerda, B. (1957/58), *Dictaten, deel III Exegese Oude Testament (Deuteronomium),* 2 series, Kampen.

Hooykaas, R. (1972), *Religion and the Rise of Modern Science.* Scottish Academic Press, Edinburgh/London.

Horwitz, L.K. & E. Tchernov (1989), Subsistence Patterns in Ancient Jerusalem.

A Study of Animal Remains from the Ophel. In: Mazar, E. & B. Mazar (eds.), *Excavations in the South of the Temple Mount: The Ophel of Biblical Jerusalem*. Qedem 29, Jerusalem.

Hostetler, John A. (1968), *Amish Society*. Revised edition, John Hopkins University Press, Baltimore.

Hough, A. (1997), *God is not 'Green'. A re-examination of Eco-Theology*. Morehouse Publishing Company

Houghton, J.T. (1989), New Ideas of Chaos in Physics. *Science and Christian Belief*, 1, 41-51.

Houston, W. (1979), "And let them have Dominium... " Biblical views of Man in Relation to the Environmental Crisis. In: Livingstone, E.A. (ed.) *Studia Biblica 1978, I. Paper on Old Testament and Related Themes. JSOT*, suppl. series, 11, Sheffield, 161-184.

Houston, W. (1993), *Purity and Monotheism. Clean and Unclean Animals in Biblical Law*. Journal for the Study of the Old Testament (*JSTO*), Suppl. series, 140, Sheffield Academic Press, Sheffield.

Houtepen, A.W.J. (1990), 'Integrity of Creation': naar een ecologische scheppingstheologie. *Tijdschrift voor Theologie*, 30, 1, 51-76.

Houtman, C. (1974), *De hemel in het Oude Testament. Een onderzoek naar de voorstellingen van het oude Israël omtrent de kosmos*. Wever, Franeker.

Houtman, C. (1980), *Inleiding in de Pentateuch*. 2 vols., Kok, Kampen.

Houtman, C. (1981), Ezra and the Law. Observations on the supposed relation between Ezra and the Pentateuch. In: A.S. van der Woude (ed.), *Remembering all the Way...A Collection of Old Testament Studies Published on the Occasion of the Fortieth Anniversary of the Oudtestamentisch Werkgezelschap in Nederland*. Oudtestamentische Studiën, Vol. XXI, E.J. Brill, Leiden, 91-115.

Houtman, C. (1982), *Wereld en tegenwereld. Mens en milieu in de Bijbel*. Ten Have, Baarn.

Houtman, C. (1984), Another Look at forbidden Mixtures. *Vetus Testamentum*, XXXIV, 2, 226-228.

Huber, P., *Die Kunstschätze der heiligen Berge*, Pattloch, 3rd edition, Weltbilt Verlag, Augsburg.

Hübner, U. (1989), Schweine, Schweineknochen und ein Speiseverbot im alten Israel. *Vetus Testamentum*, 39, 225-236.

Hueting, R. (1974), *Nieuwe schaarste en economische groei*. Agon/Elsevier, Amsterdam.

Hueting, R. & L. Reijnders (1996), Duurzaamheid is een objectief begrip. *ESB*, 81, 425-427.

Hueting, R. & L. Reijnders (1996), Duurzaamheid en preferenties. *ESB*, 81, 537-539.

Hueting, R. & L. Reijnders (1998), Sustainability is an objective Concept. *Ecological Economics* 27, 139-147.

Hueting, R. & L. Reijnders (1999), Uncertainty and Sustainability. *Ecological Economics*, 29, 9-11.

Hughes, J. Donald (1975), *Ecology in Ancient Civilizations*. University of New Mexico Press, Albuquerque.

Hughes, J. Donald (1975[a]), Ecology in Ancient Greece. *Inquiry*, 18, 115-125.

Hughes, J. Donald (1980), Early Greek and Roman Environmentalists. In: Lester J. Bilsky (ed.) *Historical Ecology Essays on Environment and Social change*. National University Publications Kennikat Press, Port Washington, New York, 45-59.

Hughes, J. Donald (1983/1996), *North American Indian Ecology*. Texas Western Press, El Paso.

Hughes, J. Donald (1994), *Pan's Travail. Environmental Problems of the Ancient Greeks and Romans*. John Hopkins University Press, Baltimore.

Hughes, J. Donald & J.C. Thirgood (1982). Deforestation in ancient Greece and Rome: A cause of collapse. *The Ecologist*, 12, 5, 196-208

Hulst, A.R. (1958), *Kol Bāsār* in der priesterlichen Fluterzählung. *OuTS*, 12, 28-68.

Hultgård, Å. (1987), The Burnt-offering in Early Jewish Religion: Sources, Practices and Purpose. In: Linders, T. & G. Nordquist (eds.), *Gifts to the Gods*. Proceedings of the Uppsala Symposium 1985. Acta Univ. Ups. Boreas, Uppsala Studies in Ancient Mediteranean and Near Eastern Civilizations 15, Uppsala, 83-93.

Hume, C.W. (1957). *The Status of Animals in the Christian Religion*. The Universities Federation for Animal Welfare: UFAW, 7a Lamb's Conduit Passage. London, W.C. 1.

Hume, C.W. (1962), *Man and Beast*. The Universities Federation for Animal Welfare: UFAW, 7a Lamb's Conduit Passage, London W.C.1.

Hunn, E. (1979), The Abominations of Leviticus Revisited: A Commentary on Anomaly in symbolic Anthropology. In: Ellen, R.F. & D. Reason (eds.), *Classifications in their Social Context*. Academic Press, London, 103-116.

Hunt, John (1969), *A World Full of Animals*. David McKay Company Inc., New York.

Huntington, S.P. (1996/1998), *The Clash of Civilizations and the Remaking of World Order*. Simon & Schuster / Touchstone, London.

Huppes, G. & R. Heijungs (1993), Integrale milieubeoordeling: ordinale versus kardinale benadering. *Milieu*, 8, 109-110.

Huxley, Th. H. (1863/1959), *Man's Place in Nature*. Introduction by Ashley Montegu, Ann Arbor Paperbacks, University of Michigan Press, Ann Arbor Michigan.

Huxham, M. & D. Sumner (2000), *Science and Environmental Decision Making*. Prentice Hall Pearson Education ltd, Harlow, Essex.

Iamblich/Porphyry (1987), *Leven en Leer van Pythagoras*. Introduction, translation and commentary by H.W.A. van Rooyen-Dijkman, Ambo, Baarn.

Isenberg, A.C. (2000), *The Destruction of the Bison*. Cambridge University Press, Cambridge.

Isidore de Séville (1986), *Etymologies*. Livre XII. Des Animaux, traduit et commenté par: Jacques André. Société d'édition 'Les Belles Lettres', Paris.

Israel, Jonathan I. (2001), *Radical Enlightenment. Philosophy and the Making of Modernity 1650-1750*. Oxford University Press, Oxford.

Janowski, B., U. Neumann-Gorsolke & U. Gleßmer (eds.) (1993), *Gefährten und Feinde des Menschen. Das Tier in der Lebenswelt des alten Israel*. Neukirchener Verlag, Neukirchen-Vluyn.

Jacobs, Milton (1977), Animals and God-Man-Nature in the Old Testament. *Jewish Journal of Sociology*, 18, 2, 141-154.

Jaenen, C.J. (1979), Review of Calvin Martin's: Keepers of the Game. *Journal of Interdisciplinary History*, X, 374-376.

Jagersma, H. (1990), *Tekst & Interpretatie. Studies over getallen, teksten, verhalen en geschiedenis in het Oude Testament*. Callenbach, Nijkerk.

Jagersma, H. (1995), *Genesis 1:1-25:11*. Callenbach, Nijkerk.

Jagersma, H & M. Vervenne (1992), *Inleiding in het Oude Testament*. J.H. Kok, Kampen.

Jenni, E. & C. Westermann (eds.) (1986), *Theologisches Handwörterbuch zum Alten Testament (THAT)*. (Vols. I + II) München, Zürich.

Jennison, George (1937), *Animals for Show and Pleasure in Ancient Rome*. Manchester University Press. Manchester, England.

Jenson, Ph.P. (1992), *Graded Holiness. A Key to the Priestly Conception of the World.* Journal for the Study of the Old Testament (*JSOT*), Suppl. series, 106, Sheffield Academic Press, Sheffield.

Janowski, B. (1990), Tempel und Schöpfung. Schöpfungstheologische Aspekte de Priesterschriftlichen Heiligtumskonzeptionen. *Jahrbuch für Biblische Theologie* (*JBTh*), Band 5, Neukirchener Verlag, Neukirchen-Vluyn, 37-69.

Jobling, D.K. (1972), *"And have Dominion..": The Interpretation of Old Testament Texts concerning Man's Rule over the Creation (Genesis 1:26,28, 9:1-2, Psalm 8:7-9) from 200 BC to the Time of the Council of Nicea.* Dissertation, Union Theological Seminary, New York.

Johns, D.M. (1990), The Relevance of Deep Ecology to the Third World: Some Preliminary Comments. *Environmental Ethics*, 12, 2, 235-252.

Joines, K.R. (1974), *Serpent Symbolism in the Old Testament. A Linguistic, Archeological and Literary Study.* Haddonfield House, Haddonfield, New Jersey.

Jones, A. (1995), *Soul Making: The Desert Way of Spirituality. Harper and Row, San Francisco*, Ca.

Jones, Gerald E. (1972), *Concern for Animals as Manifest in Five American Churches: Bible Christian, Shaker, Latter-day Saint, Christian Scientist and Seventh-day Adventist.* Dissertation Brigham Young University, Salt Lake City, Utah.

Jones, John D. (1981), Humans and Animals: Compassion and Dominion. *Anglican Theological Review*, 63, 3, 259-272.

Jónsson, Gunnlaugur A. (1988), *The Image of God. Genesis 1:26-28 in a Century of Old Testament Research.* Coniecta Biblica (*CB*) Old Testament Series 26, Almqvist & Wiksell.

JPS (The Jewish Publication Society) (1985), *TANAKH A New Translation of the Holy Scriptures According to the Traditional Hebrew Text.* The Jewish Publication Society, Philadelphia/Jerusalem.

Kabilsingh, Chatsumaran (1998), *Buddhism and Nature Conservation.* Thammasat University Press, Bangkok.

Kaiser, Jocelyn (2000), Ecologists on a Mission to save the World. *Science*, 287, 1188-1192.

Kampen, N.G. van (1988), Ten Theorems about Quantummechanical Measurements. *Physica A*, 153, 97-113.

Kaplan, Jonathan (ed.) (1984), *International Bibliography of Jewish History and Thought.* K.G. Saur, München.

Katz, E. (1985), Organism, Community and the Substitution Problem. *Environmental Ethics*, 7, 241-256.

Katz, E. (1988), Environmental Ethics, A Select Annotated Bibliography, 1983-1987. *Research in Philosophy and Technology*, 9, 251-285.

Kay, Jeanne (1985), Preconditions of Natural Resource Conservation. *Agricultural History*, 59, 124-135.

Kay, Jeanne (1988), Concepts of Nature in the Hebrew Bible. *Environmental Ethics*, 10, 4, 309-327.

Keel, O. (1978), Jahwes Entgegnung an Ijob. Einen Deutung von Ijob 38-41 vor dem Hintergrund der zeitgenössischen Bildkunst. *Forschungen zur Religion und Literatur des Alten und Neuen Testaments (FRLANT 121)*, Vandenhoeck & Ruprecht, Göttingen, 86-125.

Keel, O. (1980), *Das Böcklein in der Milch seiner Mutter und Verwandtes. Im lichte eines altorientalischen Bildmotivs.* Orbis Biblicus et Orientalis (OBO 33), Vandenhoeck & Ruprecht, Göttingen.

Keel, O. & Chr. Uehlinger (1992), *Göttinnen, Götter und Gottessymbole.* Questiones Disputatae 134 Editiones Herder, Freiburg/Basel/Vienna.

Kehl, A. (1985), Haradrius. *R.L.A.C.*, XIII, 585-593.

Keil, C.F. and F. Delitzsch (1980), *Commentary on the Old Testament in ten Volumes. Vol. 1 The Pentateuch.* W.B. Eerdmans, Publ Comp., Grand Rapids, Michigan.

Keller, Otto (1909), *Die Antike Tierwelt, I & II.* Verlag Wilhelm Engelmann, Leipzig.

Kellert, S.R. (1989), Perceptions of Animals in America. in: Hoage, R.J. (ed.), *Perceptions of Animals in American Culture,* Smithsonian Institution Press, Washington D.C., 5-24.

Kellert, S.R. (1996), *The Value of Life. Biological Diversity and Human Society.* Island Press / Shearwater Books, Washington.

Kellert, S.R. (1997), *Kinship to Mastery. Biophilia in Human Evolution and Development.* Island Press / Shearwater Books, Washinton.

Kellert, S.R. & J.K. Berry (1985), *A Bibliography of Human/Animal Relations.* Lanham, Md.

Kellert, S.R. & E.O. Wilson (eds.) (1993), *The Biophilia Hypothesis.* Island Press, Washington.

Kennedy, Des (1992), *Living Things we love to hate.* Whitecap Books, Vancouver/Toronto.

Kessler, D. (1989), *Die heiligen Tiere und der König. Teil I: Beiträge zur Organisation, Kult und Theologie der spätzeitlichen Tierfriedhöfe.* Ägypten und Altes Testament (ÄAT 16), O. Harrassowitz, Wiesbaden.

Keulartz, J. & M. Korthals (eds.) (1997), *Museum aarde. Natuur: criterium of constructie?* Boom, Amsterdam/Meppel.

Kimpel, B. (1981), *Philosophies of Life of the Ancient Greeks and Israelites. An Analysis of their Parallels.* Philosophical Library, New York.

Kinzler, A. (1902), *Biblische Naturgeschichte.* 10th Edition. Calw und Stuttgart.

Kirchhoff, H. (1987), *Sympathie für die Kreatur, Mensch und Tier in Biblischer Sicht.* Köselverlag, München.

Kitchell, K.F. (1993), The View from Deucalion's Ark: New Windows on Antiquity. *The Classical Journal,* 88, 4, 341-357.

Klaits, J. & B.G. Klaits (eds.) (1974), *Animals and Man in Historical Perspective.* Harper & Row, New York.

Klijn, A.J.F. (1995), *Het ontstaan van het Nieuwe Testament.* Uitgeverij Callenbach, Nijkerk.

Klotz, John W. (1981), Domestic Animals and Bible Symbolism. *Concordia Journal,* 7, 4, 144-149.

Knauf, E.A. (1988), Zur Herkunft und Sozialgeschichte Israels. "Das Böckchen in der Milch seiner Mutter". *Biblica,* 69, 153-169.

Knight, D.A. (1985), Cosmogony and Order in the Hebrew Tradition. In: R.W. Lovin & F.E. Reynolds (eds.), *Cosmogony and Ethical Order.* Chicago University Press, Chicago.

Koch, K. (1983), 'Gestaltet die Erde, doch heget das Leben. Einige Klarstellungen zum *dominium terrae* in Genesis 1. In: Geyer, H.G., J.M. Schmidt, W. Schneider & M. Weinrich (eds.) (1983), *Wenn nicht jetzt, wann dann?* Aufsätze für Hans-Joachim Kraus zum 65. Geburtstag, Neukirchener Verlag, Neukirchen-Vluyn, 23-36.

Kockelkoren, P. (1994), The Mouse in the Cat's Claws. In: W. Zweers & J.J. Boersema (eds.), *Ecology, Technology and Culture.* The White Horse Press, Cambridge, 99-117.

Kockelkoren, P. a.o. (1994), Environmental policy in transformation: A philosophical approach. In: CLTM (Dutch Committee for Long-Term Environmental Policy), *The Environment: Towards a Sustainable Future.* Kluwer Academic Publishers, Dordrecht/Boston, 253-272.

Koehler, L. & W. Baumgartner (1948-), *Lexicon in Veteris Testamenti Libros.* E.J. Brill, Leiden.

Kohák, E. (1984), *The Embers and the Stars. A philosophical Inquiry into the Moral Sense of Nature.* Chicago University Press, Chicago.

König, A. (1988), New and Greater Things: Re-evaluating the Biblical Message on Creation. *Stud. Originalia,* 1, University of South Africa, Pretoria.

Konvitz, M.R. (ed.) (1972), *Judaism and Human Rights,* W.W. Norton & Company inc., New York.

Koole, J.L. (1966), 'Bijbelstudie over Psalm 8'. *GTT,* 66, 1-8.

Koolschijn, G. (1990), *Het democratische beest. Plato's tegenstander.* Bert Bakker, Amsterdam.

Kooten, G.H. van (2001), *The Pauline Debate on the Cosmos. Graeco-Roman Cosmology and Jewish Eschatology in Paul and in the Pseudo-Pauline Letters to the Collossians and the Ephesians.* Thesis, Leiden University.

Kornfeld, W. (1965), Reine und unreine Tiere im Alten Testament. *Kairos,* 7, 134-147.

Korthals, M. (1994), *Duurzaamheid en democratie,* Boom, Amsterdam.

Kraak, W. (1940), *Vogeltrek in de oudheid, in het byzonder bij Aristoteles.* Dissertation, State University of Leiden, 111, N.V. Noord-Hollandsche Uitgevers Maatschappij, Amsterdam.

Krause, M. (1983), 11 Sam. 2,4 und das Konzeptionsoptimum. *ZAW,* 95, 434-437.

Kraybill, Donald B. (ed.) (1993), *The Amish and the State.* The John Hopkins University Press, Baltimore.

Kraybill, Donald B. & S.M Nolt (1995), *Amish Enterprise. From Plows to Profits.* The John Hopkins University Press, Baltimore.

Krech, Shepard III (1999), *The Ecological Indian. Myth and History.* W.W. Norton & Company, New York.

Krech, Shepard III (ed.) (1981), *Indians, Animals and the Fur Trade: A Critique of "Keepers of the Game".* University of Georgia Press, Athens.

Krebs, A. (1993), Haben wir moralische Pflichten gegenüber Tieren? Das pathozentrische Argument in der Naturethik. *Deutsche Zeitschrift für Philosophie,* 41, 6, 1009-1022.

Krebs, Ch.J. (1988), *The Message of Ecology.* Harper & Row, New York.

Kritiek (1989), De Mislukte Natuurbeheersing. Special issue of *Kritiek,* no. 17, compiled by Guy Quintelier.

Kroes, P.A. (1987), *Filosofie van de natuurkunde,* Martinus Nijhoff, Leiden.

Krolzik, U. (1979), *Umweltkrise. Folge des Christentums?* Mit einem Vorwort von G. Altner, Kreuz Verlag, Stuttgart/Berlin.

Krolzik, U. (1989), Die Wirkungsgeschichte von Genesis 1,28. In: G. Altner (ed.), *Ökologische Theologie. Perspektieven zur Orientierung.* Kreuz Verlag, Stuttgart, 149-163.

Kuik, O & H. Verbruggen (eds.) (1991), *In Search of Indicators of Sustainable Development.* Kluwer Academic Publishers, Dordrecht/Boston/London.

La Barre, W. (1954), *The Human Animal.* University of Chicago Press, Chicago.

La Barre, W. (1962), *They shall take up Serpents. Psychology of the Southern Snake-handling Cult.* University of Minnesota press, Minneapolis.

Labuschagne, C.J. (1987/1990), *Deuteronomium.* 3 dln. Ia, Ib en II, De prediking van het Oude testament, G.F. Callenbach, Nijkerk.

Labuschagne, C.J. (1990), Het bijbels scheppingsgeloof in ecologisch perspectief. *Tijdschrift voor Theologie*, 30, 1, 5-18.

Labuschagne, C.J. (1991), *Hoe heilzaam is de bijbelwetenschap? De bijdrage van de bijbel(wetenschap) aan de cultuur.* Departing speech as Professor of Old Testament Literature, Groningen State University, 17 September 1991. G.F. Callenbach, Nijkerk.

Labuschagne, C.J. (1991[a]), Creation and the Status of Humanity in the Bible. In: V. Brümmer (ed.), *Interpreting the Universe as Creation. A Dialog of Science and Religion.* Kok Pharos, Kampen, 123-131.

Labuschagne, C.J. (1992), 'You shall not boil a Kid in its Mother's Milk' A new Proposal for the Origin of the Prohibition. In: F. García Martínez *et al.* (eds.) *The Sriptures & the Scrolls.* E.J. Brill, Leiden, 8-17.

LaFreniere, G.F. (1989), Worldviews and Environmental Ethics. *The Trumpeter*, 6, 4, 144-150.

LaFreniere, G.F. (1990), Rousseau and the European Roots of Environmentalism. *Environmental History Review*, Winter, 41-72.

Lambert, W.G. (1965), A New Look at the Babylonian Background of Genesis. *JTS (N.S.)*, 16, 285-300.

Lambert, W.G. (1994), Enuma Elish. *TUAT* III, 4, 583-587.

Lamm, Norman (1971), *Faith and Doubt.* New York.

Landmann, Michael (1959), *Das Tier in der Jüdischen Weisung.* Verlag Lambert Schneider, Heidelberg.

Landsberger, B. (1934), *Die Fauna des Alten Mesopotamien nach der 14. Tafel der Serie Har-ra = Hubullu.* Des XLII. Bandes der Abhandlungen der philologisch-historischen Klasse der Sächsichen Akademie der Wissenschaften Nr. VI, Verlag von S. Hirzel, Leipzig.

Lang, B. (ed.) (1985), *Anthropological Approaches to the Old Testament.* Philadelphia, London.

Lange, F. de (1992), Leert de natuur ons mores? Het natuurbegrip in de ethiek. *Nederlands Theologisch Tijdschrift* (*NTT*), 46, 1, 34-50.

Leach, Edmund (1964), Anthropological Aspects of Language. Animal Categories and Verbal Abuse. In: Lenneberg, E.H. (ed.) *New Directions in the Study of Language.* M.I.T. Press, Cambridge, Massachusetts, 23-63.

Leathers Kuntz, M. & P.G. Kuntz (1987), *Jacob's Ladder and the Tree of Life: Concepts of Hierarchy and the great Chain of Being,* New York.

Lecky, W.E.H. (1877), *History of European Morals from Augustus to Charlemagne.* 3rd ed. Longmans, Green and Co., London.

Leder, H.G. (1963), Südenfallerzählung und Versuchungsgeschichte. Zur Interpretation von Mc 1,12f. *Zeitschrift für Neu-Testamentische Wissenschaft (ZNW)*, 54, 188-216.

Lee, Donald C. (1980), On the Marxian View of the Relationship between Man and Nature. *Environmental Ethics*, 2, 3-16.

Leeuwen, A. Th. van (1964), *Christianity in World History, the Meeting of the Faiths of East and West.* Edinburgh House Press, London.

Leiman, S.Z. (1976), *The Canonization of Hebrew Scripture: The Talmudic and Midrashic Evidence.* Archon Books, Hamden, Connecticut.

Leiss, W. (1972), *The Domination of Nature*, G. Braziller, New York.

Lemaire, T. (1986), *De Indiaan in ons Bewustzijn: de ontmoeting van de Oude met de Nieuwe Wereld.* Ambo, Baarn.

Lenneberg, Eric H. (ed.) (1964), *New Directions in the Study of Language.* M.I.T. Press, Cambridge, Massachusetts.

Leopold, Aldo (1949), *A Sand County Almanac.* Oxford University Press, Oxford.

Leroy, P. (1995), *Milieukunde als roeping en beroep*. Inaugural speech, 9 June 1995, Catholic University of Nijmegen, Nijmegen.

Leslie, J. (1996), *The End of the World. The Science and Ethics of Human Extinction*. Routledge, London.

Levin, S.I. & E.A. Boyden (1940), *The Kosher Code of the Orthodox Jew*. Minneapolis.

Lévi Strauss, C. (1962/1966), *The Savage Mind*. University of Chicago Press, Chicago, Weidenfeld and Nicolson Ltd, London (Original: 'La Pensée Sauvage', Librarie Plon, Paris 1962).

Lewinsohn, Richard (1954), *Animals, Men and Myths: A History of the Influence of Animals on Civilization and Culture*, by Morus (Richard Lewinsohn). Victor Gollancz Ltd., London.

Lexikon der Alten Welt. Artemis Verlag.

Liedke, G. (1972), Von der Ausbeutung zur Kooperation. Theologisch-Philosophische Ueberlegungen zum Problem des Umweltschutzes. In: Weizsäcker, Ernst von (ed.), *Humanökologie und Umweltschutz*, Ernst Klett Verlag Stuttgart, Kösel Verlag München, 36-65.

Liedke, G. (1979), *Im Bauch des Fisches. Ökologische Theologie*. Kreuz Verlag, Stuttgart.

Lindberg, D.C. (1992), *The Beginning of Western Science. The European Scientific Tradition in Philosophical, Religious and Institutional Context, 600 B.C. to A.D. 1450*. University of Chicago Press, Chicago.

Lindberg, D.C. & R.L. Numbers (eds.) (1986), *God and Nature: Historical Essays on the Encounter between Christianity and Science*. University of California Press, Berkeley, California.

Linden, N. ter (1996), *Het verhaal gaat*. Uitgeverij Balans, Amsterdam.

Linders, T. & G. Nordquist (eds.) (1987), *Gifts to the Gods*. Proceedings of the Uppsala Symposium 1985. Acta Univ. Ups. Boreas. Uppsala Studies in Ancient Mediterranean and Near Eastern Civilizations 15, Uppsala.

Ling, Chee Yoke (1987), *Humanity must protect Nature. What Islam, Taoism, Hinduism & Christianity Have to Say About Environmental Protection*. Third World Science Movement, Consumers' Association of Penang, Malaysia.

Linnaeus, C. (1735/1761), *Systema Naturae*. Translated by M. Houttuyn 1761.

Linzey, A. (1976), *Animal Rights: A Christian Assessment of Man's Treatment of Animals*. S.C.M. press, London.

Linzey, A. (1986), 'The Place of Animals in Creation: A Christian View. In: Regan, T. (ed.) (1986), *Animal Sacrifices. Religious Perspectives on the Use of Animals in Science*. Temple University Press, Philadelphia, 115-148.

Linzey, A. (1987), *Christianity and the Rights of Animals*. SPCK, London.

Linzey, A (1995), *Animal Theology*. University of Illinois Press, Urbana/Chicago.

Linzey, A. & D. Cohn-Sherbock (eds.) (1997), *After Noah. Animals and the Liberation of Theology*. Cassell,

Linzey, A. & T. Regan (eds.) (1989), *Animals and Christianity: A Book of Readings*. SPCK, London, Crossroads, New York.

Linzey, A. & D. Yamamoto (eds.) (1998), *Animals on the Agenda. Questions about Animals for Theology and Ethics*. SCM Press, London.

List, P.C. (1993), *Radical Environmentalism. Philosophy and Tactics*. Wadsworth Publishing Company, Belmont, California.

Livi-Bacci, M. (1992), *A concise History of World Population*. Translation by Carl Ipsen, Blackwell, Oxford.

Livingstone, E.A. (ed.) (1979), *Studia Biblica 1978, I. Paper on Old Testament and Related Themes. JSOT*, suppl. series, 11, JSOT Press, Sheffield.

Lockwood, Michael (1979), Singer on Killing and the Preference for Life. *Inquiry*, 22, 157-170.
Lohfink, N. (1974), "Macht euch die Erde untertan"? *Orientierung*, 38, 137-142.
Long, A.A. (ed.) (1971), *Problems in Stoicism*. Athlone Press, London.
Long, A.A. (1974), *Hellenistic Philosophy, Stoics, Epicureans, Sceptics*. Duckworth, London.
Long, A.A. & D.N. Sedley (1987), *The Hellenistic philosophers*. Vol. 1, *Translations of the Principal Sources with Philosophical Commentary*, Cambridge University Press, Cambridge.
Loomis, C.G. (1948), *White Magic. An Introduction to the Folklore of Christian Legend*. The Mediaeval Academy of America, Cambridge, Mass.
Lovejoy, A.O. (1936), *The Great Chain of Being. A study of the History of an Idea*. (The William James Lectures Delivered at Harvard University, 1933), Harvard University Press, Cambridge, Massachusetts.
Lovejoy, A.O. (1948), Nature as Norm in Tertullian. In: Lovejoy, A.O. (1948), *Essays in the History of Ideas*, Baltimore, 308-338.
Lovelock, J.E. (1979/1980), *GAIA, a new Look at Life on Earth*. Oxford University Press, Oxford.
Lovelock, J.E. (1988), *The Ages of GAIA. A Biography of our Living Earth*. Oxford University Press, Oxford.
Lovin, R.W. & F.E. Reynolds (eds.) (1985), *Cosmogony and Ethical Order*. Chicago University Press, Chicago.
Löwith, K. (1967), *Gott, Mensch und Welt in der Metaphysik von Descartes bis zu Nietzsche. Göttingen.*
Ludwig, D., R. Holborn & C. Waters (1993), Uncertainty, Resource Exploitation and Conservation: Lessons from History. *Science*, 260, 17, 36.
Maag, V. (1954), Sumerische und Babylonische Mythen von der Erschaffung der Menschen. *Asiatischen Studien (AS)*, 8, 85-106.
Maag, V. (1955), Alttestamentische Anthropogonie in ihrem Verhältnis zur altorientalischen Mythologie. *Asiatischen Studien (AS)*, 9, 15-44.
Maarsingh, B. (1974), *Leviticus*. De prediking van het Oude Testament, G.F. Callenbach, Nijkerk.
Machan, T.R. (1991), Do Animals have Rights? *Public Affairs Quarterly*, 5, 2, 163-173.
Macht, David I. (1953), An Experimental Pharmacological Appreciation of Leviticus XI and Deuteronomy XIV. *Bulletin of the History of Medicine*, 27, John Hopkins University Press, 444-450.
MacIntyre, A. (1967/1993), *A short History of Ethics. A History of moral Philosophy from the Homeric Age to the Twentieth Century*. Routledge, London.
MacKenzie, J.M. (1988), *The Empire of Nature: Hunting, Conservation and British Imperialism*. Manchester University Press, Manchester.
MacKenzie, J.M. (ed.) (1990), *Imperialism and the Natural World*. Manchester University Press, Manchester.
Macquarrie, J. (1971). Creation and Environment. *The Expository Times*, 83, 1, 4-9.
Magnin, Charles (1838/1981), *Origines du Théâtre*. Hachette, Paris, 1838 (Plan-de-la-Tour, Var, France: Editions d'aujourd'hui, 1981).
Magnuson, Jon (1991), Reflections of an Bow Hunter. *Christian Century*, March 13.
Mahnke, H. (1978), *Die Versuchungsgeschichte im Rahmen der synoptischen Evangelien. Ein Beitrag zur frühen Christologie*. Beiträge zur biblischen Exegese und Theologie, Band 9, Peter Lang, Frankfurt am Main.
Maimonides, Moses (± 1194), *The Guide for the Perplexed (Moreh Nebuchim)*. Trans-

lated from the Original Arabic Text by M. Friedländer, Second Edition, revised throughout 1956, Dover Publications, New York.

Manenschijn, G. (1988), *Geplunderde aarde, getergde hemel. Ontwerp voor een christelijke milieu-ethiek.* Ten Have, Baarn.

Mannion, A.M. (1997), *Global Environmental Change. A Natural and Cultural Environmental History.* 2nd edition, Longman, Harlow Essex.

Marsh, George Perkins (1864/1871/1965), *Man and Nature; Or Physical Geography as Modified by Human Action.* Charles Scribner & Co, New York (1864/1871), Belknap Press / Harvard University Press, Cambridge Massachusetts (1965).

Martin, C. (1974), The European Impact on the Culture of a North Eastern Algonquian Tribe: an Ecological Interpretation. *William and Mary Quarterly*, 3-26.

Martin, C. (1978), *Keepers of the Game. Indian-Animal Relationships and the Furtrade.* University of California Press, Berkeley.

Martin, P.S. (1973), The Discovery of America. *Science,* 179, 968-974.

Martin, P.S. & H.E. Wright (eds.) (1967), *Pleistocene Extinction: The Search for a Cause.* Yale University Press, New Haven.

Marx, K. (1971), *Grundrisse.* D. McLellan (ed.), London.

Massey, Marshall (1991), Where are Our Churches Today? A Report on the Environmental Positions of the Thirty Largest Christian Denominations in the United States. *Firmament,* 2, 4, Winter.

May, R.M. (1988), How many Species are there on Earth. *Science*, 241, 1441-1449.

Mayer-Tasch, P.C. (ed.) (1991), *Natur denken. Eine Genealogie der Oecologischen Idee.* 2 Bänder, Fisher, Frankfurt am Main.

Mayes, A.D.H. (1994), Deuteronomy 14 and the Deuteronomic Worldview. In: García Martínez, F., A. Hilhorst, J.T.A.G.M. van Ruiten & A.S. van der Woude (eds.) (1994), *Studies in Deuteronomy in honour of C.J. Labuschagne on the occasion of his 65the birthday.* Suppl. VT, vol. 53, 165-181.

McCurley, F.R. (1974), The Home of Deuteronomy Revisited: A Methodological Analysis of the Northern Theory. In: Bream H.N., R.D. Heim & C.A. Moore (eds.) *A Light unto My Path: Old Testament Studies in Honour of Jacob M. Myers.* Temple University Press, Philadelphia, 295-317.

McDaniel, Jay B. (1989), *Of God and Pelicans. A Theology of Reverence for Life.* Westminster/John Knox Press, Louisville, Kentucky.

McDaniel, Jay B. (1990), *Earth, Sky, Gods & Mortals. Developing an Ecological Spirituality.* Twenty-Third Publications, Mystic, Conn.

McDaniel, Jay B. (1995), *With Roots and Wings. Christianity in an Age of Ecology and Dialogue.* Orbis Books, Maryknoll, New York.

McDonagh, S. (1990), *The Greening of the Church.* Geoffrey Chapman, London.

McKay, H.A. (1994), *Sabbath and Synagogue. The Question of Sabbath Worship in Ancient Judaism.* E.J. Brill, Leiden.

McKenzie, John L. (1952), God and Nature in the Old Testament. *The Catholic Biblical Quarterly*, 14, 18-145.

McKibben, Bill (1989), *The End of Nature.* Random House, New York, Anchor Books, Doubleday, New York

McNeill, J.R. (1992), *The Mountains of the Mediterranean World. An Environmental History.* Cambridge University Press, Cambridge.

McPhee, J. (1989), *The Control of Nature.* Farrar Straus & Giroux, New York.

Meadows, D.H., D.L. Meadows, J.Randers & W.W. Behrens III (1972), *The Limits to Growth. A Report for the Club of Rome's Project on the Predicament of Mankind.* A Potomac Associates Book, New American Library, New York.

Meadows, D.H., D.L. Meadows & J.Randers (1992), *Beyond the Limits. Global Collapse or a Sustainable Future.* Earthscan Publications Limited, London.

Meander (1989), *Nieuwe Strategieën voor Natuurbescherming.* Stichting Meander, Jan van Arkel, Utrecht.

Medawar, P.B. & J.S. Medawar (1983), *Aristotle to Zoos. A Philosophical Dictionary of Biology.* Harvard University Press, Cambridge, Massachusetts.

Meesters, J.H. (1966), *Op zoek naar de oorsprong van de Sabbat.* Van Gorcum & Comp. N.V.- Dr. H.J. Prakke & H.M.G. Prakke, Assen.

Meigs, A.S. (1978), A Papuan Perspective on Pollution. *Man,* 13, 304-318.

Meijer, F. & O. van Nijf (1992), *Trade, Transport and Society in the Ancient World. A Sourcebook.* Routledge, London.

Meijers, L.D. (1987), The Stuctural Analysis of the Jewish Calendar and Its Political Implications. *Anthropos,* 82, 4/6, 603-610.

Melton Journal (1991), *Judaism and Ecology. Our Earth and our Tradition.* Special issue, no. 24, Spring 1991.

Mendenhall, G.E. (1974), The Shady Side of Wisdom: The Date and Purpose of Genesis 3. In: Bream H.N., R.D. Heim & C.A. Moore (eds.) *A Light unto My Path: Old Testament Studies in Honour of Jacob M. Myers.* Temple University Press, Philadelphia, 319-334.

Merchant, Carolyn (1980), *The Death of Nature, Women, Ecology and the Scientific Revolution.* Harper & Row Publishers, San Francisco.

Merchant, Carolyn (1990), Environmental Ethics and Political Conflict: A View from California. *Environmental Ethics,* 12, 1, 45-69.

Meuli, K. (1946/1975), Griechische Opfergebräuche. in: *Gesammelte Schriften.* Basel, 907-1021.

Midgley, M. (1979/1980), *Beast and Man, The Roots of Human Nature.* The Harvester Press Ltd., Brighton (Paperback, Methuen & Co Ltd., 1980, London).

Mighetto, Lisa (1991), *Wild Animals and American Environmental Ethics.* University of Arizona Press, Tucson, Arizona.

Miles, J. (1995), *God. A Biography.* Alfred A. Knopf, New York.

Mikesell, Marvin W. (1969), The Deforestation of Mount Lebanon. *The Geographical Review,* January LIX, 1, 1-28.

Milgrom, J. (1990), Ethics and Ritual: The Foundations of the Biblical Dietary Laws. in: Firmage, E.B., B.G. Weiss & J.W. Welch (eds.), *Religion and Law: Biblical-Judaic and Islamic Perspectives.* Eisenbrauns, Winona Lake, 159-191.

Milgrom, J. (1990[a]), *Numbers. The Traditional Hebrew Text with the new JPS Translation.* JPS Torah Commentary, Jewish Publication Society, Philadelphia.

Milgrom, J. (1991), *Leviticus 1-16. A new Translation with Introduction and Commentary.* The Anchor Bible, Doubleday & Company, Garden City, New York.

Milgrom, J. (1999), The Antiquity of the Priestly Source. A reply to Joseph Blenkinsopp. *ZAW,* 111, 10-22.

Milieudefensie (1997), *God & Milieu.* Special issue, 26, 1

Mill, J.S. (1962), *Utilitarianism, On Liberty, Essay on Bentham. Together with selected Writings of Jeremy Bentham and John Austin.* Edited with an Introduction by Mary Warnock, Fontana Press, William Collins Sons & Co Ltd, Glasgow.

Mill, J.S. (1904), *Three Essays on Religion.* London.

Miller, H.B. & W.H. Williams (eds.) (1983), *Ethics and Animals,* Humana press, Clifton, New Jersey.

Miller, J.E. (1991), Structure and Meaning of the Animal Discourse in the Theophany of Job (38,39-39,30). *ZAW,* 103, 3, 418-421.

Miller, M.S. & J.L. Miller (1944), *Encyclopedia of Bible Life.* Harper & Brothers, New York/London.

Miller, P.D. Jr. (1978), *Genesis 1-11. Studies in Structure & Theme. JSOT,* Suppl. 8, Sheffield Academic Press, Sheffield.

Milton, Kay (1996), *Environmentalism and Cultural Theory; Exploring the role of Anthropology in Environmental Discourse*. Routledge, London.

Mishnah (1988), A new Translation by Jacob Neusner, Yale University Press, New Haven, Connecticut.

Moldan, B., S. Billharz & R. Matravers (eds.) (1997), *Sustainability Indicators: A Report on the Project on Indicators of Sustainable Development (SCOPE 58)*. John Wiley, Chichester/New York.

Moltmann, J. (1983), Schöpfung aus nichts. In: Geyer, H.G., J.M. Schmidt, W. Schneider & M. Weinrich (eds.) (1983), *Wenn nicht jetzt, wann dann?*. Aufsätze für Hans-Joachim Kraus zum 65. Geburtstag, Neukirchener Verlag, Neukirchen-Vluyn, 259-267.

Moltmann, J. (1985), *Gott in der Schöpfung. Ökologische Schöpfungslehre*. Chr. Kaiser, München.

Moncrief, L.W. (1970), The Cultural Basis for our Environmental Crisis. *Science*, 170, 508-512.

Montaigne, M. de (1993), *Essays*. Translated by F. de Graaff, Boom, Meppel/Amsterdam.

Montefiore, H. (ed.) (1975), *Man and Nature*: William Collins Sons & Co Ltd., London.

Montefiore, C.G. & H. Loewe (eds.). (1938), *A Rabbinic Anthology*. Meridian Books Inc., Greenwich Editions, London & New York.

Moor, J.C. de (1990), *The Rise of Yahwism. The Roots of Israelite Monotheism*. Peeters, Leuven.

Moore, P.D. (1990), The Exploitation of Forest. *Science and Christian Belief*, 2, 131-140.

More, Henry (1679), Antidote Against Atheism, book 2, chapter 9, 3, In: *Opera Omnia*, vol. 2, London.

Moraux, P. (1984), *Der Aristotelismus*. Band I en II, W. de Gruyter, Berlin.

Morris, R.K. & M.W. Fox (eds.) (1978), *On the Fifth Day. Animal Rights and Human Ethics*. Acropolis Books, Washington D.C.

Moule, C.F.D. (1964), *Man and Nature in the New Testament. Some Reflections on Biblical Ecology*. University of London, The Athlone Press, London.

Mulder, H. (1995), *Back to Our Future. Physical Constraints on Sustainable Developments Paths in an Energy-based Backcasting Approach*. Ph.D. Thesis, University of Groningen.

Mulder, M.J. (ed.) (1988), *Mikra, Text, Translation, Reading and Interpretation of the Hebrew Bible in Ancient Judaism and Early Christianity*. Van Gorcum, Assen. Fortress Press, Philadelphia.

Müller, H.-P. (1988), Die sogenanten Straußenperikope in den Gottesreden des Hiobbuches. *ZAW*, 100, 90-105.

Myers, N. (1983), A Priority-Ranking Strategy for Threatened Species? *The Environmentalist*, 3, 97-120.

Nabhan, G. (1982), *The Desert Smells Like Rain*. North Point Press, San Francisco.

Naess, A. (1973), The Shallow and the Deep. Long-range Ecology Movement. *Inquiry*, 16, 95-100.

Naess, A. (1976/1989), *Ecology, Community and Lifestyle: Outline of an Ecosophy*. Translated, revised and augmented by D. Rothenberg, Cambridge University Press, Cambridge.

Naess, A. (1979), Self-realization in Mixed Communities of Humans, Bears, Sheep, and Wolves. *Inquiry*, 22, 1-2, 231-243.

Naess, A. (1986), Intrinsic Value: Will the Defenders of Nature please rise? In:

Soulé, M.E. (ed.) *Conservation Biology: the Science of Scarcity and Diversity*. Sinauer Associates, Sunderland, Massachusetts, 504-516.

Naess, A. (1988), Deep Ecology and Ultimate Premises. *The Ecologist,* 18, 4, 5, 128-131.

Naḥmanides, Moses (1194-1270, 1965), *Commentary on the Torah*. Edition C.B. Chavel, Jerusalem, 1965.

Nash, J.A. (1991), *Loving Nature. Ecological Integrity and Christian Responsibility*. The Churches' Center for Theology and Public Policy, Washington DC and Abingdon Press, Nashville.

Nash, R.F. (1982), *Wilderness and the American Mind*. 3rd ed., Yale University Press, New Haven, Connecticut.

Nash, R.F. (1989), *The Rights of Nature. A History of Environmental Ethics*. University of Wisconsin Press, Wisconsin/London.

Nasr, Seyyed Hossein (1968), *Man and Nature. The Spiritual Crisis of Modern Man*. George Allen & Unwin Ltd., London.

Nauerth, C. (1985), Hahn. *RLAC,* XIII, 99, 360-372.

Nebel, B.J. & R.T. Wright (2000), *Environmental Science*. Seventh Edition, Prentice Hall, Upper Saddle River, NJ.

Needham, J. (1961-), *Science and Civilisation in China*. Cambridge University Press, Cambridge.

Needham, J. (1970), *Clerks and Craftsman in China and the West*. Cambridge University Press, Cambridge.

Nelissen, N., J. van der Straaten & L. Klinkers (eds.) (1997), *Classics in Environmental Studies. An Overview of Classic Texts in Environmental Studies*. International Books, Utrecht.

Neusner, J. (ed.) (1973), *The Idea of Purity in Ancient Judaism*. E.J. Brill, Leiden.

Nicholas, B. (1962), An Introduction to Roman Law. *Clarendon Law Series,* ed. H.L.A. Hart, Clarendon Press, Oxford.

Nijkamp, P. (1979), *Naar een prijzenswaardig milieu?* Van Gorcum, Assen.

Nitschke, A. (1966), Tiere Und Heilige. In: R.Vierhaus & M. Botzenhart (hrsgs) *Dauer und Wandel der Geschichte. Aspekte Europäischer Vergangenheit.* Festgabe für K. von Raumer, Verlag Aschendorff, Münster, 66-100.

Nolt, S.M. (1992), *A History of the Amish*. Good Books, Intercourse, Penn.

Nolthenius, Helene (1988), *Een man uit het dal van Spoleto. Franciscus tussen zijn tijdgenoten*. Querido, Amsterdam.

Noort, E. (1997), Die Erzählungen von der grossen Flut. Bemerkungen zu Gen.-6:5—9:17 in ihrem altorientalischen Kontext. In: F. García Martínez & G. Luttikhuizen (eds.) *Interpretations of the Flood*. Kok Pharos, Kampen, 9-43.

Nordenfelt, L. (1993), *Quality of Life, Health and Happiness*. Avebury, Hampshire.

Norton, B.G. (1987), *Why preserve Natural Variety?* Princeton University Press, Princeton New Jersey.

Noske, B. (1988), *Huilen met de wolven. Een interdisciplinaire benadering van de mens-dier relatie*, Van Gennep, Amsterdam.

Noth, M. (1966), *The Laws in the Pentateuch and Other Studies*. Oliver & Boyd Ltd., Edinburgh/London.

Nriagu, J.O. (1983), *Lead and Lead poisoning in Antiquity*. New York.

NRLO (1990), *Rapport van de commissie van Advies Ethiek en Biotechnologie bij Dieren*. NRLO, Wageningen.

Numbers, R.L. (1993), *The Creationists. The Evolution of Scientific Creationism*. University of California Press, Berkeley, California.

Nussbaum, M. and A.K. Sen (eds.) (1992), *The Quality of Life*. Clarendon Press, Oxford.

NWO (1991), *Duurzaamheid en Milieukwaliteit*, Netherlands Scientific Research Council.
Oded, B. (1986), "The Table of Nations (Genesis 10)—A Socio-Cultural Approach", *ZAW*, 98, 14-31.
Oelschlaeger, M. (1991), *The Idea of Wilderness. From Prehistory to the Age of Ecology*. Yale University Press, New Haven, Connecticut.
Oelschlaeger, M. (1994), *Caring for Creation. An Ecumenical Approach to the Environmental Crisis*. Yale University Press, New Haven, Connecticut.
Oldfield, Sara (1987), *Fragments of Paradise. A Guide for Conservation action in the U.K. dependent Territories*. British Association of Nature Conservationists and the WWF, Pisces Publ., Oxford.
Olding, A. (1991), *Modern Biology and Natural Theology*. Routledge, London.
Olsen, M.E., D.G. Lodwick & R.E. Dunlap (1992), *Viewing the World Ecologically*. Westview Press, Boulder Colorado/Oxford.
Onkelos, *Targum on Pentateuch*. Edition Horeb (1928), Berlin.
Oppenheim A.L. (1978), Man and Nature in Mesopotamian Civilization. *Dictionary of Scientific Biografy*, 15, 634-666.
Opschoor, J.B. (1987), *Duurzaamheid en verandering, over ecologische inpasbaarheid van economische ontwikkelingen*. Inaugural speech, Free University of Amsterdam.
Opschoor, J.B. (1989), *Na ons geen zondvloed. Voorwaarden voor duurzaam milieugebruik*. Kok Agora, Kampen.
Opschoor, J.B. (1994), Tijdvoorkeur en milieurisico's: een reactie. *Milieu*, 1, 38-42.
Opschoor, J.B. (1995), Ecospace and the Fall and Rise of Throughput Intensity. *Ecological Economics*, 15, 137-140.
Opschoor, J.B. (1996), *Groei en duurzaamheid: mythes en mogelijkheden*. Erasmus-lezing 1996, Stichting Erasmus Liga, Den Haag.
Opschoor, J.B. (1997), Bestond de milieu-Kuznetscurve? *ESB*, 82, 474.
Opschoor, J.B. & S.W.F. van der Ploeg (1990), Duurzaamheid en kwaliteit: hoofddoelstellingen van milieubeleid. In: C.L.T.M. *Het Milieu: denkbeelden voor de 21ste eeuw*. Kerckebosch, Zeist, 81-124.
Opschoor, J.B. & G.A.J. Klaassen (1991), Economics of sustainability or the sustainability of economics: different paradigms. *Ecological Economics*, 4, 93-115.
Orians, G.H. (1990), Ecological Concepts of Sustainability. *Environment*, 32, 9, 11-15, 34-39.
O'Riordan, T. & J. Cameron (eds.) (1994), *Interpreting the Precautionary Principle*. Earthscan Publications Ltd., London.
O'Riordan, T. (ed.) (2000), *Environmental Science for Environmental Management*. Longman Scientific & Technical, Essex.
Orwell, George (1945/1961), *Animal Farm*. Penguin Books Ltd., Harmondsworth, Middlesex. J.M. Meulenhoff, Amsterdam.
Ottoson, M. (1987), Sacrifice and sacred Meals in Ancient Israel. In: Linders, T. & G. Nordquist (eds.) *Gifts to the Gods*. Acta Univ. Ups. Boreas, Uppsala, 133-137.
Ouderkirk, W. (1999), Can Nature be Evil? Rolston, Disvalue and Theodicy. *Environmental Ethics*, 21, 135-150.
Pagels, Elaine (1985), The Politics of Paradise: Augustine's Exegesis of Genesis 1-3 Versus that of John Chrysostom. *Harvard Theological Review*, 78, 1, 2, 67-95.
Pagels, Elaine (1988), *Adam, Eve and the Serpent*. Weidenfeld & Nicolson, London.
Parmelee, A. (1959), *All the Birds of the Bible; Their Stories, Identification, and Meaning*. Lutterworth Press, London.

Parsons, H.L. (ed.) (1977), *Marx and Engels on Ecology*. Greenwood Press, Westport, Connecticut.

Pascal, Bl. (1966), *Pensées*. Translated with an Introduction by A.J. Krailsheimer, Penguin Classics, London.

Passmore, J. (1974/1980), *Man's Responsibility for Nature. Ecological Problems and Western Traditions*. 2nd ed. 1980, G. Duckworth & Co., London.

Passmore, J. (1975), The Treatment of Animals. *Journal of the History of Ideas*, 36, 195-218.

Pauly, Der Kleine (1979), *Lexikon der Antike*. In fünf Bänden, Deutscher Taschenbuch Verlag.

Paz, Uzi (1986), *The Birds of Israel*. Photographs: Yossi Eshbol. Steinmatzky Ltd., Tel Aviv.

Pearce, D., A. Markandya & E.B.Barbier (1989), *Blueprint for a Green Economy*. Earthscan Publications Ltd., London.

Pearce, D. *et al.* (1993), *Blueprint 3: Measuring Sustainable Development*. Earthscan Publications Ltd., London.

Peels, H.G.L. (1994), On the Wings of the Eagle (Dtn 32,11)—An old Misunderstanding. *ZAW*, 106, 300-303.

Pelden-Cloudt, J. von (1952), Die friedvollen Königreiche des E. Hicks. *Das Kunstwerk*, 6, 15-16.

Pelikan, J. (1961), Cosmos and Creation: Science and Technology in Reformation Thought. *Proceedings of the American Philosophical Society*, 105, 5, Philadelphia, 464-469.

Pelikan, J. (1971-89), *The Christian Tradition: A History of the Development of Doctrine*. 5 Vols. University of Chicago Press, Chicago.

Pelikan, J. (1993), *Christianity and Classical Culture. The Metamorphosis of Natural Theology in the Christian Encounter with Hellenism*. Yale University Press, Newhaven, Connecticut.

Pellegrin, P. (1982/1986), *Aristotle's Classification of Animals. Biology and the conceptual Unity of the Aristotelian Corpus*. Translation by Anthony Preus, University of California Press, Berkeley, California.

Pembroke, S.G. (1971), Oikeiôsis. In: Long, A.A. (ed.) *Problems in Stoicism*, Athlone Press, London, 114-149.

Pepper, D. (1984/1990), *The Roots of modern Environmentalism*. Routledge, London/New York.

Pepper, D. (1996), *Modern Environmentalism, an Introduction*. Routledge, London/New York.

Peters, B.G. & B.W. Hogwood (1985), In Search of the Issue-attention Cycle. *Journal of Politics*, 47, 238-253.

Peters, J.P. (1911), Cosmogony and Cosmology (Hebrew). *Encyclopaedia of Religion and Ethics (ERE)*, IV, 151-155.

Peters, J.P. (1911[a]), The Wind of God. *Journal of Biblical Literature (JBL)*, 30, 44-54.

Peters, J.P. (1914[a]), The Wind of God, part II. *Journal of Biblical Literature (JBL)*, 33, 81-87.

Peters, J.P. (1914), *The Religion of the Hebrews*. Boston & London.

Physiologus (1976), Edition Franciscus Sbordone-Mediolani, Nachdr., Hildesheim (etc.), Olms.

Pinney, R. (1964), *The Animals in the Bible. The Identity and Natural History of all Animals mentioned in the Bible*. Chilton Books, Philadelphia/New York.

Plas, D. van der, B. Becking & D. Meijer (eds.) (1990), *De schepping van de wereld. Mytische voorstellingen in het Oude Nabije Oosten*. Coutinho, Muiderberg.

Pohlenz, Max (1943/1948), *Die Stoa, Geschichte einer geistigen Bewegung.* Vol. I orig. 1943, vols. I and II, 1948, Van den Hoeck & Ruprecht, Göttingen.

Pointl, F. (1989), *De kip die over de soep vloog.* Nijgh & van Ditmar, Amsterdam.

Polkinghorne, J. (1989), A note on Chaotic Dynamics. *Science and Christian Belief,* 1, 123-127.

Polkinghorne, J. (1998), *Beyond Science: The wider Human Context.* Cambridge University Press, Cambridge.

Poll, R. van (1997), *The Perceived Quality of the Urban Residential Environment. A Multiattribute Evaluation.* Dissertation, State University of Groningen.

Poll, E. van der & J. Stapert (1988), *Als het water bitter is. Evangelisch denken en de milieucrisis.* Merweboek, Sliedrecht.

Pollack, R. (2001), *The Faith of Biology and the Biology of Faith.* Columbia University Press

Pollard, John R.T. (1977), *Birds in Greek Life and Myth.* Thames and Hudson, London.

Pollard, Nigel (1984), The Israelites and their Environment. *The Ecologist,* 14, 3, 125-33.

Pollock, N.H. (1968), *The English Game Laws in the Nineteenth Century.* Dissertation John Hopkins University, Baltimore, Maryland.

Ponting, C. (1990), Historical Perspectives on sustainable Development. *Environment,* 32, 9, 4-9 and 31-33.

Ponting, C. (1991), *A Green History of the World.* Sinclair- Stevenson, London.

Popol Vuh. The Mayan Book of the Dawn of Life. (1996), translated by Dennis Tedlock, revised edition, A Touchstone Book, Simon & Schuster, New York.

Porritt, J. (1984), *Seeing Green.* Blackwell, Oxford.

Porritt, J. (1992), Sustainable Development: Panacea, Platitude or Downright Deception? in: *Energy and the Environment: the Linacre Lectures.* Oxford University Press, Oxford.

Postumus Meyjes, C.B. (1993), *Van ophouden weten. De betekenis van het sabbatsgebod voor onze tijd.* Boekencentrum, Zoetermeer.

Pot, J.H.J. van der (1985), *Die Bewertung des Technischen Fortschritts: eine systematische Uebersicht der Theorien.* 2 vols., Van Gorcum, Assen/Maastricht.

Pot, J.H.J. van der (1994), *Steward or Sorcerer's Apprentice? The Evaluation of Technical Progress: A systematic Overview of Theories and Opinions.* 2 Vols., original German language edition 1985, translated by Chris Turner, Eburon Publishers, Delft.

Potter, Charles F. (1933), *Is that in the Bible? A Classified Collection of the Odd, Amusing, Unusual and Surprising Items of Human Interest in the Bible, many of which have escaped the Attention of the Average Reader.* Garden City Publishing Company, Inc., Garden City, New York.

Prest, J. (1981), *The Garden of Eden: The Botanic Garden and the Re-creation of Paradise.* Yale University Press, New Haven, Connecticut.

Preuss, H.D. (1982), *Deuteronomium.* Erträge der Forschung Band 164, Wissenschaftliche Buchgesellschaft, Darmstadt.

Prigogine, I. & I. Stengers (1985), *Orde uit Chaos, de nieuwe dialoog tussen de mens en de natuur.* Translated by M. Franssen and M. Morreau, Bert Bakker, Amsterdam.

Prigogine, I & I. Stengers (1988/1989), *Tussen tijd en eeuwigheid.* Bert Bakker, Amsterdam.

Pritchard, J.B. (ed.) (1950/1969), *Ancient Near Eastern Texts relating to the Old Testament (ANET).* 3rd ed. 1969 with Supplementary Texts and Pictures Relating to the Old Testament, Princeton University Press, Princeton, New Jersey.

Pritchard, J.B. (ed.) (1973), *The Ancient Near East (ANET[1]). Vol.I An Anthology of Texts and Pictures. Vol II A New Anthology of Texts and Pictures.* Princeton Paperback Printing, Princeton, New Jersey.

Pritchard, J.B. (ed.) (1991), *The Times Concise Atlas of the Bible.* Times Books, London.

Pseudepigrapha see: Charlesworth, J.H. (ed.) (1985).

Pulles, M.P.J. (1990), Openbaar vervoer of eigen auto? Kiezen van milieurelevant gedrag. *Milieu,* 1, 28-29.

Pury, A. de (1993), Gemeinschaft und Differenz. Aspekte der mensch-Tier Beziehung im alten Israel. In: Janowski, B., U. Neumann-Gorsolke & U. Gleßmer (eds.) (1993), *Gefährten und Feinde des Menschen. Das Tier in der Lebenswelt des alten Israel.* Neukirchener Verlag, Neukirchen-Vluyn, 112-149.

Putnam, H. (1981), *Reason, Truth and History.* Cambridge University Press, Cambridge.

Raay, H.G.T. van & A. E. Lugo (eds.) (1974), *Man and Environment Ltd., Natural Imbalances and Social Justice.* Rotterdam University Press, Institute of Social Studies, The Hague.

Rabb, J.D. (1992), From Triangles to Tripods: Polycentrism in Environmental Ethics. *Environmental Ethics,* 14, 177-183.

Rad, G. von (1949), *Das erste Buch Mose. Genesis Kapitel 1-12.* ATD, 2/4, Göttingen.

Rad, G. von (1969), *Theologie des Alten Testaments.* 2 vols., 6th ed., Chr. Kaiser Verlag, Munich.

Radcliffe, W. (1921/1974), *Fishing from the Earliest Times.* Ares Publishers Inc., Chicago

Radice, Betty (1971), *Who's who in the Ancient World. A Handbook to the Survivors of the Greek and Roman Classics.* Penguin Books, Harmondsworth, Middlesex, England.

Radt, S.L. (1988), *Zo herkenbaar en toch ook weer niet, een mentaliteits-verschil tussen de oude Grieken en ons in de Antigone van Sopkokles.* Departing lecture, 28 June 1988, State University of Groningen, Groningen University Press.

Ramsey, G.W. (1988), Is Name-Giving an Act of Domination in Genesis 2:23 and Elsewhere? *CBQ,* 50, 24-35.

Raphael, Chaim (1982), *The Springs of Jewish Life.* Basic Books, New York.

Rappel, S. (1996), *"Macht euch die Erde untertan". Die ökologische Krise als Folge des Christentums?* Abhandlungen zur Sozialethik Band 39, Ferdinand Schöningh, Paderborn.

Rappoport, A.S. (1995), *Ancient Israel. Myths and Legends.* Two Vols., Senate/Studio Editions, London.

Rasji: *Cf* Doron.

Rathje, W. & C. Murphy (1992), *Rubbish!-The Archeology of Garbage.* Harper Collins, New York.

Rau, G. *et al.* (eds.) (1987), *Frieden in der Schöpfung.* Gütersloh, Hildesheim.

Ray, D.L. with L. Guzzo (1994), *Environmental Overkill. Whatever Happened to Common Sense?* Harper Perennial, New York.

Ray, John (1701), *The Wisdom of God Manifested in the Works of the Creation.* 3rd ed. London.

Reallexikon für Antike und Christentum (*R.L.A.C.*), Th. Klauser, E. Dassman *et al.* (eds.), A. Hiersemann, Stuttgart.

Reader, J. (1990), *Man on Earth. A Celebration of Mankind. Portraits of Human Culture in a Multitude of Environments.* Perennial Library, Harper & Row, New York.

Rees, W.E. (1990), The Ecology of Sustainable Development. *The Ecologist,* 20, 1, 18-23.

Regan, T. (1981), The Nature and Possibility of an Environmental Ethic. *Environmental Ethics,* 3, 1, 19-34.

Regan, T. (1982), *All that dwell Therein. Essays on Animal Rights and Environmental Ethics.* University of California Press, Berkeley, Los Angeles.

Regan, T. (1983), *The Case for Animal Rights,* University of California Press, Berkeley, Los Angeles.

Regan, T. (1985), Holism as Environmental Fascism. in: White, J.E. (ed.) *Contemporary Moral Problems.* St. Paul, Minn., 291 e.v.

Regan, T. (ed.) (1986), *Animal Sacrifices. Religious Perspectives on the Use of Animals in Science.* Temple University Press, Philadelphia.

Regan, T. & P. Singer (eds.) (1976), *Animal Rights and Human Obligations.* Prentice-Hall, Englewood Cliffs, New Jersey.

Reid, W.V. & K.R. Miller (1989), *Keeping Options alive: the Scientific Basis for conserving Biodiversity.* World Resources Institute, Washington.

Reinhard, J.R. (1939/1970), *Mediaeval Pageant.* Haskell House Publishers Ltd., New York.

Reijnders, L. (1984), *Pleidooi voor een duurzame relatie met het milieu.* Van Gennep, Amsterdam.

Reijnders, L. (1990), Normen voor milieuvervuiling met het oog op duurzaamheid. *Milieu,* 5, 5, 138-140.

Reijnders, L. (1992), Milieu, duurzaamheid en de grote lijnen van het beleid. *Milieu,* 7, 2, 48-51.

Reijnders, L. (1996), *Environmentally Improved Production Processes and Products.* Kluwer Academic Press, Dordrecht/Boston/London.

Reisel, M. (1966), *Genesis. Transcriptie, verklaring, vertaling.* Kruseman, Den Haag.

Reve, Karel van het (1978), *Uren met Henk Broekhuis.* G.A. van Oorschot, Amsterdam.

Ridderbos, J. (1925), *Het verloren Paradijs. Een woord met het oog op de aangaande Genesis 2 en 3 gerezen vragen.* 2nd edition, J.H. Kok, Kampen.

Ridderbos, N.H. (1958), Het Oude Testament en de geschiedenis. *Gereformeerd Theologisch Tijdschrift,* 58, 1-45.

Riede, P. (1993), Glossar der hebräischen und aramäischen Tiernamen und Tierbezeichnungen. In: Janowski, B. (u.a. Hrsg.) *Gefährten und Feinde des Menschen.* Neukirchener Verlag, Neukirchen-Vluyn, 361-376.

Rist, John M. (ed.) (1978), *The Stoics.* University of California Press, Berkeley, California.

Ritvo, H. (1987/1990), *The Animal Estate, The English and Other Creatures in the Victorian Age.* Penguin Books, Harmondsworth, Middlesex. (1987 Harvard University Press, Harvard, Massachusetts.)

RIVM (1989), *Concern for Tomorrow. A National environmental survey 1985-2010.* National Institute of Public Health and Environmental Protection. Bilthoven.

RMB (1994), *De Distance-to-Target Methode. Advies over weging van milieu-effecten.* RMB Advisory Document 94-06, Netherlands Environmental Council (RMB), The Hague.

RMB (1995), *The Support base for environmental policy.* RMB Advisory Document 95-05, RMB, The Hague.

RMB (1996a), *Naar een biologische omgevingskwaliteit.* RMB Advisory Document 96-08, RMB, The Hague.

RMB (1996), *Milieu als motor van de economie. Bundel ter gelegenheid van het slotcongres van de Raad voor het Milieubeheer.* RMB, The Hague.

RMNO (1994), *Towards environmental Performance Indicators based on the Notion of environmental Space.*), Publication 96, Advisory Council for Research on Nature and Environment (RMNO), Rijswijk.

Robertson, G. *et al.* (eds.) (1996), *Future Natural. Nature, Science, Culture.* Routledge, London/New York.

Rodman, John (1976), The other Side of Ecology in Ancient Greece: Comments on Hughes. *Inquiry*, 19, 108-112.

Rodman, John (1979), Animal Justice: The Counter-revolution in Natural Right and Law. *Inquiry*, 22, 3-22.

Röhrig, E. (ed.) (1991), *Der Gerechte erbarmt sich seines Viehs: Stimmen zur Mitgeschöpflichkeit.* Neukirchener Verlag, Neukirchen-Vluyn.

Rollin, B.E. (1981), *Animal Rights and Human Morality.* Prometheus Books, New York.

Rollin, B.E. (1990), *The Unheeded Cry, Animal Consciousness, Animal Pain and Science.* Oxford University Press, Oxford.

Rolston III, Holmes (1975), Is there an Ecological Ethics? *Ethics*, 85, 93-109.

Rolston III, Holmes (1988), *Environmental Ethics.* Temple University Press, Philadelphia.

Rolston III, Holmes (1992), Disvalues in Nature. *The Monist*, 75, 250-278.

Rolston III, Holmes (1999), *Genes, Genesis and God. Values and their origins in natural and human history.* Cambridge University Press, Cambridge.

Rose, A. (ed.) (1992), *Judaism and Ecology.* WWF/Cassell Publishers Ltd., London.

Rosenberg, N. & L.E. Birdzell (1986), *How the West grew Rich. The Economic Transformation of the Industrial World.* Basic Books inc., New York.

Rottzoll, D.U. (1998), Die Schöpfung- und Fallerzählung in Gen 2f. Teil 2: Die Schöpfungserzählung (Gen 2). *ZAW*, 110, 1-15.

Routley, R. & V. Routley (1978), Nuclear Energy and Obligations to the Future. *Inquiry*, 21, 133-79.

Routley, V. (1981), On Karl Marx as an Environmental Hero. *Environmental Ethics*, 3, 3, 237-244.

Routley, V. (1975), Critical Notice of John Passmore, Man's Responsibility for nature. *Australian Journal of Philosophy*, 53, 171-185.

Ruiten, J.T.A.G.M. (1992), Intertextual Relationship between Isa 65,25 and 11,6-9. In: F. García Martínez *et al.* (eds.) *The Scriptures & the Scrolls.* E.J. Brill, Leiden, 31-42.

Runia, D.T. (1983), *Philo of Alexandria and the Timaeus of Plato.* Dissertatie, Vrije Universiteit, Amsterdam.

Runia, D.T. (1993), *Philo in Early Christian Literature.* Van Gorcum, Assen.

Ruse, M. (1989/1993), *The Darwinian Paradigm. Essays on its History, Philosophy and Religious Implications.* Routledge, London/New York.

Ruse, M. (1996), *Monad to Man. The Concept of Progress in Evolutionary Biology.* Cambridge University Press, Cambridge.

Ruse, M. (2001), *Can a Darwinian be a Christian?* Cambridge University Press, Cambridge.

Russell, C.A. (1994), *The Earth, Humanity and God.* University College Londen Press, London.

Russell, D.S. (1967), The Jews from Alexander to Herod. in: *The New Clarendon Bible, Old Testament,* vol. V, Oxford University Press, Oxford.

Rüterswörden, U. (1993), Dominium Terrae. Studien zur Genese einer alttestamentlichen Vorstellung. *BZAW*, 215, Walter de Gruyter, Berlin/New York.

Rüthy, A.E. (1942), *Die Pflanze und ihre Teile im Biblisch-Hebräischen Sprachgebrauch.* A. Francke A-G Verlag, Bern

Ryder, R.D. (1989), *Animal Revolution, Changing Attitudes Towards Speciesism.* Basil Blackwell, Oxford.

Ryder, R.D. (1992), *Painism: Ethics, Animal Rights and Environmentalism.* Centre for Applied Ethics, University of Wales College of Cardiff, Cardiff.

Sack, D. (2001), *Whitebread Protestants. Food and Religion in American Culture.* Palgrave Publishers, Houndmills, Basingstoke.

Safrai, S. and M. Stern (eds.) (1974), *The Jewish People in the First Century. Historical*

Geography, Political History, Social, Cultural and Religious Life and Institutions. Vol. I, Van Gorcum & Comp. B.V., Assen.

Saggs, H.W.F. (1965/1987), *Every day Life in Babylonia & Assyria.* B.T. Batsford Ltd. (1965), Dorset Press, Marboro Books Corporation, (1987), New York.

Sallares, R. (1991), *The Ecology of the Ancient Greek World.* Duckworth, London.

Salmon, Sh. (1995), *Israels Gedankenwelt in der Hebräischen Bibel.* Neukirchener Verlag, Neukirchen/Vluyn.

Salonen, A. (1973), Vögel und Vogelfang im alten Mesopotamien. *Anales Academiae Scientiarum Fennicae,* ser. B., Tom. 180, Helsinki.

Sambursky, S. (1956), *The Physical World of the Greeks.* The Macmillan Company, New York.

Sambursky, S. (1959), *Physics of the Stoics.* Routledge and Kegan Paul, London.

Sambursky, S. (1962), *The Physical World of Late Antiquity.* Routledge en Kegan Paul, London.

Sandars, N.K. (ed.) (1972), *The Epic of Gilgamesh,* Penguin Books, Harmondsworth, Middlesex, England.

Sanders, E.P. (1990), *Jewish Law from Jesus to the Mishnah.* SCM press, London. Trinity Press International, Philadelphia, Pa.

Sandbach, F.H. (1975), *The Stoics.* Chatto & Windus, London.

Santmire, Paul H. (1980), St. Augustine's Theology of the biophysical World. *Dialog,* 19, 174-85.

Santmire, Paul H. (1985), *The Travail of Nature: The Ambiguous Ecological Promise of Christian Theology.* Fortress Press, Philadelphia.

Savage, Henry L. (1933), Hunting in the Middle Ages. *Speculum,* 8, 30-42.

Sawyer, J.F.A. (ed.) (1996), Reading Leviticus. A Conversation with Mary Douglas. *JSTO Supplement Series* 227, Sheffield Academic Press, Sheffield.

Schama, S. (1995), *Landscape and Memory.* Vintage Books, New York.

Schäfer, P. (1997), *Judeophobia. Attitudes toward the Jews in the Ancient World.* Harvard University Press, Cambridge, Mass.

Schaffer, Arthur (1982), The Agricultural and Ecological Symbolism of the four species of sukkot. *Tradition,* 20, 128-40.

Schapiro, D. (1928), *L'Hygiène alimentaire des Juifs devant la science moderne.* Paris.

Scharper, Stephen Bede (1997), *Redeeming the Time. A Political Theology of the Environment.*Continuum, Nwe York.

Schestow, L. (1938), *Athen und Jerusalem. Versuch einer religiosen Philosophie.* Verlag Schmidt-Dengler, Graz.

Schlitt, M. (1992), *Umweltethik. Philosophisch-ethische Reflexionen- Theologische Grundlagen- Kriterien.* Ferdinand Schöningh, Paderborn.

Schmidt, W.H. (1967), *Die Schöpfungsgeschichte der Priesterschrift.* 2., revised and expanded edition, Neukirchener Verlag, Neukirchen-Vluyn.

Schmitz, Ph. (ed.) (1981), *Macht euch die Erde untertan? Schöpfungsglaube und Umweltkrise.* Echter Verlag, Würzburg.

Schneider, E., H. Oelke & H. Groß (1988/1989), *Die Illusion der Arche Noah. Gefahren für die Arterhaltung durch Gefangenschaftszucht.* Echo Verlag, Göttingen.

Schneider, W. (1988), *Wir Neandertaler. Der abenteurliche Aufstieg des Menschen Geschlechts.* Stern Buch im Verlag Gruner & Jahr.

Schochet, E.J. (1984), *Animal Life in Jewish Tradition. Attitudes and Relationships.* KTAV Publishing House, Inc., New York.

Scholten, L.J.R. (1993), Een beoordeling van maatstaven voor duurzame ontwikkeling. *Milieu,* 8, 242-246.

Schopenhauer, A. (1848/1961), *Parerga und Paralipomena: kleine philosophische Schrif-*

ten. (Zweiter Band, Par. 177 *Ueber das Christentum.*), Sämtliche Werke, Sechster Band, Brochhaus, Wiesbaden.

Schopenhauer, A. (1841/1965), *Die beiden Grundproblemen der Ethik.* Frankfurt am Main (1841). *The Basis of Morality.* translated by E.F.J. Payne, Library of Liberal Arts, The Bobbs-Merrill Company inc., Indianapolis.

Schottroff, L. (1989), Schöpfung im Neuen Testament. in: G. Altner, *Ökologische Theologie. Persektiven zur Orientierung.* Kreuz Verlag, Stuttgart, 130-148.

Schouten, M. (2001), *De natuur als beeld in religie, filosofie en kunst.* KNNV, Utrecht.

Schouten van der Velden, A. (1992), *Dieren uit de Bijbel. Een inventarisatie en beschrijving.* G.F. Callenbach, Nijkerk.

Schreiner, S.E. (1991), *The Theater of His Glory. Nature an the Natural Order in the Thought of John Calvin.* Studies in Historical Theology, 3, The Labyrinth Press, Durham, North Carolina.

Schreiner, S.E. (1994), *Where shall Wisdom be found? Calvin's Exegesis of Job from Medieval and Modern Perspectives.* University of Chicago Press, Chicago.

Schroeder, G.L. (1990), *Genesis and the Big Bang.* Bantam Books, New York.

Schulze, W.A. (1955), Der Heiligen und die Wilde Tiere. Zur Exegese von Mc 1: 13[b]. *Zeitschr. Neutest. Wiss.*, 46, 280-283.

Schürer, E. (1901/1911), *Geschichte des Jüdischen Volkes im Zeitalter Jesu Christi.* Three vols., index, J.C. Hinrichs'sche Buchhandlung, Leipzig.

Schürer, E. (1961), *A History of the Jewish People in the Time of Jesus.* Edited by Nahum N. Glatzer, Schocken Books, New York.

Schuster, Max (1917), *Untersuchungen zu Plutarchs Dialog De Sollertia Animalium.* Inaugural Speech, Ludwig Maximilian University, Munich, Buch- und Kunstdruckerei J.P.Himmer, Augsburg.

Schwarz, M. and M. Thompson (1990), *Devided we stand: Redefining Politics Technology and Social Choice.* Harvester Wheatsheaf, New York.

Schwarzschild, S.S. (1984), The Unnatural Jew. *Environmental Ethics*, 6, 4, 347-362.

SCP (1994), *Sociaal en cultureel rapport 1994.* Netherlands Socio-Cultural Planning Office, (SCP), Rijswijk/The Hague.

SCP (1998), *Sociaal en Cultureel Rapport 1998. 25 jaar sociale verandering.* Rijswijk/The Hague

Sedley, D. (1991), 'Is Aristotle's Teleology Anthropocentic?' *Phronesis,* 36, 2, 179-196.

Selms, A. van (1984), *Genesis.* deel I, serie De Prediking van het Oude Testament, Callenbach, Nijkerk.

SER (1989), *Advies Our Common Future.* SER Advisory Document 89/06, Netherlands Socio-Economic Council (SER), The Hague.

SER (1991), *Rapport economie en milieu.* SER Advisory Document 91/18, SER, The Hague.

Serpell, J. (1986), *In the Company of Animals. A Study of Human-Animal Relationships.* Basil Blackwell, Oxford.

Serres, M. (1990/1992), *Le contrat naturel.* Editions François Bourin, Paris.

Seters, J. van (1975), *Abraham in History and Tradition.* Yale University Press, New Haven, Connecticut.

Sevenster, J.N. (1961), *Paul and Seneca.* E.J. Brill, Leiden.

Shanklin, E. (1985), Sustenance and Symbol: Anthropological Studies of Domesticated Animals. *Ann. Rev. Anthropol.,* 14, 375-403.

Shapiro, H. & E.M. Curley (1965), *Hellenistic Philosophy.* New York.

Sheldon, J.K. (1992), *Rediscovery of Creation: A Bibliographical Study of the Church's Response to the Environmental Crisis.* Scarecrow Press.

Sherrard, Ph. (1959/1992), *The Greek East and the Latin West. A Study in Christian*

Tradition. Denise Harvey & Company, Limni, Evia, Greece.

Sherrard, Ph. (1991), *The Rape of Man and Nature. An Enquiry into yhe Origins an Consequences of Modern Science.* Golgonooza Press, Ipswich.

Short, W.J. (1983), *Saints in the World of Nature. The Animal Story as Spiritual Parable in Medieval Hagiography (900-1200).* Dissertatio Pontficiae Universitatis Gregorianae, Romae.

Short, W.J. (1990), *The Franciscans.* Dominican Publ., Rome.

Shrader-Frechette, Kristin S. (ed.) (1981), *Environmental Ethics.* Boxwood Press, Pacific Grove, California.

Shulov, A. (1975), *List of Animals Mentioned in the Bible.* The Jerusalem Biblical Zoo.

Siebert, H. (1981), Ökonomischer Theorie Naturlicher Ressourcen. *Zeitschrift für Wirtsch. und Zoz. Wiss.*, 3, 267-298.

Siebert, H. (1982), Nature as a Life Support System: Renewable Resources and Environmental Disruption. *Journal of Econ.* 42, 133-142.

Sieferle, Rolf Peter (Hrg.) (1988), *Fortschritte der Naturzerstörung.* Neue Folge Band 489, Suhrkamp Verlag, Frankfurt am Main.

Simmons, I.G. (1990), *Changing the Face of the Earth. Culture, Environment, History.* Basil Blackwell Ltd., Oxford.

Simmons, I.G. (1993), *Environmental History. A Consise Introduction.* Blackwell, Oxford.

Simoons, F. J. (1961), *Eat not this Flesh. Food Avoidances in the Old World.* University of Wisconsin Press, Madison.

Simoons, F.J. (1978), Traditional Use and Avoidance of Foods of Animal Origin: A Culture Historical view. *Bioscience,* 28, 3, 178-184.

Singer, Ch. (1974), Science and Judaism. In: Finkelstein, L. (ed.), *The Jews, Their Rôle in Civilization,* vol. 3, ch. 7, Schocken Book, New York.

Singer, P. (1975/1990), *Animal Liberation. A new Ethics for our Treatment of Animals.* 2nd ed., 1990, Random House, New York.

Singer, P. (1976), All Animals are Equal. In: Regan, T. & Singer, P. (eds.), *Animal Rights and Human Obligations.* Prentice-Hall, Englewood Cliffs, New Jersey, 148-162.

Singer, P. (1979), Killing Humans and Killing Animals. *Inquiry,* 22, 145-156.

Singer, P. (1981), *The Expanding Circle: Ethics and Sociobiology.* Farrar, Straus & Giroux, New York.

Singer, P. (ed.) (1991), *A Companion to Ethics.* Basil Blackwell, Oxford.

Singer, P. (1994), *Rethinking Life and Death.* The Text Publishing Company, Melbourne.

Singer, P. (2000), *Writings on an Ethical Life.* Fourth Estate, London

Skolimowski, H. (1988), Eco-Philosophy and Deep Ecology. *The Ecologist,* 18, 4, 5, 124-127.

Smend, R. (1989/1978), *Die Entstehung des Alten Testaments.* 4th revised edition, Verlag W. Kohlhammer, Stuttgart.

Smit, H.I. de (1974), *Levende Ziel (nèfesj chàjjá). Dierenleven in Israël Dierenleven in de Schrift De sociale mens.* Priv. publ., H. Izaäk de Smit, Noorderdiep 488, Valthermond. ISBN 90 70101 017.

Smith, F. Harold (1951), *The Influence of Religions on Man's Attitude towards Animals.* The Universities for Animal Welfare (UFAW), 8 Hamilton Close, Potters Bar, Hertsfordshire, England.

Smith, J.M.P. (1928/1929), The Use of Divine Names as Superlatives. *AJSL,* 45, 212-220.

Smith, J.Z. (1978), *Map is not Territory. Studies in the History of Religions.* Studies in Judaism in Late Antiquity, vol. 23, E.J. Brill, Leiden.

Smith, J.Z. (1987), *To take Place. Toward Theory in Ritual.* University of Chicago Press, Chicago, Michigan.

Smith, J.Z. (1987[a]), The Domestication of Sacrifice. in: Hamerton-Kelly, R.G. (ed.) (1987), *Violent Origins.* Burkert, W., R. Girard & J.Z. Smith on Ritual Killing and Cultural Formation. With an Introduction by Burton Mack and a Commentary by Renato Rosaldo, Stanford University Press, Stanford, California, 191-235.

Smith, J.Z. (1990), *Drudgery Divine. On the Comparison of Early Christianities and the Religions of Late Antiquity.* School of Oriental and African Studies, University of London, London.

Smith, M.J. (ed.) (1999), *Thinking through the Environment. A Reader.* Routledge, London/New York

Smith, Ph.B. (1989), *Determinisme en de onzekerheidsrelatie.* Departing lecture, State University of Groningen, 12 December 1989, Groningen.

Smith, V.L. (1975), Primitive Hunter Culture, Pleistocene Extinction and the Rise of Agriculture. *Journal of Political Economy,* 83, 4, 727-755.

Smith, W. Robertson (1889/1894), *Lectures on the Religion of the Semites.* First Series: the fundamental Institutions, New Edition revised by the Author, Adam and Charles Black, London.

Soest, H.J. van (1996), "*Welk is het voortreffelijkste schepsel op aarde?" De interpretatie van een omstreden bijbelse voorstelling in het 19[de] en 20[ste] eeuwse Nederland.* Dissertation, Kampen Theological University, Eburon pub., Delft.

Sofisten (1986), *De mens, maat van alle dingen, fragmenten uit de Griekse Sofistiek.* Translation from the Greek and introduction by R.Bakker, Agora publ., Kok, Kampen.

Soler, J. (1979), The Dietary Prohibitions of the Hebrews. *New York Review of Books,* June 14, 24-30.

Soler, J. (1996), Les Raison de la Bible: Règles alimentaires hébraiques. In: Flandrin, J.-L. & M. Montanari (eds.) *Histoire de L'alimentation.* Fayard, Paris, 73-84.

Solow, R.M. (1986), On the Intergenerational Allocation of Natural Resources. *Scand. Journ. of Econ.*, 88, 141-149.

Soontiëns, F.J.K. (1988), *Evolutie en Finaliteit.* Dissertation, Catholic University of Nijmegen, WNS Press, Nijmegen.

Soontiëns, F.J.K. (1990), Evolutie, teleologie en toeval. *Algemeen Nederlands Tijdschrift voor Wijsbegeerte,* 82, 1, 1-14.

Sorabji, R. (1983), *Time, Creation and the Continuum. Theories in Antiquity and the Early Middle Ages.* G. Duckworth, London.

Sorabji, R. (1993), *Animal Minds and Human Morals. The Origins of the Western Debate.* Duckworth, London.

Sordi, M. (1994), *The Christians and the Roman Empire.* Routledge, London.

Sørensen, V. (1984), *Seneca, the Humanist at the Court of Nero.* Translated by W. Glyn Jones, University of Chicago Press, Chicago.

Sorrell, R.D. (1988), *St. Francis of Assisi and Nature. Tradition and Innovation in Western Christian Attitudes toward the Environment.* Oxford University Press, Oxford.

Soulé, M.E. (ed.) (1986), *Conservation Biology: The Science of Scarcity and Diversity.* Sinauer Associates, Sunderland, Massachusetts.

Speiser, E.A. (1964), *Genesis. Introduction, Translation, and Notes.* The Anchor Bible, Doubleday & Company, Garden City, New York.

Spelsberg, G. (1984/1988), *Rauchplage. Zur Geschichte der Luftverschmutzung.* Alano Verlag 1984, Kölner Volksblatt Verlag 1988.

Sperna Weiland, J. and R.H.L. Römer (ed.) (1986), *Wetenschap is mensenwerk.* Trouw/

Kwartet, PO Box 859, 1500 AW Amsterdam.
Speyer, W. (1976), Geier. *RLAC,* IX, 430-468.
Spiegel Historiael (1993), Thema: Milieugeschiedenis, 1993, 28, 6. SDU, The Hague.
Spier, F. (1992), Een oud probleem: De relaties tussen mensen en het natuurlijk milieu in een lange termijn perspectief. *De Gids,* 155, 2, 96-108.
Spinoza, B (1989), *Tractatus Theologico-Politicus* (*Theological-Political Treatise*). Transl. S. Shirley, E.J. Brill, Leiden.
Sprigge, T.L.S. (1991), Some Recent Positions in Environmental Ethics examined. *Inquiry,* 34, 107-128.
Spring, D. & E. Spring (eds.) (1974), *Ecology and Religion in History.* Harper Torchbooks, Harper & Row, New York.
Sproul, B.C. (1979), *Primal Myths. Creating the World.* Harper & Row, New York.
Stadelmann, Luis I.J. (1970), *The Hebrew Conception of the World, A Philological and Literary Study,* Analecta Biblica 39, Pontifical Biblical Institute Press, Rome.
Stanley, D.J. & A.G. Warne (1993), Nile Delta: Recent Geological Evolution and Human Impact. *Science,* 260, 628-634.
Stebbins, Eunice Burr (1929), *The Dolphin in the Literature and Art of Greece and Rome.* Dissertation John Hopkins University, The George Banta Publ. Company, Manasha, Wisconsin.
Steck, O.H. (1970), Die Paradieserzählung. Eine Auslegung von Genesis 2,4b-3,24. *Biblische Studien* (*BS*), Heft 60, Neukirchener Verlag, Neukirchen-Vluyn.
Steck, O.H. (1978), *Welt und Umwelt.* Stuttgart.
Steck, O.H. (1981), *Der Schöpfungsbericht der Priesterschrift.* FRLANT 115, Göttingen.
Steffen, Lloyd H. (1992), In Defense of Dominion. *Environmental Ethics,* 14, 63-80.
Stein, G. (1980), Das Tier in der Bibel: Der Jüdische Mensch und sein Verhältnis zum Tier. *Judaica,* 36, 14-26 and 57-72.
Stein, S. (1957), The Dietary Laws in Rabbinic & Patristic Literature. *Studia Patristica,* 64, 141-154.
Stern, M. (ed.) (1974-1984), *Greek and Latin Authors on Jews and Judaism.* With Introductions, ranslations and commentary Vols. 1-3, Publications of the Israel Academy of Sciences and Humanities, Jerusalem.
Steur, Kl. (1935), *Poimandres en Philo: Een vergelijking van Poimandres §12-§32 met Philo's uitleg van Genesis I, 26-27 en II, 7.* Dissertatiopn, Catholic University of Nijmegen.
Stone, C.D. (1972), Should Trees Have Standing? Towards Legal Rights for Natural Objects. *Southern California Law Review,* 46, 450-501.
Stone, C.D. (1985), Should Trees have Standing? Revisited: How far will Law and Morals Reach? A Pluralist Perspective. *Southern California Law Review,* 59, 1-154.
Stone, C.D. (1987), *Earth and other Ethics: The Case for Moral Pluralism.* Harper & Row, New York.
Stone, C.D. (1988), Moral Pluralism and the Course of Environmental Ethics. *Environmental Ethics,* 10, 139-154.
Stone, C.D. (1993), *The Gnat is Older than Man. Global Environment and Human Agenda.* Princeton University Press, Princeton, New Jersey.
Stone, Glenn C. (ed.) (1971), *A New Ethic for a New Earth.* Friendship Press, New York.
Stone, Michael E. (ed.) (1984), *Jewish Writings of the Second Temple Period.* Sect. 2, vol. 2, Van Gorcum, Assen, Fortress Press, Philadelphia.
Strabo (1961), *Geography,* Transl. Horace Leonard Jones, Loeb Classical Library, W. Heinemann, London, Harvard University Press, Cambridge, Massachusetts.

Strategie richting 2000 (1992), Agricultural University of Wageningen, Wageningen.

Strizower, S. (1966), Clean and Unclean. *Jewish Chronicle*, 26 August, London.

SVF: see Arnim.

Sylvan, R. (1985), A Critique of Deep Ecology. part 1 *Radical Philosophy*, Summer, 2-12; part 2 *Radical Philosophy*, Autumn, 10-22.

Sylvan, R. & D. Bennett (1988), Taoism and Deep Ecology. *The Ecologist*, 18,4, 5, 148-150.

Tainter, J.A. (1988/1990), *The Collapse of Complex Societies*. Cambridge University Press, Cambridge (paperback 1990).

Talmud (1989-), The Steinsalt Edition, 10 vols., Random House, New York.

Tambiah, S.J. (1969), Animals are Good to Think, and Good to Prohibit. *Ethnology*, 8, 4, 423-460.

Tanakh: see JPS.

Tatenhove, J. van, B. Arts & P. Leroy (2001), *Political Modernisation and the Environment. The Renewal of Environmental Policy Arrangements*. Kluwer Academic Publishers, Dordrecht/Boston/London.

Taylor, J.H. (2001), *Death and the Afterlife in Ancient Egypt*. University of Chicago Press, Chicago

Taylor, P.W. (1983), In Defense of Biocentrism. *Environmental Ethics*, 5, 237-243.

Taylor, P.W. (1984), Are Humans Superior to Animals and Plants? *Environmental Ethics*, 6, 149-160.

Taylor, P.W. (1986), *Respect for Nature. A Theory of Environmental Ethics*. Princeton University Press, Princeton New Jersey.

Taylor, Thomas (1792/1966), *A Vindication of the Rights of Brutes*. A Facsimile Reproduction with an Introduction by Louise Schutz Boas, Scholars' Facsimiles & Reprints, Gainesville, Florida.

Tchernov, E. (1984), Commensal Animals and human Sedentism in the Middle East. In: Clutton-Brock, J. & C. Grigson (eds.), Animals and Archaeology III Early Herders and their Flocks. *BAR International Series*, 202, Oxford, 91-115.

Tellegen, E. & D.K.J. Tommel (1989), Milieuactie en milieubeleid. In: J.J. Boersema *et al.*(eds.), *Basisboek Milieukunde*. 3rd ed., Boom, Meppel, 55-74.

Tennekes, H. *et al.* (1990), *De vlinder van Lorenz. De verrassende dynamica van chaos*. Aramith uitgevers, Bloemendaal.

Tennekes, H. (1994), The Limits of Science. In: W. Zweers & J.J. Boersema (eds.), *Ecology, Technology and Culture*. The White Horse Press, Cambridge, 72-88.

Terian, A. (1981), *Philonis Alexandrini De Animalibus: The Armenian Text with an Introduction, Translation, and Commentary*. Scholars Press, Chico, California.

Terian, A. (1991), The Hunting Imagery of Isaiah LI, 20A. *Vetus Testamentum* XLI, 4, 462-472.

The Desert Fathers (sections of *Vitae Patrum* compiled and published in Antwerp in 1628 by H. Rosweyde), Translation from the Latin by Helen Waddell with a Preface by Basil Pennington, Vintage Spiritual Classics, VintageBooks / Random House, New York, 1998.

Thomas, K. (1983), *Man and the Natural World. Changing Attitudes in England 1500-1800*. Allen Lane, London.

Thomas, W.L. Jr. (ed.) (1956), *Man's Rôle in Changing the Face of the Earth*. University of Chicago Press, Chicago, Illinois.

Thompson, d'Arcy Wentworth (1936), *A Glossary of Greek Birds: a new Edition*. Oxford University Press, Oxford.

Thompson, Janna (1990), A Refutation of Environmental Ethics. *Environmental Ethics*, 12, 2, 147-161.

Thompson, M. (1997), *Human Choice and Climate Change, A Cultural Perspective*. Paper, NOP Workshop 'Trend breaches in social behaviour and climate change', 17/4/1997, Utrecht.

Thompson, M., R. Ellis & A. Wildavsky (1990), *Cultural Theory*. Westview Press, Boulder, Colorado.

Thompson, Stith (1955/1958), *Motif Index of Folk Literature*. 2nd ed., 6 vols., Copenhagen.

Thompson, Th.L. (1992), *Early History of the Israelite People. From the Written and Archeological Sources*. Studies in the History of the Ancient Near East (SHCANE), vol. 4, E.J. Brill, Leiden.

Tieleman, T. (1991), Diogenes of Babylon and Stoic Embryology Ps. Plutarch, Plac. V 15.4 reconsidered. *Mnemosyne*, XLIV, 1-2, 106-125.

Tijdschrift voor Theologie (1990), 30, 1, Themanummer: De Heelheid van Gods Schepping.

Till, H.J. van (1986), *The Fourth Day: What the Bible and the Heavens are telling us about the Creation*. William B. Eerdmans Publishing Company, Grand Rapids, Michigan.

Tobin, R. (1991), *The Expendable Future. US Politics and the Protection of Biological Diversity*. Duke University Press.

Tokar, B. (1988), Social Ecology, Deep Ecology and the Future of Green Political Thought. *The Ecologist*, 18, 4, 5, 138-141.

Tolman, C. (1980), Karl Marx, Alienation and the Mastery of Nature. *Environmental Ethics*, 2, 1, 3-16.

Toorn, K. van der (1985), *Sin and Sanction in Israel and Mesopotamia: A Comparative Study*. Van Gorcum, Assen/Maastricht.

Toorn, K. van der (1988), *"De mens kan niet ten hemel klimmen, noch afdalen naar het dodenrijk"*. Inaugural speech, Utrecht State University, Utrecht theology series 3, Faculty of Religious Studies, Utrecht.

Toorn, K. van der (1994), *From Her Cradle to Her Grave. JSOT* Press, Sheffield.

Toorn, K. van der (1996), *Family Religion in Babylonia, Syria and Israel: Continuity and Change in the Forms of Religious Life*. Studies in the History and Culture of the Ancient Near East (SHCANE), Vol. 7, E.J. Brill, Leiden.

Toorn, K. van der & P.W. van der Horst (1990), Nimrod before and after the Bible. *Harvard Theological Review*, 83, 1, 1-29.

Toorn, K. van der, B. Becking & P.W. van der Horst (eds.) (1999), *Dictionary of Deities and Demons in the Bible (DDD)*. Second Edition, E.J. Brill, Leiden.

Toulmin, S. (1982), *The Return to Cosmology. Postmodern Science and the Theology of Nature*. University of California Press, Berkeley, California.

Toynbee, A.J. (1934), *A Study of History*. 8 Vols., London.

Toynbee, A.J. (1972), The Religious Backgrounds of our Environmental Crisis. *International Journal of Environmental Studies*, 3, 141-146.

Toynbee, J.M.C. (1973/1983), *Tierwelt der Antike*. Band 17 Kulturgeschichte der Antiken Welt, Verlag Philipp von Zabern, Mainz (1983). (orig. Animals in Roman Life and Art, Thames and Hudson, London, 1973).

Tov, E. (1992), *Textual Criticism of the Hebrew Bible*. Fortress Press, Philadelphia, Van Gorcum, Assen/Maastricht.

Trebolle Barrera, J.C. (1998), *The Jewish Bible and the Christian Bible. An Introduction to the History of the Bible*. E.J. Brill, Leiden / Eerdmans, Grand Rapids, Michigan.

Tristram, H.B. (1880), *The Natural History of the Bible*. S.P.C.K., London. (First edition, 1867).

Troeltsch, Ernst (1911/1960), *The Social Teaching of the Christian Churches*. Translated by Olive Wyon, 2 vols., Harper Torchbooks the Cloister Library, Harper

& Row Publishers, New York. (Orig. Die Soziallehren der Christlichen Kirchen und Gruppen, 1911).

Tuan, Yi-Fu (1968), Discrepancies between Environmental Attitude and Behaviour: Examples from Europe and China. *The Canadian Geographer,* 12, 3, 176-191.

Tuan, Yi-Fu (1970), Our Treatment of the Environment in Ideal and Actuality. *American Scientist,* 58, 244-249.

Tucker, G.M. (1997), Rain on Land where no One lives: The Hebrew Bible on the Environment. *JBL,* 116, 1, 3-17.

Tucker, G.M., D.L. Petersen & R.R. Wilson (eds.) (1988), *Canon, Theology, and Old Testament Interpretation.* Essays in Honour of Brevard S. Childs, Fortress Press, Philadelphia

Turner II, B.L. *et al.* (eds.) (1991), *The Earth as transformed by Human Action. Global and Regional Changes in the Biosphere over the Past 300 Years.* Cambridge University Press with Clark University, Cambridge.

Turner, E.S. (1964), *All Heaven in a Rage.* Michael Joseph, London.

Turner, E.S. (1974), Animals and Humanitarianism. In: Klaits, J. and Klaits, B.G. (eds.) *Animals and Man in Historical Perspective,* Harper & Row, New York, 144-169.

Turner, J. (1980), *Reckoning with the Beast- Animals, Pain and Humanity in the Victorian Mind.* John Hopkins University Press, Baltimore/London.

Turner, R.K. (ed.) (1988), *Sustainable Environmental Management. Principles and Practice.* Belhaven Press, Pinter Publishers, London, Westview Press, Boulder, Colorado.

TWAT (1973-) Theologisches Wörterbuch zum Alten Testament, G.J. Botterweck *et al.* (eds.) Stuttgart.

Tyler, A.F. (1944), *Freedom's Ferment. Phases of American social History to 1860.* University of Minnesota Press, Minneapolis.

Ucko, P.J. & G.W. Dimblely (eds.) (1969), *The Domestication and Exploitation of Plants and Animals.* G. Duckworth & Co, London.

Udo de Haes, H.A. (1987). *Persoonlijk Voordeel. Van vijand tot bondgenoot in het milieubeheer.* Inaugural Speech, State University of Leiden, Leiden.

Udo de Haes, H.A. (1997), Toch een eigen plaats voor de milieukunde. In: P. Leroy (ed.), *Verbondenheid en samenhang.* Catholic University of Nijmegen, Nijmegen, 59-72.

Udo de Haes, H.A. & R. Heijungs (1996), Milieukunde: onderzoek in dwarsdoorsneden van het maatschappij-milieu systeem. *Milieu* 11, 1, 36-42.

Uhlfelder, Myral (1966), Nature in Roman Linguistic Texts. *Transactions and Proceedings of the American Philological Association,* 97, 583-95.

UNCED (1992), *Agenda 21.* United Nations Conference on Environment and Development. Rio '92, New York.

UNEP (1999), *Cultural and Spiritual Values of Biodiversity. A Complementary Contribution to the Global Biodiversity Assessment.* (G. Addison Posey (ed.) Oxford Centre for the Envirenment, Ethics and Society), Intermediate Technology Publications, London.

United Bible Societies (1972/1980), *Fauna and Flora of the Bible.* Help for Translators, United Bible Societies, London/New York/Stuttgart.

Ussishkin, D. (1978), *Excavations at Tel Lachis, 1973-1977. Preliminary Report.* Tel Aviv.

Valen, L. van (1969), Late Pleistocene Extinctions. *Proceedings of the North American Paleontological Convention,* 469-495.

Valgaerts, E. & L. Machiels (1992), *De Keltische erfenis. Riten en symbolen in het volksgeloof.* Stichting Mens en Kultuur, Gent.

Varner, Gary E. (1991), No Holism without Pluralism. *Environmental Ethics*, 13, 2, 175-179.

Vaux, R. de (1973/1978), *Hoe het oude Israël leefde. De instellingen van het Oude Testament.* 2 vols., 3rd ed., 1978 (1st ed., 1973), translated by L. Grollenberg, Wristers, Utrecht.

Verbruggen, H. & R. Dellink (1996), Duurzaam gebruik en duurzame ontwikkeling. *ESB*, 81, 532-533.

Verkiezingsprogramma's 94 (1994) Compiled and indexed by Prof. I. Lipschits, SDU, The Hague.

Vermeersch, E. (1988), *De ogen van de panda, een milieufilosofisch essay.* Marc Van de Wiele and Stichting Leefmilieu vzw, Brugge.

Vermeersch, E. (1990), Weg van het WTK-complex: onze toekomstige samenleving. In: C.L.T.M. *Het Milieu: denkbeelden voor de 21ste eeuw,* Kerckebosch, Zeist, 17-40.

Verney, P. (1979), *Animals in Peril. Man's War against Wildlife.* Brigham Young University Press, Provo, Utah.

Vervenne, M. (1991), 'Mens, kosmos en aarde. Een exegetische reflectie over Genesis 1-3'. In: Tavernier J. de & M. Vervenne (eds.), *De mens hoeder of verrader van de schepping? Over het conciliair proces.* Nike-26, Leuven/Amersfoort, 27-62.

Vlaar, M. (ed.) (1997), *De schepping. Verhalen over het begin uit alle windstreken.* De Bezige Bij, Amsterdam.

Vogel, C.J. de (1960/1964), *Greek Philosophy: A Collection of Texts with Notes and Explanations.* 3 vols., E.J. Brill, Leiden.

Vogel, C.J. de (1967), *Theoria, studies over de Griekse wijsbegeerte.* Van Gorcum en Comp. N.V., Assen.

Vogel, C.J. de (1968), *Het Humanisme en zijn historische achtergrond.* Van Gorcum & Comp. NV, Assen.

Vries, S.Ph. (1986), *Joodse Riten en Symbolen.* 6th edition, De Arbeiderpers, Amsterdam.

Vriezen, Th.C. (1958), *An Outline of Old Testament Theology.* Oxford University Press, Oxford.

Waal, F. de (1996), *Good Natured. The Origins of Right and Wrong in Humans and other Animals.* Harvard University Press, Cambridge, Massachusetts.

Waardt, J.M.D. de (1996), *Voedselvoorschriften in boeteboeken. Motieven voor het hanteren van voedselvoorschriften in vroeg-middeleeuwse Ierse boeteboeken 500-1100.* Erasmus Publishing, Rotterdam.

Wackernagel, M. & W. Rees (1996), *Our Ecological Footprint. Reducing Human Impact on the Earth.* New Society Publishers, Gabriola Island, BC. / Philadelphia, PA.

Waddell, H. (1934), *Beasts and Saints.* Translations by Helen Waddell, Constable and Company Ltd., London.

Wagner, S. Lemma *Kābaš*. *TWAT* IV, 54-60.

Wall, D. (ed.) (1994), *Green History. A Reader in Environmental Literature, Philosophy and Politics.* Routledge, London/New York.

Wallace, C.H. (1968), *Several Animals as Symbols in the OT. An Investigation of the Impact of the Animal World upon the OT Literature.* Dissertation, Basel.

Wallace-Hadrill, D.S. (1968), *The Greek Patristic View of Nature.* Manchester University Press, Manchester.

Walstra, P. & R. Jenness (1984), *Dairy Chemistry and Physics.* John Wiley & Sons, New York.

Ward, Barbara (1966), *Spaceship Earth.* Columbia University Press, New York.

Watkins, M.G. (1885), *Gleanings from the Natural History of the Ancients.* Elliot Stock, 62 Paternoster Row, E.C., London.

Watson, G. (1971), The Natural Law and Stoicism. In: A.A. Long *Problems in Stoicism.* The Athlone Press, University of London, London, 216-238.

Watts, D. (1986), *The West Indies, Patterns of Development and Environmental Change since 1492.* Cambridge University Press, Cambridge.

WCED (World Commission on Environment and Development) (1987), *Our Common Future.* Oxford University Press, Oxford.

WCMC (World Conservation Monitoring Centre) (1992), *Global Biodiversity: Status of the Earth's Living Resources.* Chapman & Hall, London.

Weber, M. (1920), Die Protestantische Ethik und der Geist des Kapitalismus. In: *Gesammelte Aufsätze zur Religionssoziologie.* I, Verlag Mohr, Tübingen, 1-206.

Weeber, K.W. (1990/1993), *Smog über Attika. Umweltverhalten im Altertum.* Artemis Verlag, Zürich/München 1990, Rowohlt Taschenbuch Verlag, Reinbek bei Hamburg, 1993.

Weinberg, Joel P. (1981), Die Natur im Weltbild des Chronisten. *Vetus Testamentum,* XXXI, 3, 324-345.

Weinberg, W. (1969-1970), *Transliteration and Transcription of Hebrew. HUCA,* XL-XLI, 1-32.

Weippert, M. (1990), Synkretismus und Monotheismus. Religionsinterne Konfliktbewältigung im alten Israel. In: Assmann, J. & D. Harth (eds.) *Kultur und Konflikt.* Edition Suhrkamp N.F. 612, Frankfurt am main, 143-179.

Weizsäcker, E.U. von (ed.) (1972), *Humanökologie und Umweltschutz.* Stuttgart/München.

Weizsäcker, E.U. von (1990), *Erdpolitik. Oekologische Realpolitik an der Schwelle zum Jahrhundert der Umwelt.* 2nd revised edition, Wissenschaftliche Buchgesellschaft, Darmstadt.

Weizsäcker, E.U. von, Amory B. Lovins & L. Hunter Lovins (1997), *Factor Four: Doubling Wealth, Halving Resource Use. A Report to the Club of Rome.* Earthscan Publications, London.

Welbourn, F.B. (1975), Man's Dominion. *Theology,* LXXVIII, 665, 561-568.

Wellhausen, J. (1878/1905), *Prolegomena zur Geschichte Israels.* 6th edition 1905, W. de Gruyter, Berlin/Leipzig.

Wendt, Herbert (1956/1959), *Out of Noah's Ark. The Story of Man's Discovery of the Animal Kingdom.* Translated by M. Bullock, Readers Union, G. Weidenfeld and Nicolson, London. (Orig. *Auf Noahs Spuren.* G. Grote'sche verlagsbuchhandlung, 1956, Hamm, Westfalen).

Wenham, G.J. (1978), The Coherence of the Flood Narrative. *VT,* 28, 3, 336-348.

Wenham, G.J. (1979), *The Book of Leviticus.* NICOT, Eerdmans, Grand Rapids, Michigan.

Wenham, G.J. (1981), The Theology of unclean Food. *Evangelical Quarterly,* 53, 6-15.

Wenham, G.J. (1982), 'Christ's Healing Ministry and His Attitude to the Law. In: Rowden, H.H. (ed.), *Christ the Lord. Studies in Christology presented to Donald Guthry.* IVP Leicester, 115-126

Wenham, G.J. (1983), Why does Sexual Intercourse defile? (Lev. 15,18). *Z.A.W,* 95, 432-434.

Weren W. & N. Poulssen (ed.) (1986), *Bij de put van Jacob: Exegetische opstellen.* Tilburg University Press, Tilburg.

Westermann, C. (1974/1983), *Genesis 1-11.* Biblischer Kommentar, Band I Teilband 1, Genesis 1-11, 3rd Ed. 1983, Neukirchener Verlag, Neukirchen-Vluyn.

Western, D. & M. Pearl (eds.) (1989), *Conservation for the Twenty-first Century*. Oxford University Press, Oxford.

Westhoff, V. (1989), Het natuurbeeld in de schriftloze culturen en in hoger ontwikkelde oosterse wereldbeschouwingen. In: W. Achterberg (ed.), *Natuur: uitbuiting of respect?* Kok Agora, Kampen, 177-198.

Westra, L. (1989), Ecology and Animals: Is there a Joint Ethic of Respect? *Environmental Ethics*, 11, 215-230.

White, J.E. (ed.) (1985), *Contemporary Moral Problems*. St. Paul, Minnesota.

White, L. Jr. (1967), The Historical Roots of our Ecologic Crisis. *Science*, 155, 1203-1207.

White, L. Jr. (1978), *Medieval Religion and Technology*. Collected Essays University of California Press, Berkeley Ca.

White, T.H. (1954), *The Book of Beast, being a Translation from a Latin Bestiary of the Twelfth Century: Bestiarius*. Edited by T.H. White., Jonathan Cape, London.

Whitehead, A.N. (1929), Universities and Their Function. In: *The Aims of Education and other Essays*, New York.

Whitehead, A.N. (1961), *Adventures of Ideas*. Cambridge University Press, Cambridge, Massachusetts.

Whitelam, K.W. (1996), *The Invention of Ancient Israel. The silencing of Palestinian history*. Routledge, London.

Wiener, Ph.P. (ed.) (1973), *Dictionary of the History of Ideas*. New York.

Wigoder, G. (ed.) (1992), *The New Standard Jewish Encyclopedia*. 7th edition, Facts on File, New York/Oxford.

Wigoder, G. (1994), *Joodse cultuur. Oorsprong en bloei*. Translated by Tinke Davids, Ambo, Baarn.

Wildavsky, A. (1995), *But is It True? A Citizen's Guide to Environmental Health and Safety Issues*. Harvard University Press, Cambridge, Massachusetts.

Wildavsky, A. & K. Dake (1990), Theories of Risk Perception. Who Fears What and Why? *Daedalus*, 112, 41-60.

Wildberger, H. (1965), Das Abbild Gottes. *Theologische Zeitschrift*, XXI, 245-259, 481-501.

Wildiers, M. (1988), *Kosmologie in de Westerse cultuur*. De Nederlandse boekhandel/ Pelckmans, Kapellen, J.H. Kok Agora, Kampen.

Wilkinson, Loren (ed.) (1980), *Earth Keeping. Christian Stewardship of Natural Resources*. William Eerdmans publ. Company, Grand Rapids, Michigan.

Williams, Arnold (1948), *The Common Expositor, An Account of the Commentaries on Genesis 1527-1633*. University of North Carolina Press, Chapel Hill.

Williams, Daniel Day (1972), Changing Concepts of Nature. In: Ian G. Barbour (ed.), *Earth might be Fair*. Prentice-Hall inc., Englewood Cliffs, New Jersey, 48-62.

Williams, G.H. (1962), *Wilderness and Paradise in Christian Thought. The Biblical Experience of the Desert in the History of Christianity & the Paradise Theme in the Theological Idea of the University*. Harper & Brothers, New York.

Williams, Howard (1883), *The Ethics of Diet. A Catena of Authorities Deprecatory of the Practice of Flesh-eating*. F. Pitman, London.

Wilson, E.O. (ed.) (1988), *Biodiversity*. National Academy Press, Washington D.C.

Wilson, E.O. (1992), *The Diversity of Life*. W.W. Norton & Company, New York/ London.

Wiman, Bo L.B. (1992), Designing Resource Systems for Sustainability: Safe-fail versus Fail-safe Strategies. in: Svedin, U. & Aniansson, B.H. (eds.), *Society an Environment: A Swedish Research Perspective*. Kluwer Academic Publishers, Dordrecht, 23-45.

Wiman, I.M.B. (1990), Expecting the Unexpected: some Ancient Roots to Current Perceptions of Nature. *Ambio*, 19, 2, 62-69.

Winden, J.C.M. van (1973), The Early Christian Exegesis of "Heaven and Earth" in Genesis 1, 1. In: *Romanitas et Christianitas studia J.H. Waszink oblata*, North-Holland publishing Company, Amsterdam, 371-383.

Wittbecker, A.E. (1990), Metaphysical Implications from Physics and Ecology. *Environmental Ethics*, 12, 2, 275-282.

Witter, D.J. (1977), *Attitudes toward Animals and their Uses: Literature Citations and Animal Welfare Organisation Data*. Microfilm PB 273584, Dept. of Agricultural Economics, Arizona University, Tucson.

Wolde, E.J. van (1989), *A Semiotic Analysis of Genesis 2-3: A Semiotic Theory and Method of Analysis applied to the Story of the Garden of Eden*. Van Gorcum, Assen.

Wolde, E.J. van (1994), *Words become Worlds. Semantic Studies of Genesis 1-11*. E.J. Brill, Leiden.

Wolde, E.J. van (1995), *Verhalen over het begin. Genesis 1-11 en andere scheppingsverhalen*. Ambo, Baarn.

Wood, G. (1972), *Animal Facts and Feats*. Doubleday, New York.

Wood, J.G. (1882), *Bible Animals; Being a Description of every living Creature mentioned in the Scriptures from the Ape to the Coral*. Longmans Green and Co, New Edition, London.

Woodkrutch, J. (1961), *The World of Animals. A Treasury of Lore, Legend and Literature by the Great Writers and Naturalist, from the 5th Century B.C. to the Present*. Simon and Schuster, New York.

Woodkrutch, J. (1974), The civilized Animal. In: Klaits, J. & Klaits, B.G. (eds.), *Animals and Man in Historical Perspective*, Harper & Row, New York, 93-107.

World Conservation Monitoring Centre (1998), *Global Biodiversity. Status of the Earth's living Resources*. Groombridge, B. (ed.), Chapman & Hall, London.

World Resources Institute (1992), *Global Biodiversity Strategy. Guidelines for Action to save, study and use Earth's Biotic Wealth Sustainably and Equitably*. In Cooperation with: IUCN, UNEP, FAO and UNESCO.

Worster, D. (1977), *Nature's Economy. The Roots of Ecology*. Cambridge University Press, Cambridge, Sierra Club Books, San Francisco.

Worster, D. (ed.) (1988), *The Ends of Earth: Perspectives on Modern Environmental History*. Cambridge University Press, Cambridge.

Worster, D. (1990), The Ecology of Order and Chaos. *Environmental History Review*, 14, 1-18.

Woude, A.S. van der (1969), 'Genesis en Exodus. Beschouwingen over de plaats van de schepping in de oudtestamentische theologie'. *Kerk en Theologie*, 20, 1-17.

Woude, A.S. van der (1973), Lemma *šēm* in: *TWAT* 2.

Woude, A.S. van der (ed.) (1981), *Remembering all the Way...A Collection of Old Testament Studies Published on the Occasion of the Fortieth Anniversary of the Oudtestamentisch Werkgezelschap in Nederland*. Oudtestamentische Studiën, deel XXI, E.J. Brill, Leiden.

Woude, A.S. van der (1990), De rollen van de Dode Zee: ontcijferingsmethoden en nieuwe teksten. *Kerk en Theologie*, 41, 2, April 1990, 92-102.

Woude, A.S. van der (1992), Pluriformity and Uniformity. Reflections on the Transmission of the Text of the Old Testament. In: J.N. Bremmer & F. García Martínez (eds.), *Sacred History and Sacred Texts in Early Judaism*. Kok Pharos, Kampen, 151-169.

Wright, G. Ernest (1957), *Biblical Archeology*. Westminster Press, Philadelphia.

Wright Mills, C. (1972), *De sociologische visie*. Aula, Het Spectrum, Utrecht/Antwerpen.

WRR (Wetenschappelijke Raad voor het Regeringsbeleid) (1994), *Duurzame risico's: een blijvend gegeven*. Sdu Uitgeverij, Den Haag.

Wünsche, August (1906), *Die Bildersprache des Alten Testaments. Ein Beitrag zur aesthetischen Würdigung des Poetischen Schrifttums im Alten Testament*. Ed. Pfeiffer Verlag, Leipzig.

Wurff, R. van der (1997), *International Climate Change Politics. Interests and Perceptions. A Comparative Study on Climate Change Politics in Germany, the United Kingdom and the United States*. Disertation, University of Amsterdam, Amsterdam.

Wybrow, C. (1991), The Bible, Baconianism and Mastery over Nature. The Old Testament and its Modern Misreading. *American University Study Series VII, Theology and Religion*, Vol. 112. New York.

Wynne-Tyson, J. (1990), *The Extended Circle. An Anthology of Humane Thought*. A Cardinal Book, Sphere Books, Orbit House, London.

Young, E.J. (1971), *Genesis Een*. Vertaald uit het Engels door H.J. Wisselink, Uitgeverij "De Vuurbaak", Groningen.

Zanden, J.L. van & S.W. Verstegen (1993), *Groene geschiedenis van Nederland*. Het Spectrum, Utrecht.

Zatelli, I. (1991), Astrology and the Worship of Stars in the Bible. *ZAW*, 103, 1, 86-99.

Zeller, H. (1867/1872), *Bijbelsch Woordenboek voor het Christelijk Volk*. J.H. Schuurman & J.P.G. Westhoff (eds.), 2 vols., M.J. Visser, 's Gravenhage.

Zenger, E. (1983), *Gottes Bogen in den Wolken. Untersuchungen zu Komposition und Theologie der Priesterschriftlichen Urgeschichte*. Verlag Katholisches Bibelwerk Gmbh, Stuttgart.

Zeuner, Fr.E. (1963), *A History of Domesticated Animals*. Hutchinson & Co, London

Zimmerman, M.E. (1988), Quantum Theory, Intrinsic Value and Panentheism. *Environmental Ethics*, 10, 1, 3-30.

Zimmerman, M.E. *et al.* (eds.) (1993), *Environmental Philosophy. From Animal Rights to radical Ecology*. Prentice-Hall, Englewood Cliffs, New Jersey.

Zobel, H.-J. Lemma *Rādāh* in: *TWAT*, Band VII, 351-358.

Zohary, M. (1982), *Plants of the Bible*, Cambridge University Press, Cambridge.

Zohary, M. (1982), *Het plantenrijk van het land van de Bijbel*. Vertaling door Ilse Dorren uit het Hebreeuws, La Rivière & Voorhoeve, Zwolle.

Zon, H. van (1986), *Een zeer onfrisse geschiedenis. Studie over niet-industriële vervuiling in Nederland, 1850-1920*. Proefschrift Rijks Universiteit Groningen.

Zon, H. van (1994), Milieugeschiedenis. In: Boersema, J.J. (e.a. red.) (1994), *Basisboek Milieukunde*. Boom, Amsterdam/Meppel, 93-103.

Zweers, W. (ed.) (1994), Radicalism or Historical Consciousness: On Breaks and Continuity in the Discussion of Basic Attitudes. In: *Ecology, Technology and Culture. Essays in Environmental Philosophy*. White Horse Press, Cambridge, 63-71.

Zweers, W. & J.J. Boersema (eds.) (1994), *Ecology, Technology and Culture. Essays in Environmental Philosophy*. White Horse Press, Cambridge.

Zweers, W. (1995), Ecological Spirituality as Point of Departure for an Intercultural Dialogue. In: *Agriculture and Spirituality*. International Books, Utrecht, 64-83.

Zweers, W. (2000), *Participating with Nature. Outline for an Ecologization of our World-view*. International Books, Utrecht.

GENERAL INDEX

INDEX OF SCRIPTURAL REFERENCES

Apocryphal and Pseudepigraphic Books

www.ingramcontent.com/pod-product-compliance
Lightning Source LLC
LaVergne TN
LVHW102226300926
849160LV00005B/37

9789004118867